de Gruyter Expositions in Mathematics 23

de Gruyter Expositions in Mathematics

1 The Analytical and Topological Theory of Semigroups, *K. H. Hofmann, J. D. Lawson, J. S. Pym* (Eds.)

2 Combinatorial Homotopy and 4-Dimensional Complexes, *H. J. Baues*

3 The Stefan Problem, *A. M. Meirmanov*

4 Finite Soluble Groups, *K. Doerk, T. O. Hawkes*

5 The Riemann Zeta-Function, *A. A. Karatsuba, S. M. Voronin*

6 Contact Geometry and Linear Differential Equations, *V. E. Nazaikinskii, V. E. Shatalov, B. Yu. Sternin*

7 Infinite Dimensional Lie Superalgebras, *Yu. A. Bahturin, A. A. Mikhalev, V. M. Petrogradsky, M. V. Zaicev*

8 Nilpotent Groups and their Automorphisms, *E. I. Khukhro*

9 Invariant Distances and Metrics in Complex Analysis, *M. Jarnicki, P. Pflug*

10 The Link Invariants of the Chern-Simons Field Theory, *E. Guadagnini*

11 Global Affine Differential Geometry of Hypersurfaces, *A.-M. Li, U. Simon, G. Zhao*

12 Moduli Spaces of Abelian Surfaces: Compactification, Degenerations, and Theta Functions, *K. Hulek, C. Kahn, S. H. Weintraub*

13 Elliptic Problems in Domains with Piecewise Smooth Boundaries, *S. A. Nazarov, B. A. Plamenevsky*

14 Subgroup Lattices of Groups, *R. Schmidt*

15 Orthogonal Decompositions and Integral Lattices, *A. I. Kostrikin, P. H. Tiep*

16 The Adjunction Theory of Complex Projective Varieties, *M. C. Beltrametti, A. J. Sommese*

17 The Restricted 3-Body Problem: Plane Periodic Orbits, *A. D. Bruno*

18 Unitary Representation Theory of Exponential Lie Groups, *H. Leptin, J. Ludwig*

19 Blow-up in Quasilinear Parabolic Equations, *A.A. Samarskii, V.A. Galaktionov, S. P. Kurdyumov, A. P. Mikhailov*

20 Semigroups in Algebra, Geometry and Analysis, *K. H. Hofmann, J. D. Lawson, E. B. Vinberg* (Eds.)

21 Compact Projective Planes, *H. Salzmann, D. Betten, T. Grundhöfer, H. Hähl, R. Löwen, M. Stroppel*

22 An Introduction to Lorentz Surfaces, *T. Weinstein*

Lectures in Real Geometry

Editor

Fabrizio Broglia

Walter de Gruyter · Berlin · New York 1996

Editor

Fabrizio Broglia
Dipartimento di Matematica
Università di Pisa
I-56127 Pisa
Italy

1991 Mathematics Subject Classification: 14Pxx, 32C05
Keywords: Real algebraic geometry, real analytic spaces, abelian varieties, Nash functions

Printed on acid-free paper which falls within the guidelines of the ANSI to ensure permanence and durability.

Library of Congress Cataloging-in-Publication Data

Lectures in real geometry / editor, Fabrizio Broglia.
p. cm. – (De Gruyter expositions in mathematics, ISSN 0938-6572 : 23)
"Elaborated versions of the lectures given ... at the Winter School in Real Geometry, held in Universidad Complutense de Madrid, January 3-7, 1994" – Fwd.
Includes bibliographical references.
ISBN 3-11-015095-6 (Berlin : alk. paper)
1. Geometry, Analytic. 2. Geometry, Algebraic. I. Broglia, Fabrizio, 1948- . II. Series
QA551.L29 1996
516.3'5–dc20 96-31731
CIP

Die Deutsche Bibliothek – CIP-Einheitsaufnahme

Lectures in real geometry / ed. Fabrizio Broglia. – Berlin ; New York : de Gruyter, 1996
(De Gruyter expositions in mathematics ; 23)
ISBN 3-11-015095-6
NE: Broglia, Fabrizio [Hrsg.]; GT

Typeset using the authors' TEX files: I. Zimmermann, Freiburg
Printing: A. Collignon GmbH, Berlin. Binding: Lüderitz & Bauer GmbH, Berlin.
Cover design: Thomas Bonnie, Hamburg.

In memoriam
Mario Raimondo

Foreword

The texts included in this book are elaborated versions of the lectures given by the authors at the Winter School in Real Geometry, held in Universidad Complutense de Madrid, January 3–7, 1994, in memory of Mario Raimondo, a bright young Italian mathematician who died prematurely on January 11, 1992.

The School, that was proposed by a group of Mario's friends during a Scientific Workshop held at Università di Genova on the first anniversary of his death, presented a postgraduate introductory course on Real Geometry and its applications. It included global and local questions on algebraic and analytic sets as well as computational aspects. It was addressed to students and researchers from other areas interested in the field.

The School was organized by the Department of Algebra at the Universidad Complutense de Madrid, and formed part of the celebrations for the seven hundredth anniversary of its foundation. It was also sponsored by the Department of Mathematics at the University of Genoa and by the ERASMUS ICP93-G-1010/11 ("Mathematics and Fundamental Applications") and ICP93-B-1142/11 ("Galois Network"). We gratefully acknowledge the ED contract CHRX-CT94-0506, in the framework of which the book was conceived.

This book brings together a variety of rather different topics which share nonetheless a common basic theme: it reflects in a sense the state of the art in each topic; it is a useful reference for anyone wishing to have an overview of the current work in some of the more important areas of Real Geometry.

We have also included an Appendix containing the two obituary lectures given at the Genoa Workshop; not just because they present the various aspects of Mario Raimondo's scientific activity, but also because they introduce the themes predominant at the Winter School—themes chosen from Mario's principal research interests.

The production of this book involved the effort of many people, some of whom wish to remain anonymous despite the unquestionable value of their contributions. We wish to thank them all for creating this tribute to a dear friend.

Fabrizio Broglia

Introduction

It is the zero sets of polynomials or analytic functions that comprise the objects of interest in both algebraic and analytic geometry. The first of these geometries is very old and owes much to the Italian school of algebraic geometry of the 19th and 20th centuries.

Complex analytic geometry was a consequence of the development of the theory of functions of several complex variables.[1] Here, of course, the Weierstrass Preparation Theorem played a primary role. The heroic period of analytic geometry took place from the 30s to the 50s, distinguishing itself by the discovery of breathtaking structural harmonies with algebra and topology.

The notable lack of such connections in real geometry seems to have discouraged early researchers.[2] Two examples illustrate well the difference with the complex case: the theorem on the analyticity of the singular locus, and the theorem on the analyticity of the topological closure of a connected component of the regular locus. That both theorems can fail over the reals can be seen from the well-known "Whitney Umbrella," a real algebraic set. This difference stems from the lack of the Fundamental Theorem of Algebra: the real field is not algebraically closed.

It turns out that both the singular locus and the closure of a connected component of the regular locus can be represented by adding inequalities to the descriptive language: both sets are necessarily semi-analytic (semi-algebraic in the algebraic case.) This shows why semi-analytic and semi-algebraic sets play an essential role in real geometry (cf. Bochnak, Coste, Roy [3]).

For many years, it seemed that algebraic methods had nothing to do with the study of semi-analytic and semi-algebraic sets, the "broken pieces" of zero sets. Thus the discovery by M. Coste that the study of semi-analytic and semi-algebraic geometry can be placed in an algebraic framework, the theory of the real spectrum, was all the more astonishing. Since then, Bröcker, continuing this work, has found some fine applications to problems involving basic sets, the building blocks of semi-analytic and semi-algebraic sets (cf. Andradas, Bröcker, Ruiz [1], [2]).

A theorem of Remmert assures that proper complex analytic maps preserve analyticity (and algebraicity in the algebraic case.) The fundamental Tarski–Seidenberg Theorem provides a real analog: polynomial maps preserve semialgebraicity; moreover, the class of semi-algebraic sets is precisely the class of images under coordinate projection of real algebraic sets. In the real analytic case, proper images of analytic sets form a new class, that of the subanalytic sets. Subanalytic geometry was

1 As presented, in particular, in the classic monograph by Osgood.

2 Perhaps A. Commessatti was the only representative of the Italian school who devoted his investigations to real algebraic geometry.

initiated by Gabrielov and Hironaka, who approached the question by means of his desingularization theorem, one of the deepest theorems of mathematical analysis.[3]

The global aspect of real geometry has been developed by Tognoli and his school in the framework of algebraic geometry and sheaf theory. In this setting, too, fundamental results of Raimondo show that great differences with the algebraically closed case appear (cf. Tognoli [5]). Recently, several important links between real algebraic geometry and computer algebra have been developed. There are basic contributions by Raimondo in this area as well (cf. Recio, Alonso [4]).

Finally, let us emphasize once more that the strong relations with algebra and topology provide a very rich context for viewing problems of complex geometry. For this reason, the study of real geometry via complexification should prove to be very useful.

References

[1] C. Andradas, L. Bröcker, J. M. Ruiz, Constructible Sets in Real Geometry, Ergeb. Math. Grenzgeb. (3) 33, Springer-Verlag 1996.

[2] C. Andradas, J. M. Ruiz, Algebraic and analytic geometry of fans, Mem. Amer. Math. Soc. 553, 1995.

[3] J. Bochnak, M. Coste, M. F. Roy, Géométrie Algébrique Réelle, Ergeb. Math. Grenzgeb. (3) 12, Springer-Verlag 1987.

[4] T. Recio, M. E. Alonso, Mario Raimondo's contributions to computer algebra, this volume.

[5] A. Tognoli, Mario Raimondo's contributions to real geometry, this volume.

Stanislaw Łojasiewicz

3 Still another approach can be found in the seminal work of Denef and van den Dries, p-adic and real subanalytic sets, Ann. of Math. 128 (1988), 79–138) who combined the Tarski–Seidenberg Theorem with Weierstrass Preparation.

Table of Contents

Appendix

Basic algorithms in real algebraic geometry and their complexity: from Sturm's theorem to the existential theory of reals

Marie-Françoise Roy *

1. Introduction

This text is devoted to the study of algorithms solving basic problems in real algebraic geometry. The first of these problems is the real counting problem.

Problem 1. *Compute the number of real roots of a polynomial.*

The solution we shall present is based on recent improvement on Sturm's 1835 original result [26] and has several interesting features. First the computations are done in the ring of coefficients of the polynomial, so that, for example, if we start with a polynomial with integer coefficients, we know the number of its real roots by performing only integer computations. Secondly the algorithm will work in an abstract setting for polynomials with coefficients in an ordered ring D contained in a real closed field R. The definitions and first examples of real closed fields are given in Section 2. The fact that the algorithm works in this more abstract setting will be useful in the other problems we shall consider next.

Let D be an ordered domain with field of fractions K and R a real closed field containing D. The *sign* $\operatorname{sign}(a)$ of an element $a \in \mathrm{R}$ is 0 if $a = 0$, 1 if $a > 0$ and -1 if $a < 0$. Problem 1 is reformulated more precisely as the following question.

Given $P \in \mathrm{D}[X]$, *compute the number of roots of* P *in* R.

The second problem we shall consider is the following

Problem 2. *Compute the real roots of a polynomial.*

as well as the related problem

*Supported in part by POSSO, Esprit BRA 6846

Problem 3. *Evaluate the sign of a polynomial at a real root of another polynomial.*

The way we are going to solve these problems does not correspond to what is usually done, that is computing some numerical approximation of the roots and then making a numerical evaluation, which gives the sign. The first reason for that is that we want an exact answer. The second reason is that we are interested in methods working in any real closed field, not necessarily archimedean.

It turns out that it is possible to characterize a real root x of a polynomial P by the signs taken by the successive derivatives of P at x, so that Problem 2 as well as Problem 3 will be solved if we can answer the following question.

We consider P in $\mathrm{D}[X]$ and $\mathcal{Q} = [Q_1, \ldots, Q_s]$ a list of polynomials in $\mathrm{D}[X]$.

A *sign pattern* σ, is an element of $\{0, 1, -1\}^s$ and the sign pattern of $\mathcal{Q}$ at x, $\mathrm{sign}(\mathcal{Q}, x)$ is the s-vector, $(\mathrm{sign}(Q_1(x)), \ldots, \mathrm{sign}(Q_s(x))$. The *sign patterns realized* by $\mathcal{Q}$ at the roots of P in R are the sign patterns $\sigma \in \{0, 1, -1\}^s$ such that

$$R(P, \mathrm{sign}(\mathcal{Q}) = \sigma) = \{x \in \mathrm{R} \mid P(x) = 0 \text{ and } \mathrm{sign}(\mathcal{Q}(x)) = \sigma\} \neq \emptyset.$$

The *cardinal* of $\mathrm{sign}(\mathcal{Q}) = \sigma$ at the real roots of P is

$$c(P, \mathrm{sign}(\mathcal{Q}) = \sigma) = \mathrm{card}\{x \in \mathrm{R} \mid P(x) = 0 \text{ and } \mathrm{sign}(\mathcal{Q}(x)) = \sigma\}.$$

We are ready to consider the *sign determination* question.

For every P in $\mathrm{D}[X]$ and for every list $\mathcal{Q}$ of polynomials in $\mathrm{D}[X]$ compute the set of sign patterns realized by $\mathcal{Q}$ at the roots in R of P and for every sign σ realized by $\mathcal{Q}$ compute the cardinal $c(P, \mathrm{sign}(\mathcal{Q}) = \sigma)$.

The answer to the *sign determination* question is given by a combinatorial algorithm based on a refinement of the real counting problem.

Let

$$\begin{aligned} c(P, Q = 0) &= \mathrm{card}\{x \in \mathrm{R} \mid P(x) = 0,\ Q(x) = 0\} \\ c(P, Q > 0) &= \mathrm{card}\{x \in \mathrm{R} \mid P(x) = 0,\ Q(x) > 0\} \\ c(P, Q < 0) &= \mathrm{card}\{x \in \mathrm{R} \mid P(x) = 0,\ Q(x) < 0\} \end{aligned}$$

and consider the following question:

Given P and Q in $\mathrm{D}[X]$, compute

$$c(P, Q = 0), c(P, Q > 0) \text{ and } c(P, Q < 0).$$

Based on an answer to this question, we shall be able to solve the *sign determination question* and thus Problems 2 and 3.

The last problem we shall consider is the following. We take a set of polynomial equations and inequalities in many variables and we want to decide if its realization set is empty or not. This problem is known as the *existential theory of reals problem.*

Problem 4. *Given a list* $\mathcal{P} = [P_1, \ldots, P_s]$ *of polynomials in* $\mathrm{D}[X_1, \ldots, X_k]$ *and a sign pattern* $\sigma \in \{0, 1, -1\}^s$, *decide if the realization set*

$$R(\mathrm{sign}(\mathcal{P}) = \sigma) = \{x \in R^k \mid \mathrm{sign}(P_i(x)) = \sigma(i)\}$$

is empty or not.

We shall answer this problem through the answer of a more general question in Section 6.

Let $\mathcal{P} = [P_1, \ldots, P_s]$ be a list of polynomials in $\mathrm{D}[X_1, \ldots, X_k]$. The *sign patterns realized* by $\mathcal{P}$ in R^k are the signs $\sigma \in \{0, 1, -1\}^s$ such that $R(\mathrm{sign}(\mathcal{P}) = \sigma)$ is non empty.

For every list $\mathcal{P}$ *of polynomials in* $\mathrm{D}[X_1, \ldots, X_k]$ *compute the list of sign patterns realized by* $\mathcal{P}$ *and for every realizable sign pattern a sample point in every conected components of its realization.*

We shall use repeatedly as a basic subroutine the solution to the following problem: find points on the various connected components of the zero set of one polynomial. Formulated more precisely

For $P \in \mathrm{D}[X_1, \ldots X_k]$ *compute a sample point in every connected component of*

$$Z(P) = \{x \in \mathrm{R}^k \mid P(x) = 0\}.$$

This last multivariate problem will be reduced to a univariate problem using the computation of the Rational Univariate Representation which is explained in Section 6, so that all the univariate methods developed before can be used.

We shall not only be interested in finding algorithms but in finding algorithms with good theoretical complexity, and, as much as possible, good practical complexity. Basic notions and techniques about complexity of algorithms are explained in Section 4.

In particular we shall obtain for Problem 4 a complexity $s^{k+1}d^{O(k)}$ which is the best known to now, and quasi optimal.

The text presented here is derived in large part from various research papers coauthored with Mari-Emi Alonso, Saugata Basu, Eberhard Becker, Michel Coste, Felipe Cucker, Laureano Gonzalez-Vega, Hervé Lanneau, Henri Lombardi, Bud Mishra, Paul Pedersen, Richard Pollack, Tomas Recio, Aviva Szpirglas, Thorsten Worman and the corresponding references ([1], [2], [9], [8], [12], [13], [14], [21], [25], [22]) are listed in the bibliograhy. I want to thank them all for letting me use our common work. A more complete bibliography appears in these papers. I did not try to give a detailed historical information about the results presented here.

The second section is devoted to the definition and first properties of real closed fields, the third section is devoted to the real root couting problem, in the fourth section we study the complexity of basic problems and apply these techniques to the real root counting methods, in the fifth section we study sign determinations and real algebraic numbers and in the sixth section the existential theory of reals. Special thanks to Laureano Gonzalez-Vega, Henri Lombardi and Richard Pollack for their help on this text.

2. Real closed fields

In this section we give the definitions and first examples of real closed fields that we shall use in the next sections (see [4] for more details).

2.1. Definition and first examples of real closed fields

Definition 1. A *real closed field* R is a real field (i.e., where -1 is not a sum of squares), admitting a unique ordering with positive cone the squares of R, and such that every polynomial in $R[X]$ of odd degree has a root in R.

A basic theorem about real closed fields is the following.

Theorem 2.1. *A field* R *is real closed if and only if* $R[i] = R[X]/(X^2 + 1)$ *is a field which is algebraically closed.*

For example, the field of real numbers is a real closed field, as is the field of *real algebraic numbers* (real numbers satisfying an equation with integer coefficients).

A *real closure* of an ordered field K is an algebraic extension which is real closed and extends the order on K.

Theorem 2.2. *Every ordered field* K *admits a unique real closure.*

The field of real algebraic numbers is the real closure of $\mathbb{Q}$.

Real closed fields are not necessarily archimedean, as we shall see soon.

Particular cases of real closed fields with a geometrical meaning are real (algebraic) numbers and non-archimedean real closed fields which contain fields of rational functions. We develop now the important case of Puiseux series.

Definition 2. Let R be a real closed field. If ϵ is a variable, one denotes by $R\langle\epsilon\rangle$ the *field of Puiseux series in* ϵ *with coefficients in* R. Its elements are the series of the form

$$\sum_{i \geq i_0,\ i \in \mathbb{Z}} a_i \epsilon^{i/q}$$

with $i \in \mathbb{Z}$, $a_i \in R$, $q \in \mathbb{N}$, ([29] or [4]).

The motivation behind this definition is the following. In order to construct a real closed field containing $\mathrm{R}(\epsilon)$ it is necessary to consider rational exponents. For example, in order to find a root of the equation $X^3 - \epsilon^2 = 0$ one needs to have $\epsilon^{2/3}$. The following result says that the consideration of rational exponents is enough to ensure that the field is real closed.

Theorem 2.3. *The field* $\mathrm{R}\langle\epsilon\rangle$ *is real closed. Positive elements are elements whose lowest degree term has positive coefficient.*

For a proof see [29] or [17].

Let K be an ordered extension of R. An element x' of K is *infinitesimal* (with respect to R) if and only if its absolute value is strictly smaller than any positive element in R. Since $a - \epsilon$ is a square in $\mathrm{R}\langle\epsilon\rangle$ for every positive $a \in \mathrm{R}$, the element ϵ is infinitesimal and positive.

The elements of $\mathrm{R}\langle\epsilon\rangle$ bounded over R form a *valuation ring* denoted $V(\epsilon)$: the elements of $V(\epsilon)$ are Puiseux series

$$\sum_{i \in \mathbb{N}} a_i \epsilon^{i/q}$$

(so that $a_i = 0$ for $i < 0$). We denote by eval_ϵ the ring homomorphism from $V(\epsilon)$ to R which maps $\sum_{i \in \mathbb{N}} a_i \epsilon^{i/q}$ to a_0.

For a subring $\mathrm{D} \subset \mathrm{R}$ we define the order 0_+ on $\mathrm{D}[\epsilon]$ making ϵ infinitesimal and positive by saying that $P \in \mathrm{D}[\epsilon]$ is positive if and only if the tail coefficient of P (the coefficient of the lowest degree term in ϵ) is positive. Similarly for $Q \in \mathrm{R}(\epsilon)$ the order making ϵ infinitesimal and positive is defined by saying that Q is positive if and only if the tail coefficients of the numerator and denominator of Q have the same sign. It follows that $P(\epsilon) > 0$ if and only if for $t \in \mathrm{R}$ sufficiently small and positive, $P(t) > 0$.

The field of rational functions $\mathrm{R}(\epsilon)$ equipped with the order 0_+ is a subfield of the field of Puiseux series $\mathrm{R}\langle\epsilon\rangle$.

The real closure of $\mathrm{R}(\epsilon)$ equipped with the order 0_+ is the field of *algebraic Puiseux series*, that is, the subfield of elements of $\mathrm{R}\langle\epsilon\rangle$ satisfying an algebraic equation with coefficients in $\mathrm{R}[\epsilon]$. They can be interpreted geometrically as the half-branches of algebraic curves above a small open interval to the right of 0.

2.2. Cauchy index and real root counting

We shall first see that the ordinary behaviour of univariate rational functions we are used to in the field of reals are also valid in any real closed field.

Let D be a domain contained in a real closed field R and K its field of fractions.

Let P and Q be univariate polynomials with coefficients in D.

The multiplicity of a root a of P is as usual the exponent k such that $P = (X-a)^k P_1$ with P_1 a polynomial such that $P_1(a) \neq 0$.

The polar multiplicity of a in Q/P is the exponent $k \in \mathbb{Z}$ such that $Q/P = Q_1/(X-a)^k P_1$ with $X-a$ dividing neither Q_1 nor P_1. The rational function Q/P has a finite limit at a when the polar multiplicity of a in Q/P is negative. The rational function Q/P has an infinite limit at a_- (resp. a_+) if its polar muyltiplicity at a is strictly positive. The limit at a_+ is $\sigma\infty$ where $\sigma = \operatorname{sign}(Q_1(a)/P_1(a))$. The limit at a_- is is $\sigma\infty$ where $\sigma = (-1)^k \operatorname{sign}(Q_1(a)/P_1(a))$.

Let $a < b$ be elements in $\mathrm{R} \cup \{-\infty, +\infty\}$. The Cauchy index $I(Q/P;]a, b[)$ of Q/P between a and b is by definition the number of jumps of the function Q/P from $-\infty$ to $+\infty$ minus the number of jumps of the function Q/P from $+\infty$ to $-\infty$ on the open interval $]a, b[$.

The connection between the Cauchy index and the number of roots in a real closed field is given by the corollary to the next proposition. Let

$$\begin{aligned} c(P;]a, b[) &= \operatorname{card}\{x \in \mathrm{R} \mid P(x) = 0,\ a < x < b\} \\ c(P, Q = 0;]a, b[) &= \operatorname{card}\{x \in \mathrm{R} \mid P(x) = 0,\ Q(x) = 0,\ a < x < b\} \\ c(P, Q > 0;]a, b[) &= \operatorname{card}\{x \in \mathrm{R} \mid P(x) = 0,\ Q(x) > 0,\ a < x < b\} \\ c(P, Q < 0;]a, b[) &= \operatorname{card}\{x \in \mathrm{R} \mid P(x) = 0,\ Q(x) < 0,\ a < x < b\} \end{aligned}$$

Proposition 2.4. *Let $a < b$ be elements in* $\mathrm{R} \cup \{-\infty, +\infty\}$. $I(P'Q/P;]a, b[) = d(P, Q;]a, b[)$ *where* $d(P, Q;]a, b[) = c(P, Q > 0;]a, b[) - c(P, Q < 0;]a, b[)$.

Proof. We restrict our attention to the roots c of P which are not roots of Q since $P'Q/P$ at a common root of P and Q has a finite limit.

Defining k as the multiplicity of the root c in P,

$$P'Q/P = kQ(c)/(X-c) + R_c$$

with R_c having a finite value at c, so it is easy to see that

- there is a jump from $-\infty$ to $+\infty$ at c in $P'Q/P$ if $Q(c) > 0$,
- there is a jump from $+\infty$ to $-\infty$ at c in $P'Q/P$ if $Q(c) < 0$. □

Corollary 2.5. $I(P'/P;]a, b[) = c(P;]a, b[)$.

3. Real root counting

As we have seen in the introduction, one basic problem is to compute the number of real roots of a polynomial. We shall do it through the computation of the Cauchy index. This is what we study in this section.

In Subsection 3.1 we give a generalisation of Sturm's theorem which is essentially due to Sylvester and define Sylvester sequence.

In Subsection 3.2 we study the subresultant sequence.

In Subsection 3.3 we introduce the Sylvester–Habicht sequence, which comes from Habicht's work ([15]). This new sequence, obtained automatically from a subresultant sequence, has some remarkable properties:

- it gives the same information as the Sylvester sequence, and this information may be recovered by looking only at its principal coefficients,
- it can be computed by ring operations and exact divisions only, in polynomial time,
- it has good specialisation properties.

Finally in 3.4 we introduce another real counting method, based on the signature of a quadratic form due to Hermite. We then compare general methods for computing the number of real roots and explain the connection between the Sylvester–Habicht sequence and Hermite's method.

3.1. Sylvester sequence

3.1.1. Definitions and notations. Let D be a domain, and K its field of fractions. Let P be a polynomial with coefficients in D. We denote by $\deg(P)$ its *degree*, and $\mathrm{cf}_j(P)$ its *coefficient of degree* j (equal to 0 if $j > \deg(P)$). Let P and Q be two univariate polynomials with coefficients in a field K, with $\deg(Q) = q$. The *remainder* in the euclidean division of P by Q, $\mathrm{Rem}(P, Q)$ is the unique polynomial R such that $P = A \cdot Q + R$ with degree strictly less than Q. It has coefficients in K. We have the relation

$$\mathrm{Rem}(aP, bQ) = a\,\mathrm{Rem}(P, Q)$$

for any $a, b \in \mathrm{R}$ with $b \neq 0$. Note that at a root x of Q,

$$\mathrm{Rem}(P, Q)(x) = P(x).$$

The *euclidean remainder sequence* of P and Q is defined by

$$\begin{aligned} \mathrm{Rem}^0(P, Q) &= P \\ \mathrm{Rem}^1(P, Q) &= Q \\ \mathrm{Rem}^{m+1}(P, Q) &= \mathrm{Rem}(\mathrm{Rem}^{m-1}(P, Q), \mathrm{Rem}^m(P, Q)) \end{aligned}$$

The *Sylvester sequence* of P and Q is a slight modification of the euclidean remainder sequence defined as follows:

$$\begin{aligned} \mathrm{Sy}^0(P, Q) &= P \\ \mathrm{Sy}^1(P, Q) &= Q \end{aligned}$$

$$\mathrm{Sy}^{m+1}(P, Q) = -\mathrm{Rem}(\mathrm{Sy}^{m-1}(P, Q), \mathrm{Sy}^m(P, Q))$$

We denote $\mathrm{Stu}^m(P)$ for $\mathrm{Sy}^m(P, P')$, which is the classical notion of the *Sturm sequence* associated to P.

We suppose now that the field K is ordered and let R be a real closed field containing K.

The number of *sign changes* $V([a_0, \ldots, a_n])$ in a list $[a_0, \ldots, a_n]$ of elements in $\mathrm{R} \setminus \{0\}$ is defined by induction on n by

$$V([a_0]) = 0,$$

$$V([a_0, \ldots, a_{n+1}]) = \begin{cases} V([a_0, \ldots, a_n]) + 1 & \text{if } \operatorname{sign}(a_n a_{n+1}) = -1, \\ V([a_0, \ldots, a_n]) & \text{otherwise.} \end{cases}$$

This definition extends to any sequence of elements in R dropping the zeroes in the sequence considered.

Example 3.1. We have $V([1, -1, 2, 0, 0, 3, 4, -5, 0, -2, 3]) = 4$.

The *sign of a polynomial P at* $+\infty$ *(resp.* $-\infty$) is given by the sign of the leading coefficients of P and the parity of its degree. It agrees with the sign of $P(M)$ for M sufficently big (resp. small).

If $\mathcal{P} = [P_0, P_1, \ldots, P_n]$ is a sequence of polynomials and a an element of $\mathrm{R} \cup \{-\infty, +\infty\}$ then we shall call *number of sign changes* of $\mathcal{P}$ at a, denoted by $V(\mathcal{P}, a)$, the number $V([P_0(a), P_1(a), \ldots, P_n(a)])$.

We denote by $V_{\mathrm{Sy}}(P, Q; a)$ the *number of sign changes of the Sylvester sequence* of P and Q at a and define:

$$\begin{aligned} V_{\mathrm{Sy}}(P, Q;]a, b[) &= V_{\mathrm{Sy}}(P, Q; a) - V_{\mathrm{Sy}}(P, Q; b) \\ V_{\mathrm{Stu}}(P;]a, b[) &= V_{\mathrm{Sy}}(P, P';]a, b[). \end{aligned}$$

3.1.2. Sturm's theorem.

Theorem 3.2. *Let* K *be an ordered field and* R *its real closure. If P and Q are two polynomials with coefficients in* K *and a and b (with $a < b$) are elements of* $\mathrm{R} \cup \{-\infty, +\infty\}$ *which are not roots of P then*

$$V_{\mathrm{Sy}}(P, Q;]a, b[) = I(Q/P;]a, b[).$$

Proof. Let $c_1 < \cdots < c_r$ be the real roots of the polynomials $\mathrm{Sy}^j(P, Q)$ in the interval $]a, b[$, so that $a = c_0 < c_1 < \cdots < c_r < c_{r+1} = b$. For every $i \in \{1, \ldots, r\}$ choose an element d_i between c_i and c_{i+1}. If $c_0 = a$ is a root of some $\mathrm{Sy}^j(P, Q)$, choose an element d_0 between c_0 and c_1, otherwise take $d_0 = c_0 = a$. If $c_{r+1} = b$ is a root of some $\mathrm{Sy}^j(P, Q)$, choose an element d_r between c_r and c_{r+1}, otherwise take $d_r = c_{r+1} = b$.

Clearly,

$$\begin{aligned} V_{\mathrm{Sy}}(P, Q;]a, b[) &= V_{\mathrm{Sy}}(P, Q; a) - V_{\mathrm{Sy}}(P, Q; b) \\ &= V_{\mathrm{Sy}}(P, Q; c_0) - V_{\mathrm{Sy}}(P, Q; d_0) \\ &\quad + \sum_{i=1}^{r} \Big[V_{\mathrm{Sy}}(P, Q; d_{i-1}) - V_{\mathrm{Sy}}(P, Q; d_i) \Big] \\ &\quad + V_{\mathrm{Sy}}(P, Q; d_r) - V_{\mathrm{Sy}}(P, Q; c_{r+1}) \end{aligned}$$

which reduces the proof of the theorem to the study of the integers

$$\begin{aligned} &V_{\mathrm{Sy}}(P, Q; c_0) - V_{\mathrm{Sy}}(P, Q; d_0),\ V_{\mathrm{Sy}}(P, Q; d_r) - V_{\mathrm{Sy}}(P, Q; c_{r+1}) \\ \text{and}\quad &V_{\mathrm{Sy}}(P, Q; d_{i-1}) - V_{\mathrm{Sy}}(P, Q; d_i),\ \text{for every } i \in \{1, \ldots, r\}. \end{aligned}$$

If $\ell + 1$ is the length of the Sylvester sequence of P and Q then $\mathrm{Sy}^{\ell}(P, Q)$ (the last element of the sequence) is hence a greatest commom divisor of P and Q.

The proof of the theorem is obtained by proving the following lemmas.

Denote $S_m = \mathrm{Sy}^m(P, Q)$.

Lemma 3.3. *If $P(c) \neq 0$ and $S_j(c) = 0$, and d is such that there is no root of any of the S_k on $]c, d[$ (resp. $]d, c[$), then there is exactly 1 sign change in the sequences $[S_{j-1}(c), S_j(c), S_{j+1}(c)]$ and $[S_{j-1}(d), S_j(d), S_{j+1}(d)]$.*

Necessarily, $S_{j-1}(c)$ and $S_{j+1}(c)$ are different from zero, because the greatest common divisor of S_j and S_{j-1} (resp. S_j and S_{j+1}) divided P and $P(c) \neq 0$. From the definition of the Sylvester sequence S_{j+1} is the negative of the remainder of S_{j-1} divided by S_j which implies $S_{j+1}(c) = -S_{j-1}(c)$ by properties of remainders. So that the number of sign changes in $[S_{j-1}(c), S_j(c), S_{j+1}(c)]$ is 1. The signs of S_{j-1} and S_{j+1} at d agree with the signs of S_{j-1} and S_{j+1} at c. This shows that whatever the signs of $S_j(d$ and $S_j(d)$ are, the number of sign changes in $[S_{j-1}(d), S_j(d), S_{j+1}(d)]$ is 1. □

Lemma 3.4. *If $i \in \{1, \ldots, r\}$ and $P(c_i) \neq 0$ then*

$$V_{\mathrm{Sy}}(P, Q, d_{i-1}) - V_{\mathrm{Sy}}(P, Q, d_i) = 0.$$

Proof. We have seen in the preceding lemma that if c_i is a root in R of S_j with $i \neq 0$ then there is exactly 1 sign change in the sequences $[S_{j-1}(d_{i-1}), S_j(d_{i-1}), S_{j+1}(d_{i-1})]$ and $[S_{j-1}(d_i), S_j(d_i), S_{j+1}(d_i)]$. □

Lemma 3.5.

$$V_{\mathrm{Sy}}(P, Q; c_0) - V_{\mathrm{Sy}}(P, Q; d_0) = V_{\mathrm{Sy}}(P, Q; d_r) - V_{\mathrm{Sy}}(P, Q; c_{r+1}) = 0$$

Proof. It is enough to consider the case of c_0. If $c_0 = d_0$, there is nothing to prove. If $S_j(c_0) = 0$, we know by Lemma 3.3 that there is exactly 1 sign change in the sequences $[S_{j-1}(c_0), S_j(c_0), S_{j+1}(c_0)]$ and $[S_{j-1}(d_0), S_j(d_0), S_{j+1}(d_0)]$. □

Denote by $\mathcal{T}$ be the sequence defined by

$$T_m = \frac{\mathrm{Sy}^m(P, Q)}{\mathrm{Sy}^\ell(P, Q)}$$

(where $\mathrm{Sy}^\ell(P, Q)$ is the last element in the Sylvester sequence of P and Q). It is clear that if x is not a root of P then $V(\mathcal{T}, x) = V_{\mathrm{Sy}}(P, Q; x)$, and that the Cauchy index of T_0 and T_1 concides with the Cauchy index of P and Q.

Lemma 3.6. *If* $i \in \{1, \ldots, r\}$ *and* $T_0(c_i) = 0$ *then*

$$V_{\mathrm{Sy}}(P, Q; d_{i-1}) - V_{\mathrm{Sy}}(P, Q; d_i) = I(Q/P;]d_{i-1}, d_i[).$$

Proof. It is clear that c_i is not a root of T_1. If c_i is also a root of some T_j ($j \in \{2, \ldots, \ell - 1\}$) then the number of sign changes in the sequences

$$[T_{j-1}(d_{i-1}), T_j(d_{i-1}), T_{j+1}(d_{i-1})] \text{ and } [T_{j-1}(d_i), T_j(d_i), T_{j+1}(d_i)]$$

is 1 by Lemma 3.3 and the definition of $\mathcal{T}$. So whether or not c_i is a root of some T_j ($j \in \{2, \ldots, \ell - 1\}$) the number of sign changes in the sequence $[T_1(d_{i-1}), \ldots, T_\ell(d_{i-1})]$ coincdes with the number of sign changes in the sequence $[T_1(d_i), \ldots, T_\ell(d_i)]$.

We consider what happens for T_0 and T_1.

If the multiplicity of the root c_i of T_0 is even, there is no jump from $-\infty$ to $+\infty$ (resp. $+\infty$ to $-\infty$) and $V_{\mathrm{Sy}}(P, Q; d_{i-1}) - V_{\mathrm{Sy}}(P, Q; d_i) = 0$.

Suppose now that this multiplicity is odd:

— if $T_1(d_{i-1})T_0(d_{i-1}) < 0$ there is a jump from $-\infty$ to $+\infty$ at c_i in Q/P and $V_{\mathrm{Sy}}(P, Q, d_{i-1}) - V_{\mathrm{Sy}}(P, Q, d_i) = 1$,

— if $T_1(d_{i-1})T_0(d_{i-1}) > 0$ is positive, there is a jump from $+\infty$ to $-\infty$ at c_i in Q/P and $V_{\mathrm{Sy}}(P, Q, d_{i-1}) - V_{\mathrm{Sy}}(P, Q, d_i) = -1$. □

With $d(P, Q;]a, b[) = c(P, Q > 0;]a, b[) - c(P, Q < 0;]a, b[)$ we have

Corollary 3.7 (Sylvester's theorem [27]). *Let* K *be an ordered field and* R *its real closure. If* P *and* Q *are two polynomials with coefficients in* K *and* a *and* b *(with* $a < b$*) are elements of* $\mathrm{R} \cup \{-\infty, +\infty\}$ *which are not roots of* P *then*

$$V_{\mathrm{Sy}}(P, P'Q;]a, b[) = d(P, Q;]a, b[).$$

Proof. Use Theorem 3.2 and Proposition 2.4. □

Corollary 3.8 (Sturm's theorem [26]). *With the same hypothesis and notations as in Theorem 3.2, the following equalities hold:*

$$V_{\mathrm{Stu}}(P;]a, b[) = c(P;]a, b[).$$

Remark 3.9. When a or b are roots of P which are not also roots of Q, it is still possible to know from $V_{\mathrm{Sy}}(P, Q;]a, b[)$ and the signs of $Q(a)$ and $Q(b)$ the number

$I(Q/P;]a, b[)$. Define

$$\epsilon(a) = \begin{cases} 1 & \text{if } Q(a)P^m(a) < 0,\ P'(a) = \cdots = P^{m-1}(a) = 0, \text{ and } m \text{ odd},\\ 0 & \text{otherwise;} \end{cases}$$

$$\epsilon(b) = \begin{cases} 1 & \text{if } Q(b)P^m(b) > 0,\ P'(b) = \cdots = P^{m-1}(b) = 0, \text{ and } m \text{ odd},\\ 0 & \text{otherwise.} \end{cases}$$

Then it is easy to see that $V_{\mathrm{Sy}}(P, Q;]a, b[) + \epsilon(a) + \epsilon(b)$ coincides with $V_{\mathrm{Sy}}(P, Q; d_0, d_r)$ and is thus equal to $I(Q/P;]d_0, d_r[) = I(Q/P;]a, b[)$.

3.1.3. Computational problems. The bitlength of coefficients in the Sturm sequence can increase dramatically as we see in the next example.

Example 3.10. Consider the following numerical example

$$\begin{aligned} P := 9X^{13} - 18X^{11} - 33X^{10} + 102X^8 + 7X^7 - 36X^6 \\ - 122X^5 + 49X^4 + 93X^3 - 42X^2 - 18X + 9. \end{aligned}$$

The Sturm sequence of P is

$$\begin{aligned}
\mathrm{Stu}^1(P) &= \frac{1}{13}(36X^{11} + 99X^{10} - 510X^8 - 42X^7 + 252X^6 + 976X^5 - 441X^4 \\
&\quad - 930X^3 + 462X^2 + 216X - 117)\\
\mathrm{Stu}^2(P) &= \frac{1}{16}(10989X^{10} + 21240X^9 - 70746X^8 - 6054X^7 - 13932X^6 + 159044X^5 \\
&\quad - 24463X^4 - 153878X^3 + 59298X^2 + 35628X - 17019)\\
\mathrm{Stu}^3(P) &= \frac{32}{1490841}(626814X^9 - 1077918X^8 + 71130X^7 - 830472X^6 + 2259119X^5 \\
&\quad + 460844X^4 - 2552804X^3 + 668517X^2 + 632094X - 256023)\\
\mathrm{Stu}^4(P) &= \frac{165649}{38804522528}(43475160X^8 - 57842286X^7 + 5258589X^6 - 92294719X^5 \\
&\quad + 134965334X^4 + 31205119X^3 - 79186035X^2 + 5258589X + 9147321)\\
\mathrm{Stu}^5(P) &= \frac{2425282658}{543561530761725025}(1584012126X^7 - 2548299819X^6 + 984706749X^5 \\
&\quad - 3696028294X^4 + 5946032911X^3 - 713636955X^2 - 2548299819X + 984706749)\\
\mathrm{Stu}^6(P) &= \frac{543561530761725025}{6761403525275795353156967 12}(12232018869X^6 - 8633929833X^5 \\
&\quad - 28541377361X^3 + 20145836277X^2 + 12232018869X - 8633929833)\\
\mathrm{Stu}^7(P) &= -\frac{6670589018492723310272156415 9514728}{180730930229098050132455395887 1415645}(3X^5 - 7X^2 + 3).
\end{aligned}$$

There are also *specialisation problems.* Let D be a domain, K its field of fractions, P and Q polynomials in D$[X]$. Suppose that the computation of the Sylvester sequence has been done in the field K, and that the coefficients of P and Q are specialized, that is we consider a ring morphism f of D to a domain D′ and images $f(P)$ and $f(Q)$ of P and Q in the ring $D'[X]$. The Sylvester sequence associated to $f(P)$ and $f(Q)$ is not easy to compute from the Sylvester sequence of P and Q because in the euclidean division process of P by Q, elements of D appear in the denominator, and may well specialize to 0. In this case the Sylvester sequence of $f(P)$ and $f(Q)$ is not obtained by specializing the Sylvester sequence of P and Q and the degree of the polynomials in the Sylvester sequence of $f(P)$ and $f(Q)$ do not agree with the degree of the polynomials in the Sylvester sequence of P and Q.

Example 3.11. Consider the general polynomial of degree 4,

$$P = X^4 + pX^2 + qX + r.$$

The Sturm sequence of P computed in $\mathbb{Q}(p, q, r)[X]$ is

$$\begin{aligned}
\mathrm{Stu}^0(P) &= X^4 + pX^2 + qX + r \\
\mathrm{Stu}^1(P) &= 4X^3 + 2pX + q \\
\mathrm{Stu}^2(P) &= \frac{-(2pX^2 + 3qX + 4r)}{4} \\
\mathrm{Stu}^3(P) &= \frac{-((2p^3 - 8pr + 9q^2)X + p^2q + 12qr)}{p^2} \\
\mathrm{Stu}^4(P) &= \frac{p^2(16p^4r - 4p^3q^2 - 128p^2r^2 + 144pq^2r - 27q^4 + 256r^3)}{4(2p^3 - 8pr + 9q^2)^2}.
\end{aligned}$$

When we choose particular values $\tilde{p}$, $\tilde{q}$ and $\tilde{r}$ for p, q and r, the Sturm sequence of $\tilde{P} = X^4 + \tilde{p}X^2 + \tilde{q}X + \tilde{r}$ is generally obtained by replacing p, q and r by $\tilde{p}$, $\tilde{q}$ and $\tilde{r}$ in the Sturm sequence of $X^4 + pX^2 + qX + r$. But in some cases (when denominators vanish) this substitution is impossible and the computation has to be made again. For $\tilde{p} = 0$, the Sturm sequence of $\tilde{P} = X^4 + \tilde{q}X + \tilde{r}$ is

$$\begin{aligned}
\mathrm{Stu}^0(\tilde{P}) &= X^4 + \tilde{q}X + \tilde{r} \\
\mathrm{Stu}^1(\tilde{P}) &= 4X^3 + \tilde{q} \\
\mathrm{Stu}^2(\tilde{P}) &= \frac{-(3\tilde{q}X + 4\tilde{r})}{4} \\
\mathrm{Stu}^3(\tilde{P}) &= \frac{-(27\tilde{q}^4 - 256\tilde{r}^3)}{27\tilde{q}^3}.
\end{aligned}$$

3.2. Subresultants and remainders

In order to solve in Section 3.3 the computational problems we just listed, we study now the theory of subresultants. The subresultants of two polynomials P and Q are polynomials which have a very close relationship to the polynomials in the remainder sequence of P and Q and are defined through determinants associated to the coefficients of P and Q. Due to their definition through determinants, their coefficients belong to the ring generated by the coefficients of P and Q.

3.2.1. Definitions. For P and Q in D[X], let $p = \sup(\deg(P), \deg(Q)+1)$, $q = \deg(Q)$, and for $j < q$ the *j-th Sylvester matrix* of P and Q, $\mathrm{Sylv}_j(P, Q)$, is the matrix whose rows on the basis

$$X^{p+q-j-1}, \ldots, X^2, X, 1$$

are the coefficient vectors of the polynomials:

$$PX^{q-j-1}, \ldots, PX, P, QX^{p-j-1}, \ldots, QX, Q.$$

This matrix has $p+q-2j$ rows and $p+q-j$ columns. If

$$P = a_p X^p + a_{p-1}X^{p-1} + \cdots + a_0$$
$$Q = b_q X^q + b_{q-1}X^{q-1} + \cdots + b_0$$

(note that the coefficients of highest degrees of P may be 0), $\mathrm{Sylv}_j(P, Q)$ is the matrix:

$$\mathrm{Sylv}_j(P,Q) = \overbrace{\begin{pmatrix} a_p & \cdots & \cdot & a_0 & & \\ & \ddots & & \ddots & & \\ & & a_p & \cdots & \cdot & a_0 \\ b_q & \cdots & b_0 & & & \\ & \ddots & & \ddots & & \\ & & & b_q & \cdots & b_0 \end{pmatrix}}^{p+q-j} \begin{matrix} \left.\vphantom{\begin{matrix}a\\a\\a\end{matrix}}\right\} q-j \\ \left.\vphantom{\begin{matrix}a\\a\\a\end{matrix}}\right\} p-j \end{matrix}$$

Note that all the matrices $\mathrm{Sylv}_j(P, Q)$ are extracted from $\mathrm{Sylv}_0(P, Q)$.

When $\deg(P) > \deg(Q)$, the matrix $\mathrm{Sylv}_0(P, Q)$ is the classical *Sylvester matrix* of P and Q.

For every k in $\{0, \ldots, p+q-j-1\}$ let $\mathrm{Sylv}_{j,k}(P, Q)$ be the square matrix of dimension $p+q-2j$ obtained by taking the first $p+q-2j-1$ first columns and the $p+q-j-k$-th column from $\mathrm{Sylv}_j(P, Q)$.

The *subresultant sequence* is the sequence $\mathrm{Sres}_j(P, Q)_{j\in\{0,\ldots,p\}}$ with $p = \sup(\deg(P), \deg(Q)+1)$, $q = \deg(Q)$

- $\mathrm{Sres}_p(P, Q) = P$,
- $\mathrm{Sres}_{p-1}(P, Q) = Q$,
- $\mathrm{Sres}_j(P, Q) = 0$ if $q < j < p-1$,
- $\mathrm{Sres}_q(P, Q) = \mathrm{cf}_q(Q)^{p-q-1}Q$,
- $\mathrm{Sres}_j(P, Q) = \sum_{k=0}^{j} \det(\mathrm{Sylv}_{j,k}(P, Q))X^k$, if $j < q$,

and the *sequence of principal subresultant coefficients* is

- $\mathrm{sr}_p(P, Q) = 1$,
- $\mathrm{sr}_j(P, Q) = \mathrm{cf}_j(\mathrm{Sres}_j(P, Q))$ if $j < p$.

Notice that $\deg(\mathrm{Sres}_j(P, Q)) \leq j$. If $\mathrm{sr}_j(P, Q) = 0$, the subresultant $\mathrm{Sres}_j(P, Q)$ is called *defective*.

Let $\widetilde{\mathrm{Sylv}}_j(P, Q)$ be the square matrix of dimension $p+q-2j$ obtained by taking the first $p+q-2j-1$ first columns from $\mathrm{Sylv}_j(P, Q)$ and putting in the last column $X^{q-j-1}P, \ldots, XP, P, X^{p-j-1}Q, \ldots, XQ, Q$.

Proposition 3.12.

- *For* $j < q$, $\mathrm{Sres}_j(P, Q) = \det(\widetilde{\mathrm{Sylv}}_j(P, Q))$.
- *The subresultants of* P, Q *belong to the ideal generated by* P *and* Q. *More precisely, there exist polynomials* U_j *and* V_j *(defined by means of determinants) of respective degrees less than or equal to* $q-j-1$ *and* $p-j-1$ *such that* $\mathrm{Sres}_j(P, Q) = U_j P + V_j Q$.

Proof. We first remark that

$$\det(\widetilde{\mathrm{Sylv}}_j(P, Q)) = \sum_{k=0}^{p+q-j+1} \det(\mathrm{Sylv}_{j,k}(P, Q))X^k,$$

using the linearity of determinants on the last column. Since for $k > j$ the matrices $\mathrm{Sylv}_{j,k}(P, Q)$ have as last column one of the first $p+q-2j-1$ columns, $\det(\mathrm{Sylv}_{j,k}(P, Q)) = 0$ for $k \in \{j+1, \ldots, p+q-j-1\}$, so that

$$\det(\widetilde{\mathrm{Sylv}}_j(P, Q)) = \sum_{k=0}^{j} \det(\mathrm{Sylv}_{j,k}(P, Q))X^k = \mathrm{Sres}_j(P, Q).$$

For the second part, the fact that the polynomial $\mathrm{Sres}_j(P, Q)$ belongs to the ideal generated by P and Q is an easy consequence of the first property, expanding $\widetilde{\mathrm{Sylv}}_j(P, Q)$ along the last column. Moreover the coefficients with degree k of U_j, $k \in \{0, \ldots, q-j-1\}$ (resp. of V_j, $k \in \{0, \ldots, p-j-1\}$), are just the cofactors of the $q-j-k$-th (resp. $q+p-2j-k$-th) element in the last column. □

When $\deg(P) > \deg(Q)$, the subresultant $\mathrm{Sres}_0(P, Q)$ agrees with the classical *resultant* and by 3.12 belongs to the ideal generated by P and Q.

3.2.2. Properties of subresultants. The properties of subresultants are based on the following proposition which shows a precise connection between subresultants and remainders.

Proposition 3.13. *Let P and Q in* $D[X]$, *p, q and j integers with* $p = \sup(\deg(P), \deg(Q)+1)$, $q = \deg(Q)$ *and* $j < q$. *Denote by R the remainder* $\mathrm{Rem}(P, Q)$ *of the euclidean division of P by Q and by r its degree.*

We have the following equalities:

- $\mathrm{Sres}_{q-1}(P, Q) = (-1)^{p-q+1}\,\mathrm{cf}_q(Q)^{p-q+1}\,\mathrm{Sres}_{q-1}(Q, R)$.
- *If* $r < j < q-1$, $\mathrm{Sres}_j(P, Q) = \mathrm{Sres}_j(Q, R) = 0$.
- *If* $j \le r$, $\mathrm{Sres}_j(P, Q) = (-1)^{(p-j)(q-j)}\,\mathrm{cf}_q(Q)^{p-r}\,\mathrm{Sres}_j(Q, R)$.

Proof. Since R is the remainder of the division of P by Q we have $P = A \cdot Q + R$ with $A = \sum_{m=0}^{p-q} a_m X^m$ (note that the coefficients of highest degrees of A may be 0). Let M_j, $j < q$, be the matrix whose rows on the basis

$$X^{p+q-j-1}, \ldots, X^2, X, 1$$

are the coefficient vectors of the polynomials:

$$QX^{p-j-1}, \ldots, QX, Q, RX^{q-j-1}, \ldots, RX, R$$

and denote by $M_{j,k}$ the square matrix of dimension $p+q-2j$ obtained by taking the first $p+q-2j-1$ first columns and the $p+q-j-k$-th column from M_j.

We prove that

$$\det(M_{j,k}) = (-1)^{(p-j)(q-j)} \det(\mathrm{Sylv}_{j,k}(P, Q)).$$

Every row PX^ℓ in the matrix $\mathrm{Sylv}_j(P, Q)$ can be replaced by the row RX^ℓ by subtracting the sum of the rows $QX^{\ell+m}$ multiplied by a_m from the row PX^ℓ. These elementary operations do not change the determinants extracted from the matrix. Putting all rows of Q first and all rows of R next to obtain the matrices $M_{j,k}$ multiplies the determinants by $(-1)^{(p-j)(q-j)}$. When $j < r$, removing the first $p-r$ rows of Q from M_j gives exactly the matrix $\mathrm{Sylv}_j(Q, R)$, and we obtain Proposition 3.13. When $j = r$, we have $\mathrm{Sres}_r(P, Q) = (-1)^{(p-r)(q-r)}\,\mathrm{cf}_q(Q)^{p-r}\,\mathrm{cf}_r(R)^{q-r-1} R$ and $\mathrm{Sres}_r(Q, R) = \mathrm{cf}_r(R)^{q-r-1} R$ because of conventions. When $r < j < q$, it is clear that $\det(M_{j,k}) = 0$, since $M_{j,k}$ is a triangular matrix with one zero on the diagonal, so that using the definitions

$$\mathrm{Sres}_j(P, Q) = \mathrm{Sres}_j(Q, R) = 0.$$

When $j = q-1$, an easy computation shows that

$$\mathrm{Sres}_{q-1}(P, Q) = (-1)^{p-q+1}\,\mathrm{cf}_q(Q)^{p-q+1} R$$

while by the definitions $\mathrm{Sres}_{q-1}(Q, R) = R$. □

Corollary 3.14. *If* $j \le q$

$$\mathrm{Sres}_j(P, Q) = \epsilon(j)\,\mathrm{cf}_q(Q)^{e(j)}\,\mathrm{Sres}_j(Q, R)$$

$$\mathrm{sr}_j(P, Q) = \epsilon(j)\,\mathrm{cf}_q(Q)^{f(j)}\,\mathrm{sr}_j(Q, R)$$

with

$$\begin{aligned}\epsilon(j) &= (-1)^{(p-j)(q-j)},\\ e(j) &= \begin{cases} p-q-1 & \text{if } j = q,\\ p-j & \text{if } r < j < q,\\ p-r & \text{if } j \le r;\end{cases}\\ f(j) &= \begin{cases} p-q & \text{if } j = q,\\ e(j) & \text{if } j < q.\end{cases}\end{aligned}$$

Proof. It is immediate from 3.13, noticing that $\mathrm{sr}_q(Q, R) = 1$. □

We define the *pseudo-remainder* of P and Q, denoted $\mathrm{Prem}(P, Q)$ as the polynomial $\mathrm{cf}_q(Q)^{p-q+1}\,\mathrm{Rem}(P, Q)$, taking as usual $p = \sup(\deg(P), \deg(Q) + 1)$. When $\deg(P) > \deg(Q)$ this definition coincides with the usual definition of pseudo-remainder. The pseudo-remainder is a multiple of the remainder and has its coefficients in the ring of coefficients of P and Q. We have the relation

$$\mathrm{Prem}(aP, bQ) = ab^{p-q+1}\,\mathrm{Prem}(P, Q)$$

for every a, b in D with $b \neq 0$. Note that at a root x of Q,

$$\mathrm{Prem}(P, Q)(x) = \mathrm{cf}_q(Q)^{p-q+1} P(x).$$

Proposition 3.15.

$$\mathrm{Sres}_{q-1}(P, Q) = (-1)^{p-q+1}\,\mathrm{Prem}(P, Q).$$

Proof. This is nothing but Proposition 3.13 with $j = q - 1$. □

We are now ready for the Subresultant theorem.

Theorem 3.16 (Subresultant theorem [15], [18], [12]). *Let P and Q be two polynomials in* $\mathrm{D}[X]$ *with* $p = \sup(\deg(P), \deg(Q) + 1)$. *For every $j \le p$ we use the following abbreviations:*

$$S_j(P, Q) = \mathrm{Sres}_j(P, Q) \quad s_j(P, Q) = \mathrm{sr}_j(P, Q).$$

The polynomials in the subresultant sequence are either proportional to the polynomials in the remainder sequence or zero. If $s_j(P, Q) \neq 0$, and if $\deg(S_{j-1}(P, Q)) = k \le j - 1$ *then $S_{k-1}(P, Q)$ is proportional to*

$$\mathrm{Rem}(S_j(P, Q), S_{j-1}(P, Q))$$

while $S_{j-1}(P, Q)$ and $S_k(P, Q)$ are proportional (so that $S_k(P, Q)$ is not defective) and $S_\ell(P, Q) = 0$ for $\ell \in \{k+1, \dots, j-2\}$. More precisely, the following relations hold:

1. $s_j(P,Q)^{j-1-k}S_k(P,Q) = \mathrm{cf}_k(S_{j-1}(P,Q))^{j-1-k}S_{j-1}(P,Q)$.

2. *If* $k < \ell < j-1$, *then* $S_\ell(P,Q) = 0$.

3. $(-s_j(P,Q))^{j-k+1}S_{k-1}(P,Q) = \mathrm{Prem}(S_j(P,Q), S_{j-1}(P,Q))$.

Proof. The proof is by induction on r where r is the degree of the remainder $R = \mathrm{Rem}(P,Q)$.

Suppose that $r = 0$. Then replacing all rows of P by rows of R shows that for every $j < q$, $S_j(P,Q) = 0$ which proves the result.

We now suppose that the theorem is true for Q, R and use the induction hypothesis to prove it for P, Q.

For $q < j \le p$, the only case to consider is $j = p$ and $k = q$. Property 1 follows from the convention. Property 2 comes from the definition of $S_j(P,Q)$ for $q \le j < p$, and property 3 is true by the corollary of Proposition 3.13.

If $j \le q$, $\deg(S_j(Q,R)) = j$, and $\deg(S_{j-1}(Q,R)) = k \le j-1$ then the following relations hold by the induction hypothesis:

1. $s_j(Q,R)^{j-1-k}S_k(Q,R) = \mathrm{cf}_k(S_{j-1}(Q,R))^{j-1-k}S_{j-1}(Q,R)$.

2. If $k < \ell < j-1$ then $S_\ell(Q,R) = 0$.

3. $(-s_j(Q,R))^{j-k+1}S_{k-1}(Q,R) = \mathrm{Prem}(S_j(Q,R), S_{j-1}(Q,R))$.

If $j \le q$, then for every $\ell \le j$, using notations of Corollary 3.14

$$S_\ell(P,Q) = \epsilon(\ell)\,\mathrm{cf}_q(Q)^{e(\ell)}S_\ell(Q,R)$$

and the second property is immediate by Corollary 3.14 for $j < r$ and by the conventions for $r \le j \le q$. Also using several times Corollary 3.14 and conventions we have easily

$$\begin{aligned} s_j(P,Q) &= \epsilon(j)\,\mathrm{cf}_q(Q)^{f(j)}s_j(Q,R) \\ \mathrm{cf}_k(S_{j-1}(P,Q)) &= \epsilon(j-1)\,\mathrm{cf}_q(Q)^{e(j-1)}\,\mathrm{cf}_k(S_{j-1}(Q,R)) \\ \mathrm{Prem}(S_j(P,Q), S_{j-1}(P,Q)) &= \epsilon c\,\mathrm{Prem}(S_j(Q,R), S_{j-1}(Q,R)) \end{aligned}$$

with $\epsilon = \epsilon(j).(\epsilon(j-1))^{j-k+1}$, $c = \mathrm{cf}_q(Q)^{e(j)+(j-k+1)e(j-1)}$.

Notice that $\epsilon(j)^{j-1-k}.\epsilon(k) = \epsilon(j-1)^{j-k}$ for every $k < j \le q$. Notice also that $(j-1-k)f(j) + e(k) = (j-k)e(j-1)$: this is immediate from the definition of $e(j)$ when $j \le r$. When $j = q, k = r$, which is the only other case to consider, we have $(q-1-r)(p-q) + p - r = (q-r)(p-q+1)$. So from

$$s_j(Q,R)^{j-1-k}S_k(Q,R) = \mathrm{cf}_k(S_{j-1}(Q,R))^{j-1-k}S_{j-1}(Q,R)$$

one can deduce, multiplying both sides by $\epsilon(j-1)^{j-k}\mathrm{cf}_q(Q)^{(j-k)e(j-1)}$ that

$$s_j(P,Q)^{j-1-k}S_k(Q,R) = \mathrm{cf}_k(S_{j-1}(P,Q))^{j-1-k}S_{j-1}(P,Q).$$

We thus proved the first property.

Notice now that for every $k < j < q$ $\epsilon(j)^{j-k+1}.\epsilon(k-1) = \epsilon(j)\epsilon(j-1)^{j-k+1}$. Notice also that $f(j)(j-k+1) + e(k-1) = e(j) + (j-k+1)e(j-1)$: this is immediate when $j \leq r$, and when $q = j, r = k$, it is also true since $f(q)(q-r+1) + e(r-1) = (p-q)(q-r+1) + (p-r)$ and $e(q) + (q-r+1)e(q-1) = p-q-1+(q-r+1)(p-q+1)$. So from

$$(-s_j(Q, R))^{j-k+1} S_{k-1}(Q, R) = \mathrm{Prem}(S_j(Q, R), S_{j-1}(Q, R))$$

one can deduce multiplying both sides by ϵc that

$$(-s_j(P, Q))^{j-k+1} S_{k-1}(P, Q) = \mathrm{Prem}(S_j(P, Q), S_{j-1}(P, Q))$$

which is the third property we want. □

3.3. Sylvester–Habicht sequence

In Paragraph 3.1 we have defined the Sylvester sequence of P and Q. Using subresultants instead of remainders we are going to define the Sylvester–Habicht sequence: the formal analogue of the Sylvester sequence so that the sign variations (correctly defined) in the Sylvester–Habicht sequence gives – as the Sylvester sequence does – the Cauchy index.

3.3.1. Properties of the Sylvester–Habicht sequence. Let P and Q be polynomials and $p = \sup(\deg(P), \deg(Q)+1)$. The *Sylvester–Habicht sequence* associated to P and Q is the sequence of polynomials:

$$\mathrm{SyHa}_j(P, Q) = \delta_{p-j}\,\mathrm{Sres}_j(P, Q)$$

and $\delta_k = (-1)^{(k(k-1)/2)}$. The *$j$-th principal Sylvester–Habicht coefficient* of P and Q is $\mathrm{syha}_j(P, Q) := \mathrm{cf}_j(\mathrm{SyHa}_j(P, Q))$ for $j < p$ and by convention $\mathrm{syha}_p(P, Q) = 1$.

This means that it is the sequence of subresultants of P and Q modified by multiplying the first two polynomials by $+1$, the next two by -1 and so on (not taking into account the fact that some subresultants are defective or zero).

When $Q = P'$, we denote by $\mathrm{StHa}(P) = \mathrm{SyHa}(P, P')$ the *Sturm–Habicht sequence* of P.

The definition of the Sylvester–Habicht sequence through subresultants allows us to give a structure theorem for this new sequence. This theorem is used to give an algorithm for computing the Sylvester–Habicht sequence and, most important, to prove that the Sylvester–Habicht sequence can be used for real root counting as was the Sylvester sequence. But the Sylvester–Habicht sequence has nicer properties than the Sylvester sequence: nice specialization properties, good worst-case complexity bounds, etc. The next section will be devoted to the proof of these results using only the Sylvester–Habicht structure theorem.

Theorem 3.17 (Sylvester–Habicht structure theorem). *Let P, Q be polynomials in* $D[X]$ *with* $p = \sup(\deg(P), \deg(Q) + 1)$. *For every j in $\{0, \dots, p\}$ we make the following abbreviations:*

$$H_j = \mathrm{SyHa}_j(P, Q) \quad h_j = \mathrm{syha}_j(P, Q)$$

and $h_p = 1$. Then for every $j \in \{1, \dots, p\}$ such that $h_j \neq 0$ and $\deg(H_{j-1}) = k \leq j - 1$ *we have*

1. $(h_j)^{j-1-k} H_k = \delta_{j-k} \mathrm{cf}_k(H_{j-1})^{j-1-k} H_{j-1}$,
2. *if $k < \ell < j - 2$, then $H_\ell = 0$,*
3. $(h_j)^{j-k+1} H_{k-1} = \delta_{j-k+2} \,\mathrm{Prem}(H_j, H_{j-1})$.

Proof. It is straightforward using the definition of $\mathrm{SyHa}_j(P, Q)$, Subresultant theorem 3.16, paying attention to the signs introduced by the δ's and noticing that $(\delta_{p-j})^{j-1-k} \cdot \delta_{p-k} = \delta_{j-k} \cdot (\delta_{p-j+1})^{j-k}$ and $(-\delta_{p-j})^{j-k+1} \cdot \delta_{p-k+1} = \delta_{j-k+2} \cdot \delta_{p-j} \cdot (\delta_{p-j+1})^{j-k+1}$. □

3.3.2. Sylvester–Habicht sequence and Cauchy index. The principal aim of this section is to show how to use the polynomials in the Sylvester–Habicht sequence of P and Q to compute the Cauchy index. The next definitions introduce new sign counting functions that will be useful.

Definition 3. Let $\mathcal{P} = [P_0, P_1, \dots, P_n]$ be a sequence of polynomials and a be an element of $R \cup \{-\infty, +\infty\}$. Then $W(\mathcal{P}; a)$, the *modified number of sign changes* of $\mathcal{P}$ at a, is the number defined as follows:

1. Delete from $\mathcal{P}$ those polynomials which are identically 0 to obtain the list of polynomials $[Q_0, \dots, Q_s]$ in $D[X]$,
2. Define $W(\mathcal{P}; a)$ as the number of sign changes in the list $[Q_0(a), \dots, Q_s(a)]$, the usual definition being modified only for groups of two zeroes as follows:
   - count 1 sign variation for the groups: $[+, 0, 0, -]$ and $[-, 0, 0, +]$,
   - count 2 sign variations for the groups: $[+, 0, 0, +]$ and $[-, 0, 0, -]$.

We denote by $W_{\mathrm{SyHa}}(P, Q; a) = W(\mathrm{SyHa}(P, Q); a)$ the modified number of sign changes of the Sylvester–Habicht sequence at a.

Note that the *Sylvester–Habicht structure theorem* implies that if a is not a root of P, it is not possible to have more than two consecutive zeroes in the values at a of the non zero polynomials of the Sylvester–Habicht sequence.

Note also that $W_{\mathrm{SyHa}}(P, Q; +\infty)$ (resp. $W_{\mathrm{SyHa}}(P, Q; -\infty)$) coincides with the number of sign variations in the $\{\mathrm{SyHa}_j(P, Q)\}_{j=0,\dots,p}$ evaluated at $+\infty$ (resp. $-\infty$).

Definition 4. Let P and Q be polynomials in $\mathrm{D}[X]$ and a, b in $\mathrm{R} \cup \{-\infty, +\infty\}$ with $a < b$. We define

$$W_{\mathrm{SyHa}}(P, Q;]a, b[) = W_{\mathrm{SyHa}}(P, Q; a) - W_{\mathrm{SyHa}}(P, Q; b).$$

Theorem 3.18. *Let P and Q be polynomials in $\mathrm{D}[X]$ and a, b in $\mathrm{R} \cup \{-\infty, +\infty\}$ with $a < b$ and $P(a)P(b) \neq 0$. Then:*

$$W_{\mathrm{SyHa}}(P, Q;]a, b[) = I(Q/P;]a, b[).$$

Proof. Let $p = \sup(\deg(P), \deg(Q) + 1)$, $q = \deg(Q)$. If $c_1 < \cdots < c_r$ are the real roots in $]a, b]$ of the non identically zero polynomials

$$H_j = \mathrm{SyHa}_j(P, Q) \quad j \in \{0, \ldots, p\},$$

with $h_j = \mathrm{syha}_j(P, Q)$, we write $a = c_0 < c_1 < \cdots < c_r < c_{r+1} = b$. For every $i \in \{1, \ldots, r\}$ choose an element d_i between c_i and c_{i+1}. If $c_0 = a$ is a root of some $\mathrm{Sy}^j(P, Q)$, choose an element d_0 between c_0 and c_1, otherwise take $d_0 = c_0 = a$. If $c_{r+1} = b$ is a root of some $\mathrm{Sy}^j(P, Q)$, choose an element d_r between c_r and c_{r+1}, otherwise take $d_r = c_{r+1} = b$.

We have the following equalities:

$$\begin{aligned} W_{\mathrm{SyHa}}(P, Q;]a, b[) &= W_{\mathrm{SyHa}}(P, Q; a) - W_{\mathrm{SyHa}}(P, Q; b) \\ &= W_{\mathrm{SyHa}}(P, Q; c_0) - W_{\mathrm{SyHa}}(P, Q; d_0) \\ &\quad + \sum_{i=1}^{r} \Big[W_{\mathrm{SyHa}}(P, Q; d_{i-1}) - W_{\mathrm{SyHa}}(P, Q; d_i) \Big] \\ &\quad + W_{\mathrm{SyHa}}(P, Q; d_r) - W_{\mathrm{SyHa}}(P, Q; c_{r+1}) \end{aligned}$$

which reduces the proof of the theorem to the study of the integers

$$W_{\mathrm{SyHa}}(P, Q; d_{i-1}) - W_{\mathrm{SyHa}}(P, Q; d_i),\ W_{\mathrm{SyHa}}(P, Q; c_0) - W_{\mathrm{SyHa}}(P, Q; d_0)$$
$$\text{and} \quad W_{\mathrm{SyHa}}(P, Q; d_r) - W_{\mathrm{SyHa}}(P, Q; c_{r+1})$$

for every $i \in \{1, \ldots, r\}$.

We remark that those polynomials in the list $[H_j : 0 \leq j \leq p]$ which are not identically 0 do not vanish at any d_i and that H_ℓ, the last polynomial which is not identically 0 in the Sylvester–Habicht sequence, is a greatest common divisor of P and Q. So that, if x is not a root of P, $H_\ell(x) \neq 0$. The proof of the theorem is obtained by proving the following lemmas.

Lemma 3.19. *If $P(c) \neq 0$ and $H_j(c) = 0$, with $\deg(H_j) = j$ and d is such that there is no root of any of the H_k on $]c, d[$ (resp. $]d, c[$), then*

$$\mathrm{sign}(H_{j-1}(d)H_{j+1}(d)) = -1$$

so that

$$W([H_{j+1}, H_j, H_{j-1}]; c) = W([H_{j+1}, H_j, H_{j-1}]; d) = 1 \text{ if } i = 0, \ldots, r.$$

Proof. Since $H_j(c) = 0$ and $H_\ell(c) \neq 0$, we have $H_{j+1}(c) \neq 0$. So applying identity 3 in Theorem 3.17 and using properties of pseudo-remainders we get:

$$\begin{aligned} &\mathrm{cf}_{j+1}(H_{j+1})^2 H_{j-1} = \delta_3 \operatorname{Prem}(H_{j+1}, H_j) \\ \Longrightarrow\ &\mathrm{cf}_{j+1}(H_{j+1})^2 H_{j-1}(c) = -\,\mathrm{cf}_j(H_j)^2 H_{j+1}(c) \\ \Longrightarrow\ &\operatorname{sign}(H_{j-1}(c) H_{j+1}(c)) = -1 \end{aligned}$$

which implies, whatever is the sign of $H_j(d)$:

$$W([H_{j+1}, H_j, H_{j-1}]; c) = W([H_{j+1}, H_j, H_{j-1}]; d) = 1 \ \text{ if } i = 0, \ldots, r$$

□

Lemma 3.20. *If* $P(c) \neq 0$ *and* $H_j(c) = 0$, *with* $\deg(H_j) = k < j$, *and* d *is such that there is no root of any of the* H_k *on* $]c, d[$ *(resp.* $]d, c[$*), then*

$$\operatorname{sign}(H_{k-1}(d) H_k(d) H_j(d) H_{j+1}(d)) = -1$$

so that

$$W([H_{j+1}, H_j, H_k, H_{k-1}]; c) = W([H_{j+1}, H_k, H_{k-1}]; d) = 2 \ \text{ if } i = 0, \ldots, r$$

when $H_{j+1}(c) H_{k-1}(c) > 0$,

$$W([H_{j+1}, H_j, H_k, H_{k-1}]; c) = W([H_{j+1}, H_k, H_{k-1}]; d) = 1 \ \text{ if } i = 0, \ldots, r$$

when $H_{j+1}(c) H_{k-1}(c) < 0$.

Proof. Again since $H_\ell(c) \neq 0$, $H_{j+1}(c) \neq 0$. Applying identity 3 in Theorem 3.17 and the properties of pseudo-remainders, we get

$$\mathrm{cf}_{j+1}(H_{j+1})^{j-k+2} H_{k-1}(c) = \delta_{j-k+3}\, \mathrm{cf}_k(H_j)^{j-k+2} H_{j+1}(c),$$

so that $H_{k-1}(c) \neq 0$ and using identity 1 in Theorem 3.17 we see that

$$\operatorname{sign}(H_{k-1}(c) H_{j+1}(d)) = \frac{\delta_{j-k+3}}{\delta_{j-k+1}} \operatorname{sign}(H_k(d) H_j(d)).$$

Hence

$$\operatorname{sign}(H_{k-1}(d) H_k(d) H_j(d) H_{j+1}(d)) = -1$$

which implies looking at all possible cases that

$$W([H_{j+1}, H_j, H_k, H_{k-1}]; d) = \begin{cases} 2 & \text{if } H_{k-1}(c) H_{j+1}(c) > 0, \\ 1 & \text{if } H_{k-1}(c) H_{j+1}(c) < 0, \end{cases}$$

so that in all cases

$$W([H_{j+1}, H_j, H_k, H_{k-1}]; d) = W([H_{j+1}, H_j, H_k, H_{k-1}]; c).$$

□

Lemma 3.21. *If* $i \in \{1, \ldots, r\}$ *and* $P(c_i) \neq 0$ *then* $W_{\text{SyHa}}(P, Q; d_{i-1}) - W_{\text{SyHa}}(P, Q; d_i) = 0$.

Proof. We note first that since $P(c_i) \neq 0$, the signs of H_p, H_ℓ and H_k, with $H_k(c_i) \neq 0$, in d_{i-1} and d_i coincide. So, we need only know the behaviour of the polynomials which are not identically 0 in the sequence when c_i is a root of some H_j.

The only two possibilities are the following:

(i) $j \leq p-1$ with $\deg(H_j) = j$, $\deg(H_{j+1}) = j+1$ and $H_j(c_i) = 0$: according to Lemma 3.19,

$$\begin{aligned} W([H_{j+1}, H_j, H_{j-1}]; c_i) &= W([H_{j+1}, H_j, H_{j-1}]; d_{i-1}) \\ &= W([H_{j+1}, H_j, H_{j-1}]; d_i), \end{aligned}$$

(ii) $j \leq p-1$ with $k = \deg(H_j) < j$, $\deg(H_{j+1}) = j+1$ and $H_j(c_i) = 0$: in this case according to Lemma 3.20,

$$\begin{aligned} W([H_{j+1}, H_j, H_k, H_{k-1}]; c_i) &= W([H_{j+1}, H_k, H_{k-1}]; d_i) \\ &= W([H_{j+1}, H_j, H_k, H_{k-1}]; d_{i-1}). \end{aligned}$$

Thus we conclude that if c_i satisfies $P(c_i) \neq 0$, it follows that $W_{\text{SyHa}}(P, Q; d_{i-1}) - W_{\text{SyHa}}(P, Q; d_i) = 0$ which is what we wanted to show. □

Lemma 3.22.

$$W_{\text{SyHa}}(P, Q; c_0) - W_{\text{SyHa}}(P, Q; d_0) = W_{\text{SyHa}}(P, Q; d_r) - W_{\text{SyHa}}(P, Q; c_{r+1}) = 0$$

Proof. Clearly it is enough to study the case corresponding to c_0. If $c_0 = d_0$, there is nothing to prove. Given that $P(c_0) \neq 0$, it is enough to consider the two following cases:

(i) $j \leq p-1$ with $\deg(H_j) = j$, $\deg(H_{j+1}) = j+1$ and $H_j(c_0) = 0$: in this case according to Lemma 3.19,

$$W([H_{j+1}, H_j, H_{j-1}]; c_0) = W([H_{j+1}, H_j, H_{j-1}]; d_0),$$

(ii) $j \leq p-1$ with $\deg(H_j) = k < j$, $\deg(H_{j+1}) = j+1$ and $H_j(c_0) = 0$: according to Lemma 3.20, we have

$$W([H_{j+1}, H_j, H_k, H_{k-1}]; c_0) = W([H_{j+1}, H_j, H_k, H_{k-1}]; d_0).$$

The two cases (i) and (ii) allow us to conclude that:

$$W_{\text{SyHa}}(P, Q; c_0) - W_{\text{SyHa}}(P, Q; d_0) = 0$$

as we wanted to show. □

Lemma 3.23. *If* $i \in \{1, \ldots, r\}$ *and* $P(c_i) = 0$ *then*

$$W_{\text{SyHa}}(P, Q; d_{i-1}) - W_{\text{SyHa}}(P, Q; d_i) = I(Q/P;]d_{i-1}, d_i[).$$

Proof. Let $p = \sup(\deg(P), \deg(Q) + 1)$. Let ℓ be such that $H_\ell \neq 0$, $H_j = 0$ for every $j < \ell$. We define a new sequence $\mathcal{G} = [G_{p-\ell}, \ldots, G_0]$ of polynomials: let H be the monic polynomial proportional to H_ℓ and define

$$G_j = H_{j+\ell}/H \qquad j \in \{0, \ldots, p - \ell\}.$$

The first thing to observe is that the Cauchy index of $G_{p-\ell}/G_{(p-\ell)-1}$ coincides with the Cauchy index of Q/P. Clearly we also have $W([G_{p-\ell}, \ldots, G_0]; d) = W([H_p, \ldots, H_\ell]; d)$ with $d \in \{d_i, d_{i-1}\}$.

For every j in $\{0, \ldots, p - \ell\}$ we define

$$g_j = h_{j+\ell}.$$

Then for every $j \in \{0, \ldots, p - \ell\}$ such that $g_j \neq 0$ and $\deg(G_{j-1}) = k \leq j - 1$ the sequence $\mathcal{G}$ has the following properties:

1. $(g_j)^{j-1-k} G_k = \delta_{j-k} \mathrm{cf}_k(G_{j-1})^{j-1-k} G_{j-1}$,
2. if $k < \ell < j - 2$, then $G_\ell = 0$,
3. $(g_j)^{j-k+1} G_{k-1} = \delta_{j-k+2} \mathrm{Prem}(G_j, G_{j-1})$,

as an easy consequence of Theorem 3.17, given the definition of $\mathcal{G}$.

We can write

$$P(X) = (X - c_i)^e p(X) \qquad p(c_i) \neq 0$$
$$Q(X) = (X - c_i)^f q(X) \qquad q(c_i) \neq 0$$

and we only need to study two cases:

(i) $f \geq e$. In this case $G_{p-\ell}(c_i) \neq 0$ and we can proceed as in Lemma 3.21 using the properties of $\mathcal{G}$.

(ii) $f < e$. In this case $G_{p-\ell}(c_i) = 0$ and $G_{(p-\ell)-1}(c_i) \neq 0$ Proceeding as in Lemma 3.21 and using the properties of $\mathcal{G}$, we conclude that

$$W([G_{(p-\ell)-1}, \ldots, G_0]; d_{i-1}) - W([G_{(p-\ell)l-1}, \ldots, G_0]; d_i) = 0.$$

So we only need to study $W([G_{p-\ell}, G_{(p-\ell)-1}]; d_{i-1}) - W([G_{p-\ell}, G_{(p-\ell)-1}]; d_i)$ when $G_{(p-\ell)-1}(c_i) \neq 0$:

$$\begin{aligned} G_{p-\ell}(X) &= (X - c_i)^{e-f} \underline{p}(X) & \underline{p}(c_i) &\neq 0 \\ G_{(p-\ell)-1}(X) &= \underline{q}(X) & \underline{q}(c_i) &\neq 0 \end{aligned}$$

which gives looking at all possible cases:

$$\begin{aligned}W([G_{p-\ell}, G_{(p-\ell)-1}]; d_{i-1}) \ - \ & W([G_{p-\ell}, G_{(p-\ell)-1}]; d_i) \\ &= \begin{cases} 0 & \text{if } e-f \text{ is even} \\ 1 & \text{if } e-f \text{ is odd and } p(c_i)q(c_i) > 0 \\ -1 & \text{if } e-f \text{ is odd and } p(c_i)q(c_i) < 0 \end{cases} \\ &= I(Q/P;]d_{i-1}, d_i[)\end{aligned}$$

□

This ends the proof of the theorem. □

Corollary 3.24. *Let P and Q be polynomials in* D$[X]$ *and a, b in* R $\cup \{-\infty, +\infty\}$ *with $a < b$ and $P(a)P(b) \neq 0$. Then:*

$$W_{\mathrm{SyHa}}(P, P'Q;]a, b[) = d(P, Q;]a, b[)$$

where $d(P, Q;]a, b[) = c(P, Q > 0;]a, b[) - c(P, Q < 0;]a, b[)$.

Remark 3.25. Exactly as for Sylvester's sequence, when a or b are roots of P which are not also roots of Q, it is still possible to know from $W_{\mathrm{SyHa}}(P, Q;]a, b[)$ and the sign of $Q(a)$ and of $Q(b)$ the number $I(Q/P;]a, b[)$. Define

$$\epsilon(a) = \begin{cases} 1 & \text{if } Q(a)P^m(a) < 0,\ P'(a) = \cdots = P^{m-1}(a) = 0, \text{ and } m \text{ odd} \\ 0 & \text{otherwise} \end{cases}$$

$$\epsilon(b) = \begin{cases} 1 & \text{if } Q(b)P^m(b) > 0,\ P'(b) = \cdots = P^{m-1}(b) = 0, \text{ and } m \text{ odd} \\ 0 & \text{otherwise} \end{cases}$$

Then it is easy to see that $W_{\mathrm{SyHa}}(P, Q;]a, b[) + \epsilon(a) + \epsilon(b)$ coincides with $W_{\mathrm{SyHa}}(P, Q; c_0^+, c_r^+)$ and is thus equal to $I(Q/P;]c_0^+, c_r^+[) = I(Q/P;]a, b[)$.

3.3.3. Computing the Sylvester–Habicht sequence. The following algorithm is an immediate consequence of the Sylvester–Habicht structure theorem.

Algorithm SyHa

Input: The polynomials P and Q.

Output: The Sylvester–Habicht sequence

$$H_j = \mathrm{SyHa}_j(P, Q)$$

$(0 \leq j \leq \sup(\deg(P), \deg(Q) + 1))$.

Initialization:

$$p := \sup(\deg(P), \deg(Q) + 1),\ H_p := P,\ h_p := 1,\ H_{p-1} := Q,\ j := p$$

Next Step: H_j, h_j and H_{j-1} are already known with H_j and h_j non zero and $k = \deg(H_{j-1})$. The lacking H_ℓ are going to be computed up to H_{k-1}.

1. if $k < j-1$:

$$H_k = \delta_{j-k} \frac{\mathrm{cf}_k(H_{j-1})^{j-1-k} H_{j-1}}{h_j^{j-1-k}}$$

2. if $k < j-2$ and $k < \ell < j-1$:

$$H_\ell = 0$$

3.

$$H_{k-1} = \delta_{j-k+2} \frac{\mathrm{Prem}(H_j, H_{j-1})}{h_j^{j-k+1}}$$

4. $h_k = \mathrm{cf}_k(H_k)$ and $j = k$.

End: The algorithm ends when one has computed H_0 that is when $j \leq 1$.

We consider again the numerical example introduced at the end of Subsection 3.1.

$$P := 9X^{13} - 18X^{11} - 33X^{10} + 102X^8 + 7X^7 - 36X^6 \\ -122X^5 + 49X^4 + 93X^3 - 42X^2 - 18X + 9$$

The Sturm–Habicht sequence of P is

$$\mathrm{StHa}_0(P) = \mathrm{StHa}_1(P) = \mathrm{StHa}_2(P) = \mathrm{StHa}_3(P) = \mathrm{StHa}_4(P) = 0$$
$$\mathrm{StHa}_5(P) = -550392371009120750400(3X^5 - 7X^2 + 3)$$
$$\mathrm{StHa}_6(P) = -12397455648(12232018869X^6 - 8633929833X^5 - 28541377361X^3 \\ + 20145836277X^2 + 12232018869X - 8633929833)$$
$$\mathrm{StHa}_7(P) = -1377495072(1584012126X^7 - 2548299819X^6 + 984706749X^5 \\ - 3696028294X^4 + 5946032911X^3 - 713636955X^2 - 2548299819X + 984706749)$$
$$\mathrm{StHa}_8(P) = -38263752(43475160X^8 - 57842286X^7 + 5258589X^6 - 92294719X^5 \\ + 134965334X^4 + 31205119X^3 - 79186035X^2 + 5258589X + 9147321)$$
$$\mathrm{StHa}_9(P) = -1062882(626814X^9 - 1077918X^8 + 71130X^7 - 830472X^6 \\ + 2259119X^5 + 460844X^4 - 2552804X^3 + 668517X^2 + 632094X - 256023)$$
$$\mathrm{StHa}_{10}(P) = -6561(10989X^{10} + 21240X^9 - 70746X^8 - 6054X^7 - 13932X^6 \\ + 159044X^5 - 24463X^4 - 153878X^3 + 59298X^2 + 35628X - 17019)$$
$$\mathrm{StHa}_{11}(P) = 1053(36X^{11} + 99X^{10} - 510X^8 - 42X^7 + 252X^6 976X^5 - 441X^4 \\ - 930X^3 + 462X^2 + 216X - 117)$$
$$\mathrm{StHa}_{12}(P) = P'$$
$$\mathrm{StHa}_{13}(P) = P.$$

Given a specialisation, i.e., a ring morphism $f\colon \mathrm{D} \to \mathrm{D}'$, we are interested in computing easily the Sylvester–Habicht sequence of $f(P)$ and $f(Q)$ when the Sylvester–

Habicht sequence of P and Q is known. The reason why we are interested in the behavior of Sylvester–Habicht sequence under specialization it is that arithmetic operations could be easier in $\mathrm{D}[X]$ than in $\mathrm{D}'[X]$ (in particular the exact divisions appearing in the algorithm SyHa).

Proposition 3.26. *Let* $f\colon \mathrm{D} \to \mathrm{D}'$ *be a ring homomorphim such that* $p = \sup(\deg(P), \deg(Q)+1)$, $q = \deg(Q) = \deg(f(Q))$, *then for all* $j \leq p$

$$\mathrm{SyHa}_j(f(P), f(Q)) = f(\mathrm{SyHa}_j(P, Q)).$$

Proof. Clear from the definition of subresultants as determinants and of the Sylvester–Habicht sequence. □

A more detailed study of the specialization properties of the Sylvester–Habicht sequence can be found in [13].

We consider again the general polynomial of degree 4:

$$P = X^4 + pX^2 + qX + r.$$

The Sturm–Habicht sequence of P is formed by the polynomials (belonging to $\mathbb{Z}[p,q,r][X]$):

$$\begin{aligned}
\mathrm{StHa}_4(P) &= X^4 + pX^2 + qX + r \\
\mathrm{StHa}_3(P) &= 4X^3 + 2pX + q \\
\mathrm{StHa}_2(P) &= -4(2pX^2 + 3qX + 4r) \\
\mathrm{StHa}_1(P) &= -4((2p^3 - 8pr + 9q^2)X + p^2q + 12qr) \\
\mathrm{StHa}_0(P) &= 16p^4r - 4p^3q^2 - 128p^2r^2 + 144pq^2r - 27q^4 + 256r^3
\end{aligned}$$

which agrees, up to squares in $\mathbb{Q}(p,q,r)$, with the generic Sturm sequence for P (see Example 3.10).

If $p = 0$, the Sturm–Habicht sequence of the polynomial $P = X^4 + qX + r$ is

$$\begin{aligned}
\mathrm{StHa}_4(P) &= X^4 + qX + r \\
\mathrm{StHa}_3(P) &= 4X^3 + q \\
\mathrm{StHa}_2(P) &= -4(3qX + 4r) \\
\mathrm{StHa}_1(P) &= -12q(3qX + 4r) \\
\mathrm{StHa}_0(P) &= -27q^4 + 256r^3
\end{aligned}$$

which is the specialisation of the Sturm–Habicht sequence of P with $p = 0$. Comparing with Example 3.10 we observe that the polynomials in the Sturm–Habicht sequence are multiple of polynomials of the Sturm sequence, with some sign changes and repetitions.

3.4. Quadratic forms, Hankel matrices and real roots

We have already studied in Sections 3.1 and 3.3 Sturm and Sturm–Habicht methods solving the real root counting problem. Another method dealing with the same problem, due to Hermite, uses the signature of a quadratic form. In this section we shall describe Hermite's method and the relations between these methods, the Sylvester–Habicht sequence being the concept which will allow to connect them.

In all this Section 3.4 the polynomial P will be assumed monic.

3.4.1. Hermite's quadratic form. Let D be a domain and K its fraction field. We consider two polynomials P and Q in $\mathrm{D}[X]$, with P monic, of degree p:

$$P = X^p + a_{p-1}X^{p-1} + \cdots + a_1X + a_0,$$
$$Q = b_qX^q + b_{q-1}X^{q-1} + \cdots + b_1X + b_0.$$

If $x_1, \ldots, x_p$ are the roots of P in an algebraic closure C of K, counted with multiplicities, then we define the quadratic form $B(P, Q)$ of the p variables $f_0, \ldots, f_{p-1}$ in the following way:

$$B(P, Q) = \sum_{i=1}^{p} Q(x_i)(f_0 + f_1x_i + \cdots + f_{p-1}x_i^{p-1})^2.$$

The expression of $B(P, Q)$ is symetric in the x_i's and since P is monic then the quadratic form $B(P, Q)$ has coefficients in D.

If $\leq$ is an order on K and R is the real closure of K for the order $\leq$, the quadratic form $B(P, Q)$ has a signature in the field R (this signature depends on the order $\leq$ considered on K). We consider $\mathrm{C} = \mathrm{R}[i]$ (with $i^2 = -1$). The roots of P in R will be called real roots and the roots in $\mathrm{C} \setminus \mathrm{R}$, complex roots.

Theorem 3.27 (Hermite's method [16]). *With the above notations:*
— *the rank of $B(P, Q)$ is the number of roots x in C of P such that $Q(x) \neq 0$,*
— *the signature of $B(P, Q)$ is equal to $d(P, Q) = c(P, Q, > 0) - c(P, Q, < 0)$.*

Proof. Let $y_1, \ldots, y_r$ be the distinct real roots of P, $m_1, \ldots, m_r$ their multiplicities, $z_1, \overline{z_1}, \ldots, z_t, \overline{z_t}$ the complex distinct roots of P and $n_1, \ldots, n_t$ their multiplicities.

For $x \in \mathrm{C}$, let L be the linear form on C^n defined by

$$L(x, f) = f_0 + f_1x + \cdots + f_{p-1}x^{p-1},$$

and $b(x, f) = L(x, f)^2$. The quadratic form $B(P, Q)$ is equal to

$$B(P, Q) = \sum_{j=1}^{n} m_j Q(y_j)b(y_j, f) + \sum_{h=1}^{t} n_h(Q(z_h)b(z_h, f) + Q(\overline{z_h})b(\overline{z_h}, f)).$$

The linear forms $L(y_j, f)$, $L(z_h, f)$ and $L(\overline{z_h}, f)$ are linearly independent (the roots are distinct and it is sufficient to consider a Vandermonde determinant). This gives the proof for (i). Writing $Q(z_h) = {d_h}^2$ and decomposing $d_h b(z_h, f)$ under the form $P_h + iQ_h$ with P_h and Q_h real linear forms, it is clear that $Q(z_h)b(z_h, f) + Q(\overline{z_h})b(\overline{z_h}, f)$ is the difference of two squares of real linear forms.

So the signature of $B(P, Q)$ is equal to $d(P, Q) = c(P, Q > 0) - c(P, Q < 0)$.□

Corollary 3.28. *The signature of $B(P)$ is equal to the number of real roots of P.*

We are now going to relate the coefficients of $B(P, Q)$ with the rational function $P'Q/P$, which gives a method for computing $B(P, Q)$. We have

$$B(P, Q) = \sum_{k=0}^{p-1}\sum_{j=0}^{p-1}\sum_{i=1}^{p} Q(x_i)x_i^{k+j} f_k f_j.$$

When $Q = 1$, denoting by $B(P)$ the quadratic form $B(P, 1)$ we get

$$B(P) = \sum_{i=1}^{p}(f_0 + f_1 x_i + \cdots + f_{p-1}{x_i}^{p-1})^2,$$

so that

$$B(P) = \sum_{k=0}^{p-1}\sum_{j=0}^{p-1}\sum_{i=1}^{p} {x_i}^{k+j} f_k f_j.$$

The formula

$$P = \prod_{i=1}^{p}(X - x_i)$$

allows to write

$$P'/P = \sum_{i=1}^{p} 1/(X - x_i).$$

Considering the euclidean division of Q by $X - x_i$ and taking A_i as the quotient

$$P'Q = Q(x_i) + (X - x_i)A_i$$

we derive the following expression for the rational function $P'Q/P$:

$$P'Q/P = \sum_{i=1}^{p} A_i + \sum_{i=1}^{p} Q(x_i)/(X - x_i).$$

So, the coefficient of $1/X^{j+1}$ in the development of $P'Q/P$ in powers of $1/X$ is hence, with the preceding notations, $\sum_{i=1}^{p} Q(x_i)x_i^j$.

We are going to associate now to any rational function Q/P a quadratic form $C(P, Q)$ generalizing Hermite's quadratic form whose signature will be the Cauchy index of Q/P. Precisely we shall have $C(P, P'Q) = B(P, Q)$

Let $Q/P = \sum_{i=i_0}^{\infty} c_i/X^{i+1}$ (with $i_0 \in \mathbb{Z}$) and define

$$C(P, Q) = \sum_{k=0}^{p-1}\sum_{j=0}^{p-1} c_{k+j} f_k f_j.$$

It is clear that $C(P, P'Q) = B(P, Q)$ (with the convention $c_j = 0$ if $0 \leq j < i_0$).

Proposition 3.29. *The signature of $C(P, Q)$ is the Cauchy index of Q/P between $-\infty$ and $+\infty$.*

Proof. Consider the quotient ring $A = K[X]/P$, with linear basis $1, X, \ldots, X^{p-1}$, and the linear form ℓ defined over A by

$$\ell(1) = \cdots = \ell(X^{p-2}) = 0, \; \ell(X^{p-1}) = 1.$$

Lemma 3.30. *For every rational function $Q/P = \sum_{i=0}^{\infty} c_i/X^{i+1}$ we have $\ell(QX^i) = c_i$, where QX^i is identified with its remainder modulo P.*

Proof. The formula

$$Q/P = \sum_{i=i_0}^{\infty} c_i/X^{i+1}$$

gives, by identification, the following relations ($\star$) between the coefficients of $P = X^p + a_{p-1}X^{p-1} + \cdots + a_0$, the coefficients of $Q = b_{p-1}X^{p-1} + \cdots + b_0$ (b_{p-1} being perhaps zero) and the c_i, $i \geq 0$:

$$\begin{aligned} c_0 &= b_{p-1} \\ c_1 + a_{p-1}c_0 &= b_{p-2} \\ c_2 + a_{p-1}c_1 + a_{p-2}c_0 &= b_{p-3} \\ &\vdots \\ c_{p-1} + \cdots + a_1c_0 &= b_0 \\ c_{n-1} + \cdots + a_1c_{n-p} &= 0 \quad \text{for } n > p. \end{aligned} \qquad (\star)$$

One obtains precisely the equalities $\ell(QX^i) = c_i$ by computing $\ell(Q) = b_{p-1}$, $\ell(XQ)$, $\ldots$, $\ell(X^iQ)$, knowing $\ell(X^jP) = 0$. □

After the lemma, the quadratic form $C(P, Q)$ is identified to the quadratic form associating to $f = f_0 + \cdots + f_{p-1}X^{p-1} \in A$ the expression $\ell(Qf^2)$.

Let $[(L_{x,i} = P(X)/(X-x)^i)_{x\in Z(P), 1\leq i\leq \mu(x)}]$ be the Lagrange basis of A, where $Z(P)$ is the set of roots of P in C and $\mu(x)$ is the multiplicity of the root x. It is clear that $\ell(L_{x,1}) = 1$, $\ell(L_{x,i}) = 0$, $i > 1$.

We compute the matrix of $C(P, Q)$ in the Lagrange basis of A. Its coefficients are the $\ell(QL_{x,i}L_{y,j})$ with $i \le \mu(x)$, $j \le \mu(y)$. It is easy to see that for $x \neq y$, $L_{x,i}L_{y,j} = 0$, so that the quadratic form $C(P, Q)$ is decomposed into a sum of orthogonal quadratic forms $C(P, Q)_x$ in the Lagrange basis. The part corresponding to two complex conjugate roots gives a real expression with signature 0. If x is a root of P of multiplicity m in R (and a root of Q of multiplicity r, $r \ge 0$), $Q = (Q^{(r)}(x)/r!)(X-x)^r + (X-x)^{r+1}q(X)$. Then for $i+j-r < m+1$, $\ell(QL_{x,i}L_{x,j}) = 0$ and for $i+j+r = m+1$, $\ell(QL_{x,i}L_{x,j}) = \ell(L_{x,1}L_{x,m}Q(X)/(X-x)^r) = (P^m(x)/m!)(Q^{(r)}(x)/r!)$. It is easy to compute the corresponding signature, thanks to the following lemma.

Lemma 3.31. *The signature of the quadratic form of the d variables* $f_1, \ldots, f_d$ $\phi = \sum_{j+k=d+r} f_j f_k + \sum_{j+k>d+r} a_{j,k} f_j f_k$ *is 0 if* $d+r$ *is even and* 1 *if* $d+r$ *is odd.*

Proof. It is obviously true if every $a_{j,k}$ is zero, and since the quadratic form $\phi_t = \sum_{j+k=d+r} f_j f_k + \sum_{j+k>d+r} a_{j,k} t f_j f_k$ is of rank $d-r$ for every t, its signature is constant. □

So that, using the lemma, the signature is 0 if $m-r$ is even, 1 if $m-r$ is odd and $P^{\mu(x)}(x)Q^{(r)}(x) > 0$ (resp. -1 if $m-r$ is odd and $P^{\mu(x)}(x)Q^{(r)}(x) < 0$).

This ends the proof of the proposition since the number of jumps from $-\infty$ to $+\infty$ or to $+\infty$ to $-\infty$ of Q/P at x is 0 if $m-r$ is even, the number of jumps from $-\infty$ to $+\infty$ in Q/P at x is 1 if $m-r$ is odd and $P^{\mu(x)}(x)Q^{(r)}(x) > 0$, (resp. the number of jumps from $+\infty$ to $-\infty$ in Q/P at x is 1 if $m-r$ is odd and $P^{\mu(x)}(x)Q^{(r)}(x) < 0$). □

Quadratic forms of the type

$$A = \sum_{k=0}^{p-1}\sum_{l=0}^{p-1} c_{k+j} f_k f_l$$

are called Hankel forms. To a rational function Q/P we have associated a Hankel form $C(P, Q)$. Conversely, given P of degree d, we can associate to a Hankel form

$$C = \sum_{k=0}^{p-1}\sum_{l=0}^{p-1} c_{k+j} f_k f_l$$

a polynomial of degree $< p$, Q, such that

$$Q/P = \sum_{i=0}^{\infty} c_i / X^{i+1}$$

and thus $C(P, Q) = C$.

The first p of the identities ($\star$) give the coefficients of Q as a function of $c_0, \ldots, c_{p-1}$, so that the Hankel quadratic form C determines uniquely Q of degree $< p$ such that $C(P, Q) = C$.

An attractive method for computing the signature of a quadratic form is the following one. The signature of a quadratic form is equal to the difference between the number of positive eigenvalues and negative eigenvalues of its associated symmetric matrix. Since the characteristic polynomial of the matrix has all its roots real, its number of positive real roots and negative real roots counted with multiplicities are obtained by an easy application of Descartes' rule (see [20]). More efficient methods to compute the signature can be found in [24].

However we shall be able to design an alternative method for computing the signature of a Hankel quadratic form by relating them to Sylvester–Habicht sequences. This approach will provide the best method known solving the real root counting problem.

3.4.2. Bezoutians and principal Sylvester–Habicht coefficients.

Definition 5. The *principal minors of a matrix*

$$A = \begin{pmatrix} a_{1,1} & \cdots & a_{1,p} \\ \vdots & & \vdots \\ a_{p,1} & \cdots & a_{p,p} \end{pmatrix}$$

are, by definition, the determinants $\det(A_k)$ of the matrices

$$A_k = \begin{pmatrix} a_{1,1} & \cdots & a_{1,k} \\ \vdots & & \vdots \\ a_{k,1} & \cdots & a_{k,k} \end{pmatrix}$$

with $k \in \{1, \ldots, p\}$.

The *bezoutians, $C_k(P, Q)$, associated to P and Q* are the principal minors of the matrix

$$\begin{pmatrix} c_0 & \cdots & c_{p-1} \\ \vdots & & \vdots \\ c_{p-1} & \cdots & c_{2p-2} \end{pmatrix}$$

associated to the quadratic form $C(P, Q)$.

Proposition 3.32. *Let P and Q be two polynomials with*

$$P = X^p + a_{p-1}X^{p-1} + \cdots + a_0,$$
$$Q = b_{p-1}X^{p-1} + b_{p-2}X^{p-2} + \cdots + b_0.$$

Denoting by h_j the coefficient of degree j of $\mathrm{SyHa}_j(P, Q)$ *then for all $k \in \{1, \ldots, p\}$ we have*

$$h_{p-k} = C_k(P, Q).$$

Proof. Denote by $s_j = \mathrm{sr}_j(P, Q)$ the coefficients of degree j of $\mathrm{Sres}_j(P, Q)$. From the definition of the Sylvester–Habicht sequence

$$h_{p-k} = (-1)^{k(k-1)/2} s_{p-k}.$$

By definition, we have:

$$s_{p-k} = \begin{vmatrix} 1 & a_{p-1} & . & . & . & . & . & . & a_{p-2k+2} \\ 0 & 1 & a_{p-1} & . & . & . & . & . & a_{p-2k+3} \\ . & . & . & . & . & . & . & . & . \\ . & . & . & . & . & . & . & . & . \\ 0 & . & . & 0 & 1 & a_{p-1} & . & . & a_{p-k} \\ b_{p-1} & . & . & . & . & . & . & . & b_{p-2k+1} \\ 0 & b_{p-1} & . & . & . & . & . & . & b_{p-2k+2} \\ . & . & . & . & . & . & . & . & . \\ . & . & . & . & . & . & . & . & . \\ 0 & . & . & 0 & 0 & b_{p-1} & . & . & b_{p-1-k} \end{vmatrix}.$$

From the relations (⋆) we deduce $s_{p-k} = D \times D'$ with

$$D = \begin{vmatrix} 1 & 0 & . & . & . & . & . & . & 0 \\ 0 & 1 & 0 & . & . & . & . & . & 0 \\ . & . & . & . & . & . & . & . & . \\ . & . & . & . & . & . & . & . & . \\ 0 & . & . & 0 & 1 & 0 & . & . & 0 \\ c_0 & c_1 & . & . & . & c_{k-1} & . & . & c_{2k-2} \\ 0 & c_0 & c_1 & . & . & . & . & . & c_{2k-3} \\ . & . & . & . & . & . & . & . & . \\ . & . & . & . & . & . & . & . & . \\ 0 & . & . & 0 & 0 & c_0 & c_1 & . & c_{k-1} \end{vmatrix},$$

$$D' = \begin{vmatrix} 1 & a_{p-1} & . & . & . & . & . & . & a_{p-2k+2} \\ 0 & 1 & a_{p-1} & . & . & . & . & . & a_{p-2k+3} \\ . & . & . & . & . & . & . & . & . \\ . & . & . & . & . & . & . & . & . \\ 0 & . & . & 0 & 1 & a_{p-1} & . & . & a_{p-k} \\ 0 & . & . & . & 0 & 1 & a_{p-1} & . & a_{p-k+1} \\ . & . & . & . & . & . & . & . & . \\ . & . & . & . & . & . & . & . & . \\ . & . & . & . & . & . & . & . & . \\ 0 & . & . & . & . & . & . & . & 1 \end{vmatrix}.$$

This last equality allows to write

$$s_{p-k} = \begin{vmatrix} c_{k-1} & . & . & . & c_{2k-2} \\ . & . & . & . & . \\ . & . & . & . & . \\ . & . & . & . & . \\ c_0 & c_1 & . & . & c_{k-1} \end{vmatrix}$$

and, after some line permutation,

$$s_{p-k} = (-1)^{k(k-1)/2} C_k(P, Q)$$

as we wanted to show. □

We indicate now how to compute the Cauchy index from the signs of the h_{p-k}. We have seen right above that these principal Sylvester–Habicht coefficients coincide with the Bezoutians.

Given $[a_0, \ldots, a_n]$, a sequence of non zero elements of an ordered field K, we denote by $D([a_0, \ldots, a_n])$ the difference between the number of sign permanences and the number of sign changes in $[a_0, \ldots, a_n]$.

If A is a symmetric matrix with coefficients in a real closed field then it is easy to prove, when no $\det(A_k)$ vanishes, that the signature of the matrix A is equal to $D([1, \det(A_1), \ldots, \det(A_p)])$. This easy result is known as Jacobi's rule.

If some principal minor of the matrix vanishes then it is no more true that the principal minors determine the signature of the quadratic form (for this see [11], volume I, chap 10). Nevertheless it is posssible to generalize Jacobi's rule and to determine the signature, by looking only at the principal minors, in the particular case of Hankel forms.

We generalize the number $D([a_0, \ldots, a_n])$ to the case of a sequence $[a_0, \ldots, a_n]$ of elements of R with $a_0 \neq 0$ including zeroes, in the following way:

1. Consider the zero elements in $a_0, \ldots, a_n$, define $k(0) = -1$ and
$$[a_0, \ldots, a_n] = [a_0, \ldots, a_{j(1)}, 0, \ldots, 0, a_{k(1)}, \ldots, a_{j(2)}, 0, \ldots, 0, \ldots a_{k(t-1)}, \ldots, a_{j(t)}, 0, \ldots, 0]$$
(every element a_j, such that $k(\ell - 1) < j \leq j(\ell)$, $\ell \in \{1, \ldots, t\}$ is non zero).

2. Define
$$D([a_0, \ldots, a_n]) = \sum_{\ell=1}^{t} D([a_{k(\ell-1)}, \ldots, a_{j(\ell)}]) + \sum_{\ell=1}^{t} \epsilon_\ell$$
where
$$\epsilon_\ell = \begin{cases} 0 & \text{if } j(\ell) - k(\ell) \text{ is even} \\ (-1)^{(j(\ell)-k(\ell)-1)/2} \operatorname{sign}(a_{k(\ell)} \cdot a_{j(\ell)}) & \text{if } j(\ell) - k(\ell) \text{ is odd} \end{cases}.$$

Proposition 3.33. *The number $D([\mathrm{syha}_k(P, Q)]_{k=p,\ldots,0})$ is equal to the Cauchy index of Q/P between $-\infty$ and $+\infty$.*

Proof. We denote $H_j = \mathrm{SyHa}_j(P, Q)$ and $h_j = \mathrm{syha}_j(P, Q)$. If all the polynomials in the Sylvester–Habicht sequence associated to P and Q are regular (their index in the Sylvester–Habicht sequence is equal to their degree) then it is clear that

$$I(Q/P;]-\infty, +\infty[) = \mathrm{W}_{\mathrm{SyHa}}(P, Q, -\infty) - \mathrm{W}_{\mathrm{SyHa}}(P, Q, +\infty)$$

$$= \mathrm{V}([(-1)^p, (-1)^{p-1}h_{p-1}, \dots, -h_1, h_0]) - \mathrm{V}([1, h_{p-1}, \dots, h_1, h_0])$$
$$= D([h_p, \dots, h_0]).$$

Problems arise when in the Sylvester–Habicht sequence there are defective polynomials, i.e., there appears some h_{j-1} which is equal to zero. In this case, the principal Sylvester–Habicht coefficient is not the leading coefficient of H_{j-1}. Denote as usual by k the degree of H_{j-1}. The situation is as follows:

$$h_j \neq 0,\ h_{j-1} = \cdots = h_{k+1} = 0,\ h_k \neq 0$$

and we need to prove the following equality:

$$\tau \overset{\text{def}}{=} \mathrm{V}([H_j, H_{j-1}, H_k]; -\infty)) - \mathrm{V}([H_j, H_{j-1}, H_k]; +\infty))$$
$$= \sigma \overset{\text{def}}{=} \begin{cases} 0 & \text{if } j-k \text{ is even,} \\ (-1)^{j-k-1/2}\operatorname{sign}(h_k h_j) & \text{if } j-k \text{ is odd.} \end{cases}$$

Applying the Sylvester–Habicht structure theorem to this situation we first notice that H_{j-1} and H_k are proportional, so that $\operatorname{sign}(H_{j-1}H_k, +\infty) = \operatorname{sign}(H_{j-1}H_k, -\infty)$. More precisely, denoting by c the leading coefficient of H_{j-1} we get the following relations:

$$(h_j)^{j-k-1}h_k = \delta_{j-k}c^{j-k}. \tag{$\star\star$}$$

We denote by σ_j, σ_{j-1} and σ_k respectively $\operatorname{sign}(H_j, +\infty)$, $\operatorname{sign}(H_{j-1}, +\infty)$ and $\operatorname{sign}(H_k, +\infty)$. When $j-k$ is even, the sequence of signs of $[H_j, H_{j-1}, H_k]$ at $+\infty$ is $[\sigma_j, \sigma_{j-1}, \sigma_k]$ and at $-\infty$ is $[\sigma_j, \sigma_{j-1}, \sigma_k]$ when j is even (resp. $[-\sigma_j, \sigma_{j-1}, -\sigma_k]$ when j is odd), which implies that $\tau = 0$ when $j-k$ is even. When $j-k$ is odd, the sequence of signs of $[H_j, H_{j-1}, H_k]$ at $+\infty$ is $[\sigma_j, \sigma_{j-1}, \sigma_k]$ and at $-\infty$ is $[\sigma_j, -\sigma_{j-1}, -\sigma_k]$ when j is even (resp. $[-\sigma_j, \sigma_{j-1}, \sigma_k]$ when j is odd). Also using ($\star\star$) we see that $\delta_{j-k} = \sigma_{j-1}\sigma_k$ so that $\sigma = \operatorname{sign}(h_k h_j)$, when $(j-k-1)/2$ is even, because $\delta_{j-k} = 1$, and $\sigma = -\operatorname{sign}(h_k h_j)$ when $(j-k-1)/2$ is odd, because $\delta_{j-k} = -1$. The equality between σ and τ is now obtained in all cases:

- if $j-k$ is even then $\tau = \sigma = 0$,
- if $j-k$ is odd and $(j-k-1)/2$ even then $\tau = \sigma = \operatorname{sign}(h_k h_j)$,
- if $j-k$ is odd and $(j-k-1)/2$ odd then $\tau = \sigma = -\operatorname{sign}(h_k h_j)$.

□

We now give a proof of the following theorem due to Frobenius [10].

Proposition 3.34. *With the preceding notations the signature of a Hankel form with associated symmetric matrix C, is equal to* $D([1, \det(C_1), \dots, \det(C_p)])$.

Proof. Apply the preceding propositions. □

3.5. Summary and discussion

We sum up our results.

Theorem 3.35.

1. *The numbers* $\mathrm{V}_{\mathrm{Sy}}(P, Q;]a, b[)$ *and* $\mathrm{W}_{\mathrm{SyHa}}(P, Q;]a, b[)$ *coincide and they are equal to* $I(Q/P;]a, b[)$.

2. *The numbers* $\mathrm{V}_{\mathrm{Sy}}(P, Q)$, $\mathrm{V}_{\mathrm{SyHa}}(P, Q)$ *and* $D([\mathrm{syha}_{p-k}(P, Q)]_{k=0,\dots,p})$ *coincide, and are equal to the signature of* $C(P, Q)$, *when* P *is monic. They are equal to* $I(Q/P;)\infty, +\infty[)$.

Which is the best method for computing $d(P, Q) = c(P, Q > 0) - c(P, Q < 0)$? We have seen that the Sylvester–Habicht method is better than the Sylvester method. The signature of Hermite's quadratic form could be performed directly or by the bezoutians method. But it can be observed that the Sylvester–Habicht method, using Algorithm SyHa, is quicker than the bezoutians method and computes the whole Sylvester–Habicht sequence.

So the best strategy to compute the number $d(P, Q)$ is:

- compute the Sylvester–Habicht sequence of P and $P'Q$ using Algorithm SyHa,
- compute $d(P, Q)$ from the Sylvester–Habicht principal coefficients.

4. Complexity of algorithms

We have compared in the preceding section several methods for counting real root and seen in examples that their efficiency is not the same.

We are going to make these notions more precise.

In order to study the complexity of an algorithm we need to choose

- parameters of complexity describing the size of the input,
- basic operations which we consider as the unit time of computation.

The *complexity* of a given algorithm is a function associating to an input size t the maximum number of basic operations performed by the algorithm when it runs over all possible inputs of size bounded by t.

The *parallel complexity* of a given algorithm is a function associating to an input size t the maximum depth of the tree describing the basic operations performed by the algorithm when it runs over all possible inputs of size bounded by t.

We give some examples that will be useful in the sequel.

- If we consider a univariate polynomial with coefficients in $\mathbb{Z}$, the parameters of complexity will be the degree and the bitlength of its integer coefficients, and the basic operations will be the bit operations.
- If we consider a univariate polynomial with coefficients in an ordered field K, the complexity parameter will be the degree, and the basic operations will be the arithmetic operations and sign evaluations in K.
- If we consider a univariate polynomial with coefficients in an ordered domain D, the complexity parameter will be the degree, and the basic operations will be the arithmetic operations and sign evaluations in D.
- If we consider a list of univariate polynomials with coefficients in an ordered domain D, the complexity parameter will be the degree and the number of polynomials, and the basic operations will be the arithmetic operations and sign evaluations in D.
- If we consider a multivariate polynomial with coefficients in an ordered domain D, the complexity parameters are the number of variables and the total degree of the polynomial.
- If we consider a list of multivariate polynomials with coefficients in an ordered domain D, the complexity parameters are the number of variables, the total degree of the polynomials and the number of polynomials.

Sometimes in order to have a better understanding of the complexity of algorithms, we shall consider as basic operation the call to a blackbox solving a subproblem.

We give now some useful results on the computation of the euclidean remainder sequence that we defined in Section 2.

Proposition 4.1. *Let P and Q be two univariate polynomials with coefficients in a field* K. *The complexity of the computation of the euclidean remainder sequence of P and Q by the usual division method is $O(pq)$ arithmetic operations on* K.

Proof. It will be based on the following obvious lemma.

Lemma 4.2. *The number of arithmetic operations in* K *in the euclidean division of P by Q is* $\mathrm{O}((p-q)q)$.

We define $r_i = \deg(\mathrm{Rem}^i(P, Q))$. The number of arithmetic operations in order to compute $\mathrm{Rem}^{i+1}(P, Q)$ knowing $\mathrm{Rem}^i(P, Q)$ and $\mathrm{Rem}^{i-1}(P, Q)$ is $\mathrm{O}((r_{i-1} - r_i)(r_i))$ hence making the sum over all i and estimating r_i by q we get $\mathrm{O}(pq)$. □

The number of arithmetic operations in the field K to get the Sylvester sequence of P and Q of degrees p and q is also $\mathrm{O}(pq)$.

But if we take in consideration also the growth of the coefficients, the situation with the Sylvester sequence is not well controlled and we use rather the Sylvester–Habicht sequence.

In order to study the bitlength of the coefficients in the Sylvester–Habicht sequence, we need to study some easy properties of determinants.

Proposition 4.3. *Let A be an $n \times n$ matrix with integer coefficients. Then the determinant of A is bounded by the product of the euclidean norm of the columns of A.*

Proof. Clear by Gram–Schmidt orthogonalization. □

Corollary 4.4. *Let A be an $n \times n$ matrix with integer coefficients of bitlength t. Then the bitlength of the determinant of A is bounded by $(t + \log n) \times n$.*

Proof. Clear by the preceding proposition. □

Proposition 4.5. *Let A be an $n \times n$ matrix with coefficients polynomials in $T_1, \ldots, T_c$ of degrees t. Then the degree with respect to $T_1, \ldots, T_c$ of the determinant of A is bounded by $t \times n$.*

Proof. It is immediate. □

Proposition 4.6. *Let $P(X)$ and $Q(X)$ be two polynomials in $\mathrm{D}[X]$ of degree p and q. Then the complexity of the computation of the Sylvester–Habicht sequence of P and Q is in $\mathrm{O}(pq)$ arithmetic operations in* D.

Proof. Using the SyHa algorithm, the number of arithmetic operations in D is in $\mathrm{O}(pq)$ as in 4.1. □

Proposition 4.7. *Let $P(X)$ and $Q(X)$ be two polynomials in $\mathbb{Z}[X]$ of degree p and q with respective bitlength of coefficients t and t'. Then the bitlength of the coefficients in the Sylvester–Habicht sequence is $\mathrm{O}((t+t'+\log(p+q))(p+q))$ and the complexity of the computation of the Sylvester–Habicht sequence of P and Q is in $\mathrm{O}(pq)$ arithmetic operations over $\mathbb{Z}$ and $\mathrm{O}(pq(t+t'+\log(p+q))^2)$ bit operations using classical arithmetic.*

Proof. Use the preceding proposition and the bound on the coefficients of the Sylvester–Habicht sequence of $\mathrm{O}((t+t'+\log(p+q))(p+q))$ given by Proposition 4.3. □

Proposition 4.8. *Let $P(X)$ and $Q(X)$ be two polynomials in $\mathrm{D}[T_1, \ldots, T_c][X]$ of degree p and q with total degree d with respect to $[T_1, \ldots, T_c]$. Then the degrees of the polynomials in the Sylvester–Habicht sequence with respect to $[T_1, \ldots, T_c]$ is $(p+q) \times d$ and the complexity of the computation of the Sylvester–Habicht sequence of P and Q is in $(\mathrm{O}(pq)d)^{\mathrm{O}(c)}$ arithmetic operations over* D.

5. Sign determinations

In this section, we study polynomial time formal methods to deal with inequalities and real algebraic numbers. In the first paragraph we design a basic algorithm to deal with simultaneous inequalities. In subsection 5.2 we prove Thom's lemma and use it to define a method of computation with real algebraic numbers. We follow [8] and [25].

5.1. Simultaneous inequalities

Consider a polynomial P with coefficients in D and define

$$Z = \{x \in \mathrm{R} \mid P(x) = 0\}.$$

Let $\mathcal{Q} = [Q_1, \ldots, Q_s]$ be a list of polynomials with coefficients in D and let $\sigma = [\sigma_1, \ldots, \sigma_s]$ be a sign pattern in $\{0, 1, -1\}^s$. The subject of this paragraph is the computation of the number elements of Z where $\mathcal{Q}$ has the sign pattern σ. Given $\mathcal{Q} = [Q_1, \ldots, Q_s] \in \mathrm{D}[X]^s$, and $\sigma \in \{O, 1, -1\}^s$, the *realization of the sign-pattern σ at the roots of P* is

$$R(P, \operatorname{sign}(\mathcal{Q}) = \sigma) = \{x \in Z \mid (\forall i)\ Q_i(x)\sigma_i\}$$

and $c(P, \operatorname{sign}(\mathcal{Q}) = \sigma) = \operatorname{card}(R(P, \operatorname{sign}(\mathcal{Q}) = \sigma))$ is the cardinal of the elements of Z giving to $\mathcal{Q}$ the sign pattern σ.

Our algorithms are based on the fact that it is possible to compute $d(P, Q) = c(P, Q > 0) - c(P, Q < 0)$, as we have seen in the last section. We call $d(P, Q)$ a *Sturm-query* and consider its computation as a basic blackbox.

5.1.1. Preliminaries.

Definition 6. A *pseudo-partition* of Z is a list $C = [C_1, \ldots, C_n]$ of n subsets of Z whose union is Z and whose intersections two by two are empty (the difference with a partition is that some C_i may be empty). A list of polynomials $H = [H_1, \ldots, H_m]$ is *adapted* to a pseudo-partition $C = [C_1, \ldots, C_n]$ of Z if on each subset of C the signs of the polynomials of H are fixed. If H is adapted to C, the *matrix of signs of H on C*, $A(C, H)$ is the $n \times n$ matrix whose (i, j)'s entry is the sign of H_j on C_i. We denote by $c(C)$ the list of cardinals of the list C, by $\operatorname{sign}(H(x))$ the list of signs of the list H at x and by $d(H)$ the list of Sturm queries of the list H.

For example $[1, Q, Q^2]$ is adapted to

$$[R(P, Q = 0), R(P, Q > 0) \text{ and } R(P, Q < 0)]$$

and the matrix

$$A = \begin{bmatrix} 1 & 1 & 1 \\ 0 & 1 & -1 \\ 0 & 1 & 1 \end{bmatrix}$$

is the matrix of signs of $[1, Q, Q^2]$ on $[R(P, Q = 0), R(P, Q > 0), R(P, Q < 0)]$.

Note that $\{O, 1, -1\}^s$ is totally ordered by the lexicographical ordering defined by $0 < 1 < -1$. Lists of sign patterns will be always ordered with the lexicographical ordering.

A list Σ of sign conditions is *complete for* $\mathcal{Q}$ over Z if $Z = \cup_{\sigma \in \Sigma} R_\sigma(P, \mathcal{Q})$. If Σ is an ordered list of sign conditions complete for $\mathcal{Q}$ over Z, we denote by $R_\Sigma(\mathcal{Q})$ the pseudo-partition of Z $[R_\sigma(\mathcal{Q})_{\sigma\in\Sigma}]$ and by $c_\Sigma(\mathcal{Q})$ the ordered list of corresponding cardinals. We number elements of $\{0, 1, 2\}^s$ by elements of $\{0, \ldots, 3^s - 1\}$ using basis 3, so that the list of polynomials $[(Q_1)^{\nu_1} \cdots (Q_s)^{\nu_s}]$, with $\nu_i \in \{0, 1, 2\}$ gets totally ordered by considering exponents. We denote by $H(\mathcal{Q})$ this ordered list. Then it is clear that $H(\mathcal{Q})$ is adapted to $R_\Sigma(\mathcal{Q})$.

Proposition 5.1. *Given a pseudo-partition C of Z, and a list of polynomials H adapted to C,*

$$A(C, H) \cdot c(C) = d(P, H).$$

Proof. It is obvious since the j-the row of $A(C, H)$ is the list of signs of H_j on the list C. □

Corollary 5.2. *Given $Q \in \mathrm{D}[X]$,*

$$A \cdot \begin{bmatrix} c(P, Q = 0) \\ c(P, Q > 0) \\ c(P, Q < 0) \end{bmatrix} = \begin{bmatrix} d(P, 1) \\ d(P, Q) \\ d(P, Q^2) \end{bmatrix}.$$

Proof. We have seen above that A is the matrix of signs of $[1, Q, Q^2]$ on $[R(P, Q = 0), R(P, Q > 0), R(P, Q < 0)]$. □

If $\sigma = [\sigma_1, \ldots, \sigma_n]$ and $\sigma' = [\sigma'_1, \ldots, \sigma'_\nu]$ are two lists of sign patterns belonging to $\{0, 1, -1\}^s$ and $\{0, 1, -1\}^t$, denote by $\sigma \wedge \sigma'$ the list of sign patterns belonging to $\{0, 1, -1\}^{s+t}$ defined by

$$[(\sigma_1, \sigma'_1), \ldots, (\sigma_n, \sigma'_1), \ldots, (\sigma_1, \sigma'_\nu), \ldots, (\sigma_n, \sigma'_\nu)].$$

If $C = [C_1, \ldots, C_n]$ and $C' = [C'_1, \ldots, C'_\nu]$ are two pseudo-partitions of Z denote by $C \cap C'$ the pseudo-partition of Z defined by

$$[C_1 \cap C'_1, \ldots, C_n \cap C'_1, \ldots, C_1 \cap C'_\nu, \ldots, C_n \cap C'_\nu].$$

If $H = [H_1, \ldots, H_m]$ and $H' = [H'_1, \ldots, H'_\mu]$ are two lists of polynomials denote by $H \cdot H'$ the list of $m\mu$ polynomials

$$[H_1 \cdot H'_1, \ldots, H_m \cdot H'_1, \ldots, H_1 \cdot H'_\mu, \ldots, H_m \cdot H'_\mu].$$

If A and A' are two $n \times m$ and $\nu \times \mu$ matrices, denote by $A \otimes A'$ the $n\nu \times m\mu$ matrix obtained in replacing the entry $a'_{i,j}$ of A' by the matrix $a'_{i,j}A$.

Proposition 5.3. *Consider two pesudo-partitions C and C' of Z and two lists of polynomials H and H' such that H (resp. H') is adapted to C (resp. C'). Then*

$$A(C, H) \otimes A(C', H') = A(C \cap C', H \cdot H').$$

Proof. This follows immediately from the definitions. □

Corollary 5.4. *Consider two pseudo-partitions C and C' of Z, and two lists of polynomials H and H' such that H (resp. H') is adapted to C (resp. C'). Then*

$$\big(A(C, H) \otimes A(C', H')\big)c(C \cap C') = d(P, H \cdot H').$$

Proof. Immediate from the preceding propositions. □

Let $\Sigma(\mathcal{Q})$ be the ordered list of non empty sign condtions of $\mathcal{Q}$ over Z, $R(\mathcal{Q})$ the corresponding partition of Z and $c(\mathcal{Q})$ the corresponding list of cardinals of $R_\sigma(\mathcal{Q})$ for σ in $\Sigma(\mathcal{Q})$. We define $K(\mathcal{Q})$ as the smallest sublist of $H(\mathcal{Q})$ (according to the lexicographical ordering) such that the matrix $A(R(\mathcal{Q}), K(\mathcal{Q})$ is invertible.

We need the following definition: given h and k in $H(\mathcal{Q})$ h *precedes* k if $h \in H(\mathcal{Q}')$ with $\mathcal{Q}' \subset \mathcal{Q}$ and there exists a polynomial $h' \in H(\mathcal{Q} \setminus \mathcal{Q}')$ with $k = h \cdot h'$.

Proposition 5.5. *Let $\mathcal{Q}' \subset \mathcal{Q}$, $h \in H(\mathcal{Q}')$ and $k \in H(\mathcal{Q} \setminus \mathcal{Q}')$. If $h \cdot k \in K(\mathcal{Q})$, then $h \in K(\mathcal{Q})$.*

Proof. Given $h \in H(\mathcal{Q})$, denote by $\operatorname{sign}(h)$ the vector of signs of h on $R(\mathcal{Q})$. The vector $\operatorname{sign}(h)$ is a row of the matrix $A(R(\mathcal{Q}), H(\mathcal{Q}))$. Suppose that $\operatorname{sign}(h)$ is linear combination of rows above it in the matrix $A(R(\mathcal{Q}), H(\mathcal{Q}))$ so that $\operatorname{sign}(h) = \sum_{h' \in H} \lambda_{h'} \operatorname{sign}(h')$ with all $h' \in H$ being before h in $H(\mathcal{Q})$. Then it is clear that $\operatorname{sign}(h \cdot k) = \sum_{h' \in H} \lambda_{h'} \operatorname{sign}(h' \cdot k)$ and given the construction of the list $H(\mathcal{Q})$, if h' is before h in $H(\mathcal{Q})$, $h' \cdot k$ is before $h \cdot k$ in $H(\mathcal{Q})$. □

Corollary 5.6. *If r is the number of elements of Z, and $\mathcal{P} = (P_1, \ldots, P_s)$, the elements of $K(\mathcal{Q})$ are products of at most $\log_2(r)$ polynomials among*

$$\{1, P_1, P_1^2, \ldots, P_s, P_s^2\}.$$

Proof. The number of distinct sign conditions in $\Sigma(\mathcal{Q})$ is not greater than r, so the length of $K(\mathcal{Q})$ is not greater than r as well. If h is a product of ℓ polynomials of the form $\{1, P_1, P_1^2, \ldots, P_s, P_s^2\}$, there are at least 2^ℓ polynomials in $H(\mathcal{Q})$ preceeding h. □

Note also that if $\mathcal{Q}$ is the list $(Q_1, \ldots, Q_s)$ and $\mathcal{Q}' = (Q_1, \ldots, Q_\ell)$ with $1 \leq \ell \leq s$, $K(\mathcal{Q}) \subset K(\mathcal{Q}') \cdot K(\mathcal{Q} \setminus \mathcal{Q}')$.

5.1.2. Simultaneous inequalities. In this paragraph, P will denote a polynomial with coefficients in D of degree p, r the number of roots of P in R, $r \leq p$, and

$\mathcal{Q} = [Q_1, \ldots, Q_s]$ a list of polynomials with coefficients in D.

The aim of the algorithm we shall describe is to determine the number of roots of P in R giving fixed signs to $\mathcal{Q} = [Q_1, \ldots, Q_s]$.

We describe the algorithm corresponding to the case $s = 1$. The *input* is P, Q, two polynomials with coefficients in D. The *output* is $c(P, Q = 0)$, $c(P, Q > 0)$ and $c(P, Q < 0)$.

Algorithm SIstart

Step 1: Compute $d(P, 1)$, $d(P, Q)$ and $d(P, Q^2)$.

Step 2: From these values and

$$\begin{bmatrix} 1 & 1 & 1 \\ 0 & 1 & -1 \\ 0 & 1 & 1 \end{bmatrix} \begin{bmatrix} c(P, Q = 0) \\ c(P, Q > 0) \\ c(P, Q < 0) \end{bmatrix} = \begin{bmatrix} \mathrm{d}(P, 1) \\ \mathrm{d}(P, Q) \\ \mathrm{d}(P, Q^2) \end{bmatrix}$$

compute $c(P, Q = 0)$, $c(P, Q > 0)$ and $c(P, Q < 0)$.

The idea for $s \neq 1$ is to compute the Sturm-queries $d(P, H)$ of polynomials of the form $H = (Q_1)^{\nu_1} \ldots (Q_s)^{\nu_s}$, with $\nu_i \in \{0, 1, 2\}$ and deduce the values of all the $c(P, \operatorname{sign}(\mathcal{Q}) = \sigma)$ by inverting a linear system.

We number elements of $\{0, 1, -1\}^s$ by elements of $\{0, \ldots, 3^s - 1\}$ using basis 3, and identifying -1 and 2, so that the set of sign-patterns gets totally ordered. We denote by $C(\mathcal{Q}) = [R(P, \operatorname{sign}(\mathcal{Q}) = \sigma)]_{\sigma=0,\ldots,3^s-1}$ and by $c(\mathcal{Q})$ the list of corresponding cardinals that we want to compute.

Similarly, we number elements of $\{0, 1, 2\}^s$ by elements of $\{0, \ldots, 3^s - 1\}$ using basis 3, so that the list of polynomials $[(Q_1)^{\nu_1} \ldots (Q_s)^{\nu_s}]$ with $\nu_i \in \{0, 1, 2, \}$ gets totally ordered by considering exponents. We denote by $H(\mathcal{Q})$ this ordered list, and by $d(\mathcal{Q})$ the list of corresponding Sturm queries.

We get the following naive sign determination algorithm.

The *input* of the algorithm consists of P and $\mathcal{Q} = [Q_1, \ldots, Q_s]$, and the *output* is the list of cardinals $c(\mathcal{Q})$.

Algorithm Naive SI

Step 1: Compute the Sturm-queries $d(\mathcal{Q}) = [d(P, H(\mathcal{Q}))]$.

Step 2: Take the $3^s \times 3^s$ matrix $A(s)$ defined inductively by

$$A(1) = A, A(\ell + 1) = A(\ell) \otimes A.$$

Notice that $A(s)$ is the matrix of signs of $H(\mathcal{Q})$ on $C(\mathcal{Q})$. Compute the list $c(\mathcal{Q})$ from the equality $A(s) \cdot c(\mathcal{Q}) = d(\mathcal{Q})$ by inverting $A(s)$.

The complexity of the preceding algorithm is exponential in s, since we have 3^s polynomials, hence 3^s Sylvester–Habicht sequences to compute, to get the $c(P, \operatorname{sign}(\mathcal{Q}) = \sigma)$ for the 3^s sign conditions σ.

To avoid this exponential growth, the idea in [3] is to remark that, for any $\mathcal{Q}$, the number $r(\mathcal{Q})$ of distinct sign conditions realized by $\mathcal{Q}$ at the roots of P in R is less than the number r of roots of P in R, so that the number of non empty sign conditions can never exceed r.

Precisely one wants to determine

- the list of sign patterns $\Sigma(\mathcal{Q}) = [\sigma(\mathcal{Q})_1, \ldots, \sigma(\mathcal{Q})_{r(\mathcal{Q})}]$, with $r(\mathcal{Q}) \leq r$, realized by $\mathcal{Q}$ at the roots of P in R, in the order induced by the order already defined over sign patterns,
- the list $c(\mathcal{Q})$ of cardinals of the list of corresponding non empty realizations $R(\mathcal{Q})$.

The *input* of the algorithm consists of $P, \mathcal{Q} = [Q_1, \ldots, Q_s]$. The *output* **SIout**$(P, \mathcal{Q})$ of the algorithm is the following:

- $(1_\mathcal{Q})$ the list $\Sigma(\mathcal{Q})$,
- $(2_\mathcal{Q})$ the list $c(\mathcal{Q})$,
- $(3_\mathcal{Q})$ the list of polynomials $K(\mathcal{Q})$ (see Subection 5.1.1) and the list $d(\mathcal{Q}) = [d(P, K(\mathcal{Q}))]$. The list $K(\mathcal{Q})$ is adapted to $R(\mathcal{Q})$, where $R(\mathcal{Q})$ is the list of realizations of sign patterns in $\Sigma(\mathcal{Q})$. The list $K(\mathcal{Q})$ is a sublist of $H(\mathcal{Q})$,
- $(4_\mathcal{Q})$ an invertible matrix $B(\mathcal{Q})$ which is the matrix of signs of $K(\mathcal{Q})$ over $R(\mathcal{Q})$.

Suppose that $\mathcal{Q} = [\mathcal{Q}_1, \mathcal{Q}_2]$. We are going to describe first a subroutine **SIcomb** describing how to combine **SIout**$(P, \mathcal{Q}_1)$ and **SIout**$(P, \mathcal{Q}_2)$, to get **SIout**$(P, \mathcal{Q})$.

Algorithm SIcomb

Input SIout$(P, \mathcal{Q}_1)$, and **SIout**$(P, \mathcal{Q}_2)$.

Step 1: Define a list of $r' = r(\mathcal{Q}_1)r(\mathcal{Q}_2)$ polynomials

$$K' = K(\mathcal{Q}_1) \cdot K(\mathcal{Q}_2)$$

and compute the list of Sturm-queries $d' = d(P, K')$.

Step 2: Take the $r' \times r'$ matrix $B'(\mathcal{Q}) = B(\mathcal{Q}_1) \otimes B(\mathcal{Q}_2)$. Note that $B'(\mathcal{Q})$ is the matrix of signs of K' on $R' = R(\mathcal{Q}_1) \cap R(\mathcal{Q}_2)$. Compute the list $c' = c(R')$ from the equality

$$B'(\mathcal{Q}) \cdot c' = d'$$

by inverting $B'(\mathcal{Q})$.

Step 3: Define $r(\mathcal{Q})$ as the number ($\leq r$) of non zero elements in c'. Remove from c' the zero elements, and get $c(\mathcal{Q})$. Remove from $\Sigma' = \Sigma(\mathcal{Q}_1) \wedge \Sigma(\mathcal{Q}_2)$ the sign conditions corresponding to elements equal to 0 in c' and get $\Sigma(\mathcal{Q})$. Remove from $B'(\mathcal{Q})$ the columns corresponding to elements equal to 0 in c', which gives an $r' \times r(\mathcal{Q})$ matrix B", the matrix of signs of K' over $R(\mathcal{Q})$.

Step 4: Extract an invertible matrix $B(\mathcal{Q})$, of dimension $r(\mathcal{Q})$ from B'' by taking the $r(\mathcal{Q})$ first independent rows of B''. Define $K(\mathcal{Q})$ by extracting from the list K' the corresponding polynomials. It is clear that $B(\mathcal{Q})$ is the matrix of signs of $K(\mathcal{Q})$ over $R(\mathcal{Q})$.

It is easy to verify $(1_{\mathcal{Q}}), \ldots, (4_{\mathcal{Q}})$.

We explain now how to compute **SIout**$(P, \mathcal{Q})$. When the list $\mathcal{Q}$ has a single element, we call algorithm **SIstart**, followed by steps 3 and 4 of **SIcomb**. Next there are two options of the sign determinations algorithm:

- Algorithm **SSI**, for sequential simultaneous inequalities, which consists in applying inductively algorithm **SIcomb** to $\mathcal{Q}_{\ell-1} = [Q_1, \ldots, Q_{\ell-1}]$, and Q_ℓ,
- Algorithm **PSI**, for parallel simultaneous inequalities, which consists in applying inductively algorithm **SIcomb** to $[Q_1, \ldots, Q_t]$, and $[Q_{t+1}, \ldots, Q_s]$, where $t = 2^{u-1}$, with $t < s < 2^u$.

In terms of total amount of computations, algorithm **SSI** is better, and its number of inductive steps is s, while algorithm **PSI** is better for parallel complexity since its number of inductive steps is $\log_2(s)$.

Example. We explain on an example how **SSI** works. We consider

$$P = (X^3 - 1)(X^2 - 9),\ Q_1 = X,\ Q_2 = X + 1,\ Q_3 = X - 2.$$

We have $d(P, 1) = 3$: so P has 3 real roots.

Also $d(P, Q_1) = 1$ and $d(P, Q_1^2) = 3$, hence

$$\begin{bmatrix} 1 & 1 & 1 \\ 0 & 1 & -1 \\ 0 & 1 & 1 \end{bmatrix} \cdot \begin{bmatrix} c(P, Q_1 = 0) \\ c(P, Q_1 > 0) \\ c(P, Q_1 < 0) \end{bmatrix} = \begin{bmatrix} 3 \\ 1 \\ 3 \end{bmatrix}$$

which means, solving the system, that P has

$$\begin{cases} 0 \text{ root with } Q_1 = 0 \\ 2 \text{ roots with } Q_1 > 0 \\ 1 \text{ root with } Q_1 < 0. \end{cases}$$

So we have $r(\mathcal{Q}_1) = 2$; the list of polynomials $K(\mathcal{Q}_1)$ is $[1, Q_1]$ and

$$B(\mathcal{Q}_1) = \begin{bmatrix} 1 & 1 \\ 1 & -1 \end{bmatrix}.$$

We consider now Q_2: $d(P, Q_2) = 1, d(P, {Q_2}^2) = 3$, hence $c(P, Q_2 = 0) = 0$. We need to compute only $d(P, Q_1 Q_2)$ which is equal to 3 hence we have

$$\begin{bmatrix} 1 & 1 & 1 & 1 \\ 1 & -1 & 1 & -1 \\ 1 & 1 & -1 & -1 \\ 1 & -1 & -1 & 1 \end{bmatrix} \begin{bmatrix} c(P, \operatorname{sign}([Q_1, Q_2]) = [1, 1]) \\ c(P, \operatorname{sign}([Q_1, Q_2]) = [-1, 1]) \\ c(P, \operatorname{sign}([Q_1, Q_2]) = [1, -1]) \\ c(P, \operatorname{sign}([Q_1, Q_2]) = [-1, -1]) \end{bmatrix} = \begin{bmatrix} 3 \\ 1 \\ 1 \\ 3 \end{bmatrix}$$

Solving the system we find that P has

$$\begin{cases} 2 \text{ roots with } Q_1 > 0 \text{ and } Q_2 > 0 \\ 1 \text{ root with } Q_1 < 0 \text{ and } Q_2 < 0. \end{cases}$$

So we have $r(\mathcal{Q}_2) = r(\mathcal{Q}_1)$, the list of polynomials $K(\mathcal{Q}_2)$ is $[1, Q_1]$ and $B(\mathcal{Q}_2) = B(\mathcal{Q}_1)$.

We add Q_3: $d(P, Q_3) = -1, d(P, {Q_3}^2) = 3$, hence $c(P, Q_3 = 0) = 0$.

We need only to compute $d(P, Q_1Q_3)$ which is equal to 1 hence, inverting again a 4×4 matrix, P has

$$\begin{cases} 1 \text{ root with } Q_1 > 0 \text{ and } Q_2 > 0 \text{ and } Q_3 > 0 \\ 1 \text{ root with } Q_1 > 0 \text{ and } Q_2 > 0 \text{ and } Q_3 < 0 \\ 1 \text{ root with } Q_1 < 0 \text{ and } Q_2 < 0 \text{ and } Q_3 < 0. \end{cases}$$

So we have $r(\mathcal{Q}) = 3$, the list of polynomials is $[1, Q_1, Q_3]$ and

$$B(\mathcal{Q}) = \begin{bmatrix} 1 & 1 & 1 \\ 1 & 1 & -1 \\ 1 & -1 & -1 \end{bmatrix}.$$

To summarize $r(\mathcal{Q}) = 3$ and the sign conditions realized by $\mathcal{Q} = [Q_1, Q_2, Q_3]$ at the three real roots of P are the three sign conditions

$$\begin{aligned} \sigma(\mathcal{Q})_1 &= (>0, >0, >0), \\ \sigma(\mathcal{Q})_2 &= (>0, >0, <0), \\ \sigma(\mathcal{Q})_3 &= (<0, <0, <0). \end{aligned}$$

5.1.3. Complexity of the computation. We take now the following notations:

- P will denote a polynomial with coefficients in D,
- r the number of roots of P in R, and d its degree,
- $\mathcal{Q} = [Q_1, \ldots, Q_s]$ polynomials with coefficients in D,
- q an integer greater than or equal to $\mathrm{d}(Q_i)$.

We shall evaluate the combinatorial complexity of our sequential algorithms by counting the number of calls to the Sturm-query blackbox. The complexity of the parallel algorithms is not detailed (see [9]).

Proposition 5.7. *The combinatorial complexity of* **SSI** *is linear in s and polynomial in r. The calls to Sturm query blackbox is done for polynomials of degrees at most* $\mathrm{O}(q \log r)$.

Proof. In **SIcomb**, which is done s times, the number of calls to the Sturm-query blackbox is bounded by $2r$. We have seen in the last proposition that these calls are made for polynomials which are products of at most $\log r$ products of Q_i or Q_i^2. □

Proposition 5.8. *The number of arithmetic operations in* D *for* **SSI** *when we use the Sylvester–Habicht sequence to answer to Sturm-queries is linear in s and polynomial in p and q. More precisely the number of arithmetic operations is* $O(sr(p(\log(r)q+p)))$. *If moreover all coefficients are integers of bit size bounded by* τ, *the bit complexity is linear in s and polynomial in p, q and* τ.

Proof. In **SIcomb**, which is done s times, the number of Sylvester–Habicht sequences to compute is bounded by $2r$. Each computation of Sylvester–Habicht sequence takes $O(p(\log(r)q + p))$ arithmetical operations, since the polynomials in the $K(\mathcal{Q})$ are of degree $2\log(r)q$. So, the total number in **SSI** of arithmetic operations is in $O(sr(p(\log(r)q + p)))$. Using Proposition 4.3 it is clear that the bitlength of integers in intermediate computations, which are determinants extracted from the Sylvester matrix, is in $O(\tau(\log(r)q + p))$. □

5.2. Thom's lemma and its consequences

Thom's lemma, a very simple and basic result in real algebraic geometry, has interesting computational consequences.

One of them is the fact that a root $x \in \mathrm{R}$ of a polynomial P of degree p with coefficients in R may be distinguished from the other roots of P in R by the signs of the derivatives $P^{(i)}$ of P at x, $i = 1, \ldots, p-1$. This offers a new possibility for the coding of real algebraic numbers and the computation with these numbers (see [8], [25]).

5.2.1. Thom's lemma. A *sign condition* is > 0, < 0 or $= 0$. A *generalized sign condition* is > 0 or < 0 or $= 0$ or ≥ 0 or ≤ 0.

If $\sigma = (\sigma(i))_{i=0,\ldots,d-1}$ is a d-uple of generalized sign conditions, $\bar{\sigma}$ is the d-uple obtained by relaxing the strict inequalities of σ, that is by replacing > 0 (resp. < 0) by ≥ 0 (resp ≤ 0).

Proposition 5.9 (Thom's lemma). *Let P be a polynomial of degree p with coefficients in* R, $P', \ldots, P^{(p-1)}$ *its derivatives and* $\sigma = (\sigma(i))_{i=0,\ldots,p-1}$ *a p-uple of generalized sign conditions.*

Let

$$R(\sigma) = \{x \in \mathrm{R} \mid P^{(i)}(x)\sigma(i), i = 0, \ldots, p-1\}$$

Then

(i) $R(\sigma)$ *is either empty or a point or a (closed, open or semi open) interval,*

(ii) *if* $R(\sigma)$ *is non empty, the closure of* $R(\sigma)$ *is* $R(\bar{\sigma})$.

Proof. By induction on the degree p of P. There is nothing to prove if $p = 0$. Let σ be a function from $\{1, \ldots, p+1\}$ into $\{> 0, < 0, = 0, \geq 0, \leq 0\}$, τ its restriction to

$\{1, \ldots, p\}$. If

$$R(\tau) = \{x \in \mathrm{R} \mid P^{(i)}(x)\sigma'(i),\ i = 1, \ldots, p-1\}$$

is empty or a point then

$$R(\sigma) = \{x \in \mathrm{R} \mid P^{(i)}(x)\sigma(i),\ i = 0, \ldots, p\}$$

verifies (i) and (ii). If $R(\tau)$ is an interval, the derivative P' has a fixed constant sign on the interior of the interval, hence P is increasing or decreasing on the whole interval, which gives the properties (i) and (ii) for $R(\sigma)$. □

One does not obtain in general the closure of a set defined by strict inequalities by relaxing these inequalities, as is shown by the example of $\{x \in \mathrm{R} \mid x^3 - x^2 > 0\}$.

Classical algorithms proposed to work on real algebraic numbers are semi-numerical: one characterizes a real algebraic number x by means of a squarefree defining polynomial P with integer coefficients and an interval that isolates x from all other roots of P ([7]).

The approach we propose here is purely formal and relies on Thom's lemma.

Proposition 5.10. *Let P be a polynomial of degree p with coefficients in* D. *Let x and x' be two elements of* R. *Suppose the signs $\sigma(i)$ and $\sigma'(i)$ of $P^{(i)}(x)$ and $P^{(i)}(x')$, $i = 0, \ldots, p-1$, are given. Then:*

(i) *If $\sigma = \sigma'$ with $\sigma(0) = \sigma'(0) = 0$ then $x = x'$.*

(ii) *If $\sigma \neq \sigma'$ one can decide whether $x < x'$ or $x > x'$ as follows. Let k be the smallest integer such that $\sigma(p-k)$ and $\sigma'(p-k)$ are different.*

Then

(a) $\sigma(p-k+1) = \sigma'(p-k+1) \neq 0$,

(b) *if $\sigma(p-k+1) = \sigma'(p-k+1)$ is > 0, $x > x'$ if and only if $P^{(p-k)}(x) \geq P^{(p-k)}(x')$ (which is known by looking at $\sigma(p-k)$ and $\sigma'(p-k)$),*

(c) *if $\sigma(p-k+1) = \sigma'(p-k+1)$ is < 0, $x > x'$ if and only if $P^{(p-k)}(x) \leq P^{(p-k)}(x')$ (which is known by looking at $\sigma(p-k)$ and $\sigma'(p-k)$).*

Proof. (i) and (ii) are consequences of Thom's lemma.

(ii)(a) because of point (i) in Thom's lemma applied to $P^{(p-k+1)}$, (ii)(b) and (ii)(c) because the set $\{x \in \mathrm{R} \mid P^{(i)}(x)\sigma(i),\ i = 1, \ldots, p-k-1\}$, is connected by Thom's lemma applied to $P^{(p-k+1)}$, and that, on an interval, the sign of the derivative of a polynomial gives its variation. □

5.2.2. Coding of real algebraic numbers. In this paragraph, we denote by P a polynomial with coefficients in D of degree p, r the number of roots of P in R, $r \leq p$.

Remark that P does not need to be irreducible.

As seen in Proposition 5.10, a root x of P in R is characterized by the sequence of signs it gives to the derivatives $P^{(i)}$ of P. So a possible coding of x is this sequence of signs . We simply define algorithm **RAN** as **SSI** applied to P and $[P^{(p-1)}, \ldots, P']$. In fact it is possible in some cases to save some computations as we can see in the next example. This improvements are detailed in [25].

We characterize the real roots of the polynomial

$$P = 16X^6 - 48X^5 - 200X^4 + 320X^3 + 541X^2 - 71X - 50.$$

The polynomial P is square-free so that the sign of P' at two consecutive roots of P changes. So here it is necessary to compute **SSI** up to $P^{(5)}$ and $P^{(4)}$, and then the roots may be characterized by the signs of P'.

The final result is:

$$\begin{aligned} x_1 &= [P = 0, P^{(5)} > 0, P^{(4)} > 0, P' < 0], \\ x_2 &= [P = 0, P^{(5)} > 0, P^{(4)} > 0, P' > 0], \\ x_3 &= [P = 0, P^{(5)} < 0, P^{(4)} > 0, P' < 0], \\ x_4 &= [P = 0, P^{(5)} < 0, P^{(4)} > 0, P' > 0], \\ x_5 &= [P = 0, P^{(5)} < 0, P^{(4)} < 0, P' < 0], \\ x_6 &= [P = 0, P^{(5)} < 0, P^{(4)} < 0, P' > 0]. \end{aligned}$$

It is also clear that

$$x_1 < x_2 < x_3 < x_4 < x_5 < x_6.$$

Denote by

- P a polynomial with coefficients in D (resp. $\mathbb{Z}$) of degree p,
- τ an integer greater than the bitlength of the coefficients of P (in case of integer coefficients).

Proposition 5.11. *The complexity of algorithm* **RAN** *counting arithmetic operations in* D *(resp. bit operations in* $\mathbb{Z}$*) is polynomial in* p *(resp. polynomial in* p *and* τ*) if we use Sylvester–Habicht sequence for the Sturm-queries.*

5.2.3. Sign determination at real algebraic numbers. Denote by Q a polynomial with coefficients in D. We want to compute the signs taken by Q at any root of P in R. We suppose that the roots of P are already coded by **RAN** which is **SSI** applied to P and its derivatives and we apply **SIcomb** to Q. Since the real roots of P are already individually coded, the output of this call of **SIcomb** does not modify further the list K and the matrix B in the output of **RAN**.

Denote by $\mathcal{Q}$ a list of polynomials with coefficients in D. We want to compute the signs taken by $\mathcal{Q}$ at any root of P in R. This algorithm **RANSI** consists only in s applications of **SIcomb** to the various Q_i in the list $\mathcal{Q}$.

If we want to compare two elements x_1 and x_2 in R coded by Thom's lemma as roots in R of the polynomial P_1 of degree p_1 and P_2 of degree p_2, we apply **RANSI** to P_1 and the derivatives of $P_1 P_2$, then **RANSI** to P_2 and the derivatives of $P_1 P_2$.

Then, the coding of x_1 and x_2 as roots of $P_1 P_2$ in R is known and those two algebraic numbers can be compared by Proposition 3 as roots of $P_1 P_2$ in R.

6. Existential theory of reals

We now study the multivariate case and give an algorithm deciding if a combination of equations and inequations has a solution or not, which is known as the existential theory of reals problem.

In the first subsection, we study multivariate polynomials systems. In the second subsection we describe some geometric properties of semi-algebraic sets. In the third subsection we describe an algorithm for finding points on algebraic hypersurfaces and in the last section we solve the existential theory of reals with precise complexity bounds.

6.1. Solving multivariate polynomial systems

Let K be a field of characteristic zero and $\bar{\mathrm{K}}$ an algebraically closed field containing it. Let $A = \mathrm{K}[X_1, \ldots, X_k]/(P_1, \ldots, P_s)$ be a finitely generated K algebra. We define $\bar{A} = A \otimes_{\mathrm{K}} \bar{\mathrm{K}}$. The set of common zeroes of $P_1, \ldots, P_s$ in a field L extension of K is denoted by $\mathrm{Z}_{\mathrm{L}}(A)$.

6.1.1. Preliminaries about finite-dimensional algebras. We are going to give proofs of the following well known results:

Theorem 6.1. *The* K*-algebra* $A = \mathrm{K}[X_1, \ldots, X_k]/(P_1, \ldots, P_s)$ *is a finite dimensional vector space if and only if* $\mathrm{Z}_{\bar{\mathrm{K}}}(A)$ *is finite. Moreover the cardinal of* $\mathrm{Z}_{\bar{\mathrm{K}}}(A)$ *is always smaller than the dimension of A as a* K*-vector space.*

Theorem 6.2. *For every* $\alpha \in \mathrm{Z}_{\bar{\mathrm{K}}}(A)$ *there exists a local ring* $\bar{A}_\alpha$ *such that,* $\bar{A} \cong \prod_{\alpha \in Z_{\bar{\mathrm{K}}}(A)} \bar{A}_\alpha$.

We denote by μ_α the dimension of $\bar{A}_\alpha$ as a $\bar{K}$-vector space. We call μ_α, the *multiplicity* of the zero $\alpha \in \mathrm{Z}_{\bar{\mathrm{K}}}(A)$. If every μ_α is 1 A is said to be *regular*. It is the case if and only if $\bar{A}$ is a product of a finite number of copies of $\bar{\mathrm{K}}$.

The next result is less well known but extremely useful.

Theorem 6.3 (Stickelberger's theorem). *Let $f \in A$ and L_f be the linear endomorphim of multiplication by f. Then $L_f(\bar{A}_\alpha) \subset \bar{A}_\alpha$. The restriction of L_f to $\bar{A}_\alpha$, $L_{f,\alpha}$, has only one eigenvalue $f(\alpha)$ with multiplicity μ_α.*

All the proofs in this section rely on the famous Hilbert Nullstellensatz that we prove for completeness. We first give it in its weak form. The proof we present, due to Michel Coste, is a simplification of Van der Waerden's proof [28].

Theorem 6.4 (weak form of Hilbert's Nullstellensatz). *Let K be an algebraically closed field and let $P_1, \dots, P_s$ be polynomials of $\mathrm{K}[X_1, \dots, X_k]$ without common zeroes in K^k. There exist polynomials $Q_1, \dots, Q_s$ of $\mathrm{K}[X_1, \dots, X_k]$ such that $1 = P_1Q_1 + \cdots + P_sQ_s$.*

Proof. The proof is by induction on k. When $k = 1$, the ideal generated by $P_1, \dots, P_s$ in $\mathrm{K}[X_1]$ is principal, and generated by Q. If Q is not constant, it has a zero in K since K is algebraically closed, and this zero is common to all P_i.

Suppose now $k > 1$, and the theorem true for $k - 1$. Since K is infinite, one can suppose that the polynomial P_1 is monic in X_k, after a linear change of variables. Take a new indeterminate U, and put

$$Q(U, X_1, \dots, X_k) = P_2 + UP_3 + \cdots + U^{m-2}P_s.$$

The resultant of P_1 and Q with respect to X_k belongs to $\mathrm{K}[U, X_1, \dots, X_{k-1}]$, and is written

$$\mathrm{Res}_{X_k}(P_1, Q) = D_\ell(X_1, \dots, X_{k-1})U^\ell + \cdots + D_0(X_1, \dots, X_{k-1}).$$

This resultant belongs to the ideal generated by P_1 and Q, and there are polynomials Λ and Θ of $\mathrm{K}[U, X_1, \dots, X_k]$ such that

$$\mathrm{Res}_{X_k}(P_1, Q) = \Lambda P_1 + \Theta Q.$$

Identifying the coefficients in this equality between polynomials in U, one sees that the $D_0, \dots, D_\ell$ belong to the ideal generated by $P_1, \dots, P_s$.

Suppose now that $D_0, \dots, D_\ell$ have a common zero x' in K^{k-1}. For every $a \in \mathrm{K}$, $\mathrm{Res}_{X_k}(P_1, Q)(a, x') = 0$. Since P_1 is monic in X_k, its leading coefficient in X_k never vanishes, and so the annulation of the resultant implies that for every $a \in \mathrm{K}$ the polynomials $P_1(x', X_k)$ et $Q(a, x', X_k)$ have a common root in K. Since $P_1(x', X_k)$ has a finite number of roots in K, one of them, say α, is a root of $Q(a, x', X_k)$ for infinitely many $a \in \mathrm{K}$. Chosing $s - 1$ such distinct elements $a_1, \dots, a_{s-1}$, one gets a system of $s - 1$ homogeneous linear equations in the $P_j(x', \alpha)$ for $j = 2, \dots, s$

$$P_2(x', \alpha) + a_i P_2(x', \alpha) + \cdots + a_i^{s-1} P_s(x', \alpha) = 0.$$

The determinant of this system is a non zero Vandermonde determinant, and so one has $P_2(x', \alpha) = \cdots = P_s(x', \alpha) = 0$. Hence (x', α) is a common zero to $P_1, \dots, P_s$, which is contrary to the hypothesis.

One knows thus that $D_0, \dots, D_\ell$ have no common zeroes in K^{k-1}. By induction hypothesis, 1 belongs to the ideal generated by $D_0, \dots, D_\ell$ in $\mathrm{K}[X_1, \dots, X_{k-1}]$. As we have seen that $D_0, \dots, D_\ell$ are in the ideal generated by $P_1, \dots, P_s$ in $\mathrm{K}[X_1, \dots, X_k]$, one concludes that 1 belongs to this ideal too, which means that there exist polynomials $Q_1, \dots, Q_s$ of $\mathrm{K}[X_1, \dots, X_k]$ such that $1 = P_1 Q_1 + \cdots + P_s Q_s$. □

Hilbert's Nullstellensatz is as usual derived from the weak form using Rabinovitch's trick.

Theorem 6.5 (Hilbert's Nullstellensatz). *Let* K *be an algebraically closed field and* $P_1, \dots, P_s, P$ *be polynomials with coefficients in* K. *If a polynomial* P *vanishes on the common zeroes of* $P_1, \dots, P_s$ *in* K, *a power of* P *belongs to the ideal* $(P_1, \dots, P_s)$.

Proof. The set of polynomials $(P_1, \dots, P_s, TP - 1)$ has no common zeroes in $\bar{K}^{k+1}$, so according to the weak Hilbert Nullstellensatz we can find polynomials

$$Q_1(X_1, \dots, X_k, T), \dots, Q_s(X_1, \dots, X_k, T), Q(X_1, \dots, X_k, T)$$

such that $1 = P_1 Q_1 + \cdots + P_s Q_s + (TP - 1)Q$. Replacing everywhere T by $1/P$ and multiplying by a convenient power of P we get a power of P in the ideal $(P_1, \dots, P_s)$. □

Proof ot Theorem 6.1. If A is a finite dimensional vector space of dimension N, the powers $1, X_1, \dots, X_1^N$ of the variable X_1 are necessarily linearly dependent in A, which gives a polynomial $p_1(X_1)$ in the ideal $(P_1, \dots, P_s)$. It means that the first coordinate of any common zero of $P_1, \dots, P_s$ is a zero of p_1. Doing the same for all the variables, we see that $Z_{\bar{\mathrm{K}}}(A)$ is finite. Conversely, if $Z_{\bar{\mathrm{K}}}(A)$ is finite, take a polynomial $p_1(X_1)$ whose zeroes are the first coordinates of the elements of $Z_{\bar{\mathrm{K}}}(A)$. According to the corollary of Hilbert's Nullstellensatz a power of p_1 belongs to the ideal $(P_1, \dots, P_s)$. Doing the same for all variables we get polynomials of degree d_i in X_i in the ideal $(P_1, \dots, P_s)$. It means that any monomial of multidegree greater than $(d_1, \dots, d_k)$ is a linear combination in $\bar{A}$ of monomials of respective degree in X_i smaller than d_i, so that A itself is finite dimensional.

We define an element u of A as *separating* if two distinct zeroes of $Z_{\bar{\mathrm{K}}}(A)$ have different images in $\bar{\mathrm{K}}$ by u.

In order to prove the last assertion we use the following lemma:

Lemma 6.6. *If* $Z_{\bar{\mathrm{K}}}(A)$ *has* N *points, then one at least among the* $u_i = X_1 + iX_2 + \dots + i^{k-1}X_k$ *for* $0 \le i \le (k-1)\binom{N}{2}$ *is separating.*

Proof. Consider a couple $(x, y) = ((x_1, \dots, x_k), (y_1, \dots, y_k))$ of distinct points of $Z_{\bar{\mathrm{K}}}(A)$ and let $\ell(x, y)$ be the number of i, $0 \le i \le (k-1)\binom{N}{2}$, such that $u_i(x) = u_i(y)$. Since the polynomial $(x_1 - y_1) + (x_2 - y_2)t + \cdots + (x_k - y_k)t^{k-1}$ which is not identically zero has no more than $k - 1$ distinct roots, the number $\ell(x, y)$ is always smaller than

$k-1$. Since the total number of couples of distinct points of $Z_{\bar{K}}(A)$ is less than $\binom{N}{2}$, this completes the proof. □

If u is separating, the minimal polynomial of the endomorphism L_u is of degree at least N, since it must have N distinct roots, hence $1, u, \ldots, u^{N-1}$ are linearly independent elements of A. □

Proof of Theorem 6.2. The proof is based on the following properties:

Lemma 6.7. *If $A = \mathrm{K}[X_1, \ldots, X_k]/(P_1, \ldots, P_s)$ has a finite number of points in $\bar{\mathrm{K}}$, then there exist elements s_α of $\bar{A}$ with $s_\alpha(\alpha) = 1$, $s_\alpha(\beta) = 0$ for every $\beta \neq \alpha \in Z_{\bar{K}}(A)$.*

Proof. We can without loss of generality suppose that the variable X_1 is separating. The classical Lagrange interpolation gives polynomials in X_1 with the required properties. □

Lemma 6.8. *If $Z_{\bar{K}}(A)$ is finite, then there exist for every $\alpha \in Z_{\bar{K}}(A)$ elements e_α of $\bar{A}$ with $\sum_{\alpha \in Z_{\bar{K}}(A)} e_\alpha = 1$, $e_\alpha^2 = e_\alpha$, $e_\alpha(\alpha) = 1$, $e_\alpha e_\beta = 0$ for $\beta \neq \alpha$ $\beta, \alpha \in Z_{\bar{K}}(A)$.*

Proof. Since $s_\alpha s_\beta$ vanishes on every common zero of $P_1, \ldots, P_s$, according to Hilbert's Nullstellensatz there exist powers of s_α, denoted by t_α, such that $t_\alpha t_\beta = 0$ in $\bar{\mathrm{A}}$, $t_\alpha(\alpha) = 1$. The family of polynomials $P_1, \ldots, P_s, t_\alpha$ has no common zeroes so, according to the Hibert Nullstellensatz, there exist polynomials r_α such that $\sum t_\alpha r_\alpha = 1$ in $\bar{A}$. Take as $e_\alpha = t_\alpha r_\alpha$. It is easy to verify the claimed properties. □

Denote by $\bar{A}_\alpha = L_{e_\alpha}(\bar{A})$ the image of $\bar{A}$ under the multiplication by e_α. It is clear that $\bar{A} \cong \prod_{\alpha \in Z_{\bar{K}}(A)} \bar{A}_\alpha$. □

The element e_α is called the *idempotent* attached to α and the canonical surjection of $\bar{A}$ onto $\bar{A}_\alpha$ coincides with the multiplication by e_α.

Proof of Stickelberger's theorem 6.3. It is clear that $L_f(\bar{A}_\alpha) \subset \bar{A}_\alpha$ since $\bar{A}_\alpha$ is the image of $\bar{A}$ under the multiplication by e_α. Since $e_\alpha(f - f(\alpha))$ vanishes on the common zeroes of $P_1, \ldots, P_s$ and according to Hilbert's Nullstellensatz, there exists m such that $e_\alpha(f - f(\alpha))^m = 0$, which means that $L_{f,\alpha} - f(\alpha)\,\mathrm{Id}$ is nilpotent. □

Corollary 6.9.

- *The trace of L_f is $\sum_{\alpha \in Z_{\bar{K}}(A)} f(\alpha)$.*
- *The determinant of L_f is $\prod_{\alpha \in Z_{\bar{K}}(A)} f(\alpha)$.*

This gives us a method, known as Rational Univariate Representation [1], for solving polynomial systems of equations. For any element $u \in A$, let $\chi(u, T)$ be the characteristic polynomial for the linear transformation, L_u. Then, according to the last results,

$$\chi(u, T) = \prod_{\alpha \in Z_{\bar{K}(A)}} (T - u(\alpha))^{\mu_\alpha}.$$

We will see that provided u is separating, the points of the variety $Z_{\bar{K}}(A)$, can be expressed as rational functions of the roots of $\chi(u, t)$.

We introduce now a new variable s, and consider the polynomial $\chi(u + sv, T) \in K[s, T]$, for some $v \in A$. If u is a separating element in A, then so is $u + sv$ for almost all values of s.

Let $g(u, v, T) = \dfrac{\partial \chi(u + vs, T)}{\partial s}|_{s=0}$ and $g(u, t) = \chi'(u, T)$. Substituting $T = u(\alpha)$ in the expression for $g^{(\mu-1)}(u, T)$, we have

$$g^{(\mu-1)}(u, u(\alpha)) = \mu! \prod_{\beta \in Z(I), \beta \neq \alpha} (u(\alpha) - u(\beta))^{\mu_\beta}.$$

Similarly, computing $g^{(\mu-1)}(u, v, T)$ and substituting $t = u(\alpha)$ we have

$$g^{(\mu-1)}(u, v, u(\alpha)) = -v(\alpha)\mu! \prod_{\beta \in Z(I), \beta \neq \alpha} (u(\alpha) - u(\beta))^{\mu_\beta},$$

(here $g^{(i)}(u, v, T)$, $g^{(i)}(u, T)$ are the ith derivatives with respect to t of $g(u, v, t)$ and $g(u, t)$ resp.). So we get:

Proposition 6.10. *Let $Z_\mu = \{\alpha \in Z_{\bar{K}}(A) \mid \mu_\alpha = \mu\}$. The following equality holds for every α in Z_μ:*

$$v(\alpha) = -\frac{g^{(\mu-1)}(u, v, u(\alpha))}{g^{(\mu-1)}(u, u(\alpha))}.$$

Thus, taking $v = X_i$, for $i = 1, \ldots, k$, we can express the coordinates of the points $\alpha \in Z_\mu$, in terms of rational functions of $u(\alpha)$, which in turn are roots of the polynomial $\chi(u, t)$.

We introduce now some terminology. A *univariate representation* is a $(k + 2)$-tuples of univariate polynomials, $(f, g_0, \ldots, g_k)$. A point α is *associated* to a univariate representation $(f, g_0, \ldots, g_k)$ if there exists a root β of f such that

$$\alpha = (\frac{g_1(\beta)}{g_0(\beta)}, \ldots, \frac{g_k(\beta)}{g_0(\beta)}).$$

The tuple $(f, g_0, \ldots, g_k)$ is a *univariate representation* of α.

The last proposition says that for every multiplicity μ and every α in Z_μ, α is associated to the univariate representation

$$(f(T), g_0(T), g_1(T), \ldots, g_k(T)),$$

with $f(T) = \chi(u, T)$, $g_0(T) = g^{(\mu-1)}(u, T)$, $g_i(T) = -g^{(\mu-1)}(u, X_i, T)$.

6.1.2. Univariate Representation Subroutine. In this paragraph, basic definitions and properties of Gröbner bases are considered as known. We start with a finite dimensional algebra $A = \mathrm{K}[X_1, \ldots, X_k]/(P_1, \ldots, P_s)$ and define $I = (P_1, \ldots, P_s)$.

The subroutine we are going to describe takes as input a Gröbner basis of I and outputs a set consisting of univariate representations such that $Z_{\bar{\mathrm{K}}}(A)$ is a subset of the points associated to these univariate representations.

Let N be the dimension of the vector space A. Then as we have seen in Lemma 6.6 among the family $\{X_1 + iX_2 + i^2X_3 + \cdots + i^{k-1}X_k \mid 0 \le i \le (k-1)\binom{N}{2}\}$, there is necessarily a separating element for A. The output of the Univariate Representation Subroutine is obtained by taking all characteristic polynomials $f(u,t) = \chi(u,t)$ for all

$$u \in \{X_1 + iX_2 + i^2X_3 + \cdots + i^{k-1}X_k \mid 1 \le i \le k\binom{N}{2} + 1\}$$

and constructing the $(k+2)$-tuples of polynomials

$$(f(u,t), g^{(\mu-1)}(u,t), -g^{(\mu-1)}(u, X_1, t), \ldots, -g^{(\mu-1)}(u, X_k, t)),$$

for $1 \le \mu \le \mathrm{degree}(f)$. It is clear from the results of the last subsection that $Z_{\bar{\mathrm{K}}}(A)$ is a subset of the points associated to these univariate representations.

Note that since a Gröbner basis is given in the input, the multiplication table in A, and hence the characteristic polynomials of any element u can be computed in a time polynomial in N. For $O(kN^2)$ choices of the polynomial u, we produce $\deg(f) = O(N)$ tuples of polynomials. Thus, the set of tuples is of size $O(kN^3)$. Moreover, the degree of each polynomial in the output is bounded by $O(N)$.

6.2. Some real algebraic geometry

We shall need some properties of semi-algebraically connected components and paths in general real closed fields. A full discussion of these can be found in [4] but we offer a brief summary below.

The order relation on a real closed field R defines, as usual, the euclidean topology on R^k. Semi-algebraic sets are finite unions of sets defined by a finite number of polynomial equalities and inequalities, and semi-algebraic homeomorphisms are homeomorphims whose graph is semi-algebraic.

In particular we have the following elementary properties of semi-algebraic sets over a real closed field R (see also [4])

- The projection of a semi-algebraic set of $\mathrm{R}^{k+\ell}$ on R^k is semi-algebraic. A subset of R^k defined through a first order formula of the language of ordered fields is semi-algebraic (this is the famous Tarski–Seidenberg principle).
- A semi-algebraic set S is *semi-algebraically connected* if it is not the disjoint union of two non-empty closed semi-algebraic sets in S. A *cell* is a semi-

algebraically connected component of a semi- algebraic set S: a maximal semi-algebraically connected subset of S. A semi-algebraic set has a finite number of cells.

- A *semi-algebraic path* between x and x' in R^k is a semi-algebraic subset γ, semi-algebraically homeomorphic to the unit interval of R through a semi-algebraic homeomorphism f_γ with $f_\gamma(0) = x$ and $f_\gamma(1) = x'$. A semi-algebraic path γ is *defined over* D if the graph of f_γ is described by polynomials with coefficients in D. A semi-algebraic set is semi-algebraically connected if and only if it is semi-algebraically path connected.

- In the case of real numbers, cells of semi- algebraic sets are ordinary connected components.

- Given a point x belonging to the closure of a semi-algebraic set $S \subset \mathrm{R}^k$, there exists a semi-algebraic function ϕ from $[0, 1]$ to R^k with $\phi(0) = x$, $\phi(]0, 1]) \subset S$ (this is the curve selection lemma).

- Given a semi-algebraic set S in R^k, and K a real closed field containing R, S_{K}, the *extension* of S to K, is the subset of K^k defined by the same boolean combination of inequalities that defines S. The semi-algebraically connected components of S_{K} are the extensions to K of the semi-algebraically connected components of S.

Let D be a subring of the real closed field R. A semi-algebraic set S is *defined over* D if it can be described by polynomials with coefficients in D.

A Puiseux series $f(\epsilon) \in \mathrm{R}\langle\epsilon\rangle$ is algebraic over $\mathrm{D}[\epsilon]$ if it is a root of a polynomial $P_\epsilon(X)$ with coefficients in $\mathrm{D}[\epsilon]$. The root $f(\epsilon)$ is distinguished among the other roots of P by its Thom encoding (see Section 5.1). For t sufficiently small and positive, we may replace ϵ by t in the polynomials which are the coefficients of $P_\epsilon(X)$ to obtain the polynomial $P_t(X)$ and (since inequalities in $\mathrm{D}[\epsilon]$ will agree with the corresponding inequalities in $\mathrm{D}[t]$ for t positive and small enough) $P_t(X)$ will have the same number of roots in R as $P_\epsilon(X)$ has in $\mathrm{R}\langle\epsilon\rangle$, with the corresponding Thom encodings and the algebraic Puiseux series $f(\epsilon)$ thus defines a semi-algebraic function $f(t)$ for t small enough.

Let $S(\epsilon)$ be a semi-algebraic set in $\mathrm{R}\langle\epsilon\rangle^k$ defined over $\mathrm{D}[\epsilon]$ and for $t \in \mathrm{R}$ let $S(t)$ be the semi-algebraic set in R^n obtained by substituting t for ϵ. Let $P(S(\epsilon))$ be a property of the semi-algebraic set $S(\epsilon)$ in K^n defined over $\mathrm{D}[\epsilon]$ which is expressible by a first order formula $\Phi(\epsilon)(x_1, \ldots, x_k)$ with parameters in $\mathrm{D}[\epsilon]$. Then for t positive and small enough $P(S(t))$ is satisfied.

This is an easy consequence of the Tarski–Seidenberg principle and of the preceding lemma on the ways signs are computed in $\mathrm{D}[\epsilon]$.

We will use this property to replace ϵ by t in a path defined over $\mathrm{D}[\epsilon]$ to obtain a "neighboring" path defined over $\mathrm{D}[t]$.

We shall also need the following lemma:

Lemma 6.11. *Let $f \in D[\epsilon, \frac{1}{\epsilon}][t]$, where D is an ordered domain contained in a real closed field* R, *and let $\alpha \in \mathrm{R}\langle\epsilon\rangle$ be a real root of f bounded over* R. *Then, $\beta = \mathrm{eval}_\epsilon(\alpha)$ is a root of $f_0(t)$ where $f_0(t)$ is the coefficient of the lowest power of ϵ occurring in f, when f is expressed in powers of ϵ.*

Proof. Without loss of generality assume that,

$$f(t, \epsilon) = f_0(t) + f_1(t)\epsilon + \cdots + f_d(t)\epsilon^d.$$

Since, α is bounded over R, its order as a Puiseux series in ϵ is non-negative. Let, $\alpha = \sum_{i \geq i_0} \alpha_i \epsilon^{\frac{i}{q}}$, where $i_0 \geq 0$ and $q > 0$. We consider two cases, $i_0 = 0$ and $i_0 > 0$. If $i_0 = 0$, then $\beta = \mathrm{eval}_\epsilon(\alpha) = \alpha_0$. If $f_0(\alpha_0)$ is not 0, substituting α for t in f we see that $f_0(\alpha_0)$ is the initial term of the Puiseux series $f(\alpha, \epsilon)$. This is impossible, since α is a root of f. Thus $f_0(\alpha_0) = 0$. If $i_0 > 0$, then $\beta = \mathrm{eval}_\epsilon(\alpha) = 0$. In this case we prove that the constant term in the polynomial $f_0(t)$ is 0. Let

$$f_0(t) = a_0 + a_1 t + \cdots + a_n t^n.$$

If a_0 is not 0, substituting α for t in f, we see that $f(\alpha, \epsilon)$ has initial term a_0. This is impossible since α is a root of f.

Thus in both cases $\beta = \mathrm{eval}_\epsilon(\alpha)$ is a root of $f_0(t)$. □

6.3. Finding points on hypersurfaces

We are going to prove the following result:

Theorem 6.12. *Let P be a polynomial in k variables of degree at most d with coefficients in an ordered domain* D *contained in a real closed field* R. *There exists an algorithm finding a point in every cell of $\mathrm{Z_K}(P)$ in $d^{O(k)}$ arithmetic operations in* D, *where* K *is a real closed field extending* R.

The idea ([2]) is to modify the polynomial P so that the zero set of the modified polynomial is a smooth, bounded hypersurface and has only a finite number of critical points of the X_1 function. We then apply the Univariate Representation Subroutine to this finite number of points.

6.3.1. Preliminaries. Given $Q \in \mathrm{R}[X_1, \ldots, X_k]$, we write $Z(Q)$ to be the set of real zeroes of Q inside R^k.

Lemma 6.13. *For $Q \in \mathrm{R}[X_1, \ldots, X_k]$ of degree d, we let*

$$Q_1 = Q^2 + ({X_1}^2 + \cdots + {X_{k+1}}^2 - \Omega^2)^2,$$

with Ω an infinitely large positive variable. Then the algebraic set $Z(Q_1) \subset (\mathrm{R}\langle 1/\Omega\rangle)^{k+1}$ is contained in the open ball of center 0 and radius $\Omega + 1$. Moreover, *the extension of every semi-algebraically connected component of* $Z(Q)$ *to* $\mathrm{R}\langle 1/\Omega\rangle$, *contains the projection onto* $(\mathrm{R}\langle 1/\Omega\rangle)^k$ *of a cell of* $Z(Q_1) \subset (\mathrm{R}\langle 1/\Omega\rangle)^{k+1}$.

Proof. The first part of the lemma is obvious from the definition of Q_1. The second part follows from the fact, that the extension of a cell, C of $Z(Q)$, to $(\mathrm{R}\langle 1/\Omega\rangle)^k$, will contain at least one point, say x, that is bounded over R. Then the projection of the component of $Z(Q_1)$ containing the point, $(x, (\Omega^2 - |x|^2)^{1/2})$ is contained in $C_{\mathrm{R}\langle 1/\Omega\rangle}$. □

Given a polynomial $Q \in \mathrm{R}[X_1, \ldots, X_k]$ we define the *total degree of* Q *in* X_i as the maximal total degree of the monomials in Q containing the variable X_i.

Proposition 6.14. *For a real closed field* R, *let* $Q \in \mathrm{R}[X_1, \ldots, X_k]$ *be a non-zero polynomial, non-negative over* R^k, *whose degree is bounded by* d, *such that* $Z(Q) \subset \mathrm{R}^k$ *is contained in the open ball with center* 0 *and radius* r, *for some* $r \in \mathrm{R}$. *Let* $d_1, d_2, \ldots, d_k$ *be the total degrees of* Q *in* $X_1, X_2, \ldots, X_k$, *respectively (and, without loss of generality,* $d_1 \geq d_2 \cdots \geq d_k$*). Let*

$$Q_1 = (1-\zeta)Q + \zeta(X_1^{2(d_1+1)} + \cdots + X_k^{2(d_k+1)} - k(r^{2(d_1+1)}))$$

where ζ *is a positive infinitesimal. Then the following holds:*

1. *The algebraic set* $Z(Q_1) \subset (\mathrm{R}\langle\zeta\rangle)^k$ *is bounded and smooth.*

2. *The polynomials*

$$Q_1, \frac{\partial Q_1}{\partial X_2}, \ldots, \frac{\partial Q_1}{\partial X_k}$$

   whose zero set defines K, *form a Gröbner basis for the degree lexicographical ordering with* $X_1 > \cdots > X_k$.

3. *The set* $K \subset Z(Q_1)$ *of critical points of* $Z(Q_1)$ *(over* $\bar{\mathrm{R}}\langle\zeta\rangle$*) of the projection map onto the* X_1 *coordinate is finite (where* $\bar{\mathrm{R}} = \mathrm{R}[i]$, *is the algebraic closure of* R*).*

4. *Let* K' *be the set of real critical points of this projection map (with coordinates in* $\mathrm{R}\langle\zeta\rangle$*). For every cell,* C, *of* $Z(Q)$, *there is a point* $p \in K'$, *such that* $\mathrm{eval}_\zeta(p)$ *belongs to* C.

Proof. Since Q is assumed to be non-negative over R^k, any zero of Q_1 satisfies the inequality

$$X_1^{2(d_1+1)} + \cdots + X_k^{2(d_k+1)} - k(r^{2(d_1+1)}) \leq 0,$$

which shows that the zeroes of Q_1 lie inside a bounded ball with center 0.

To prove that $Z(Q_1)$ is smooth, consider the family of hypersurfaces

$$Q_{1,t} = (1-t)Q + t(X_1^{2(d_1+1)} + \cdots + X_k^{2(d_k+1)} - k(r^{2(d_1+1)})).$$

The variety $Z(Q_{1,t})$ is smooth if and only if the set of solutions to the system of equations,

$$Q_{1,t} = \frac{\partial Q_{1,t}}{\partial X_1} = \cdots = \frac{\partial Q_{1,t}}{\partial X_k} = 0$$

is empty. The set of t's for which this system has no solutions is constructible in the Zariski topology, open, and contains $t = 1$. Hence it contains ζ which is transcendental and thus $Z(Q_1)$ is smooth.

We consider the ideal I defined by the polynomials

$$Q_1, \frac{\partial Q_1}{\partial X_2}, \ldots, \frac{\partial Q_1}{\partial X_k}.$$

These polynomials

$$Q_1, \frac{\partial Q_1}{\partial X_2}, \ldots, \frac{\partial Q_1}{\partial X_k}$$

form a Gröbner basis of the ideal with respect to the degree lexicographical ordering with $X_1 > \cdots > X_k$. This follows immediately from Buchberger's algorithm ([5], [19]), and the following lemma which we state after introducing the necessary notation. We are given an admissible total ordering on monomials (a total order compatible with multiplication) such that $X_1 > \cdots > X_k$. Given a polynomial P we write $\ell(P)$ for its leading monomial with respect to this order, $c(m, P)$ for the coefficient of the monomial m in the polynomial P. Thus $c(\ell(P), P)$, the leading coefficient of P is the coefficient of the leading monomial in P. Given two polynomials P_1, P_2 with leading coefficients 1, the *S-polynomial* of P_1, P_2 is defined by

$$S(P_1, P_2) = m \times P_1 - n \times P_2,$$

where $m = \ell(P_2)/\gcd(\ell(P_1), \ell(P_2))$, $n = \ell(P_1)/\gcd(\ell(P_1), \ell(P_2))$. Given a finite set of polynomials, G, with leading coefficients 1, we say that P *is reduced* to P_1 modulo G if there is a $Q \in G$ and a monomial m occurring in P such that, $\ell(Q)|m$, and $P_1 = P - c(m, P)\frac{m}{\ell(Q)}Q$. Moreover, we say that P is *reducible* to P_1 modulo G if there is a finite sequence of reductions modulo G going from P to P_1. According to Buchberger's algorithm ([5], [19]) an ideal basis, G, is a Gröbner basis if all the S-polynomials for all pairs of polynomials in the basis G, can be reduced to zero modulo G.

Lemma 6.15. *If* $c(\ell(Q_1), Q_1) = c(\ell(Q_2), Q_2) = 1$ *and* $\gcd(\ell(Q_1), \ell(Q_2)) = 1$ *then the S-polynomial of* Q_1 *and* Q_2, $S(Q_1, Q_2) = \ell(Q_2) \times Q_1 - \ell(Q_1) \times Q_2$ *is reducible to* 0 *modulo* Q_1 *and* Q_2.

Proof. Let $R_1 = Q_1 - \ell(Q_1)$, $R_2 = Q_2 - \ell(Q_2)$. Then $S(Q_1, Q_2) = \ell(Q_2)R_1 - \ell(Q_1)R_2$ and there is no monomial of $\ell(Q_2)R_1$ appearing in $\ell(Q_1)R_2$ (and vice versa). This is because all monomials in R_1 and R_2 are smaller (in the given ordering) than $\ell(Q_1)$ and $\ell(Q_2)$ respectively, and $\gcd(\ell(Q_1), \ell(Q_2)) = 1$.

We successively reduce every monomial of $S(Q_1, Q_2)$ coming from $\ell(Q_2)R_1$ using Q_2 and the result is $-\ell(Q_1)R_2 - R_1R_2$ and the monomials in $-\ell(Q_1)R_2$ are distinct from the monomials in R_1R_2. Then we reduce successively every monomial of $-\ell(Q_1)R_2 - R_1R_2$ coming from $-\ell(Q_1)R_2$ using Q_1 and the result is 0.

Thus $S(Q_1, Q_2)$ is reducible to 0 modulo Q_1 and Q_2. □

We now prove Part 3. The quotient ring $A = \mathrm{R}[X_1, \ldots X_k]/I$ is a vector space, which – since the polynomials

$$Q_1, \frac{\partial Q_1}{\partial X_2}, \ldots, \frac{\partial Q_1}{\partial X_k}$$

form a Gröbner basis of the ideal with respect to the degree lexicographical ordering with $X_1 > \cdots > X_k$ – is spanned by the power products not occurring in the ideal generated by the leading terms of the polynomials. Hence, the quotient ring is spanned by power products

$$X_1^{e_1} X_2^{e_2} \ldots X_k^{e_k},\ e_1 < 2(d_1+1), e_i < 2d_i + 1,\ 2 \le i \le k.$$

Thus, the quotient ring is a finite dimensional vector space, and hence the number of zeroes of the system (over $\bar{\mathrm{R}}\langle\zeta\rangle$) is finite.

We now prove Part 4 of the Proposition. Since the image of a bounded connected semi-algebraic set under eval_ζ is again semi-algebraically connected, it is enough to prove that every point y in $Z(Q)$ belongs to the image of $\mathrm{eval}_\zeta(Z(Q_1))$.

Let $y \in Z(Q)$. Then $Q_1(y)$ is a strictly negative element of $\mathrm{R}\langle\zeta\rangle$. Now since Q is everywhere non-negative, and 0 on a subset of codimension at least 1, fixing a ball B, centered at y, of radius r_1 (r_1 in R) there exists a point y' in B with $Q(y') \in \mathrm{R}$, $Q(y') > 0$, hence $Q_1(y') > 0$ in $(\mathrm{R}\langle\zeta\rangle)$ (because ζ is infinitesimal). This implies that the sign of Q_1 changes inside the ball $B_{\mathrm{R}\langle\zeta\rangle}$ and so there is a z with coordinates in $\mathrm{R}\langle\zeta\rangle$ such that $Q_1(z) = 0$ in every ball $B_{\mathrm{R}\langle\zeta\rangle}$. Hence, the point of $Z(Q_1)$ with minimal distance to y is infinitesimally close to y and is sent by eval_ζ to y. □

The set of critical points so obtained after perturbation is finite, and we need to solve the system

$$Q_1, \frac{\partial Q_1}{\partial X_2}, \ldots, \frac{\partial Q_1}{\partial X_k}.$$

in order to compute the coordinates of these critical points.

6.3.2. Cell Representatives Subroutine. This subroutine takes as input a polynomial Q and outputs points in every cell of the zeroes of Q in $\mathrm{K} = \mathrm{R}\langle 1/\Omega\rangle$. A similar algorithm is implicit in the approaches of Canny [6] and Renegar [23]. We follow closely the strategy in [2].

Given a polynomial Q, with coefficients in any ordered domain D, denote by $d_1, \ldots, d_k$ its total degrees in $X_1, \ldots, X_k$ (with $d_1 \ge \cdots \ge d_k$). The subroutine computes a set of univariate representations, of size $(d_1 \ldots d_k)^{O(1)}$. The set of points associated to these univariate representations intersect every cell of the set $Z(Q)$. Each univariate polynomial, in the tuples $(f(t), g_0(t), \ldots, g_k(t))$, has degree bounded by $O(d_1) \ldots O(d_k)$. The subroutine uses $(d_1 \ldots d_k)^{O(1)}$ arithmetic operations in D.

In the subroutine we will introduce extra variables Ω and ζ. In order to prove the correctness of our subroutine using Proposition 6.14 we will interpret Ω to be infinitely big, and ζ to be a positive infinitesimal with the ordering $1/\Omega \gg \zeta$. However, the Cell Representatives Subroutine itself, treats Ω and ζ as just extra variables. Since we do not perform any sign determination in this subroutine, we do not make use of the ordering in the field of coefficients.

We first introduce two new variables X_{k+1} and Ω, and replace Q by

$$Q^2 + (X_1^2 + \cdots + X_k^2 + X_{k+1}^2 - \Omega^2)^2.$$

We then introduce another new variable ζ and define

$$Q_1 = (1-\zeta)Q + \zeta(X_1^{2(d_1+1)} + \cdots + X_k^{2(d_k+1)} + X_{k+1}^6 - (k+1)(\Omega+1)^{2(d+1)}).$$

We next apply the k-representation subroutine to

$$Q_1, \frac{\partial Q_1}{\partial X_2}, \ldots, \frac{\partial Q_1}{\partial X_{k+1}}$$

to get a set of $(k+3)$-tuples $(f, g_0, \ldots, g_{k+2})$. Notice that according to Proposition 6.14 the above set of polynomials forms a Gröbner basis, G, for the ideal they generate with the ordering being the degree lexicographical ordering. Moreover, its quotient ring is spanned by the power-products not in $(LT(G))$ where $(LT(G))$ is the ideal generated by the leading terms (in the degree lexicographical ordering) of the polynomials in G. This is the set of all power products $X_1^{e_1} \ldots X_{k+1}^{e_{k+1}}$, with $e_1 < 2(d_1+1), e_i < 2d_i + 1,\ 2 \le i \le k,\ e_{k+1} < 5$.

Thus, the quotient ring is a finite dimensional vector space of dimension $5(2(d_1+1))(2d_2+1)\ldots(2d_k+1)$. The multiplication table for this basis is of size $(d_1 \ldots d_k)^{O(1)}$, and the characteristic polynomial $f(t) = \chi(u,t)$ is a polynomial in t of degree at most $O(d_1)\ldots O(d_k)$.

Notice that the construction of the multiplication table for the algebra A involves reductions by polynomials in the Gröbner basis G. Moreover, the leading terms of the polynomials in G are, $\zeta X_1^{2(d_1+1)}$, $\zeta X_i^{2d_i+1}$, for $2 \le i \le k$, and ζX_{k+1}^5. Thus, the product of two basis monomials in the multiplication table will be a polynomial $\in D[\Omega, \zeta, 1/\zeta, X_1, \ldots, X_{k+1}]$. Hence, the polynomials in the tuples produced by the k-representation subroutine, $f, g_0, \ldots, g_k$ are in $D[\Omega, \zeta, 1/\zeta][t]$.

In the next step, we let ζ go to 0 and retain only those points which do not go to infinity in the process. We do this purely algebraically. However, if we interpret ζ to be an infinitesimal, then this has the same effect as applying the eval_ζ map to the points (which are bounded) represented by the tuples $(f, g_0, \ldots, g_{k+1})$, produced in the previous step.

We describe this more precisely. Given a non-zero polynomial $h \in D[\Omega, \zeta, 1/\zeta, t]$

we write it as

$$h(\Omega, \zeta, 1/\zeta, t) = \sum_{i \geq \nu(h)} h^{[i]}(\Omega, t)\zeta^i,$$

$h^{[\nu(h)]} \neq 0$. We call $\nu(h)$ the order of h with respect to ζ.

For every tuple $(f, g_0, \dots, g_k)$ produced in the previous step, we only consider those for which $\nu(f) = \nu(g_0)$, and $\nu(g_i) \geq \nu(g_0)$, and ignore the rest.

For each tuple $(f, g_0, \dots, g_k)$ retained in the previous step, we replace it by the tuple

$$\left(f^{[\nu(f)]}, g_0^{[\nu(f)]}, \dots, g_{k+1}^{[\nu(f)]}\right).$$

It follows from Lemma 6.11 that the set of points associated to the tuples obtained as above, includes the image of the set of points bounded over $\mathrm{R}\langle 1/\Omega\rangle$, associated to the original set of tuples, under the eval_ζ map.

We now project the points constructed above onto their first k coordinates. Note that the points we obtained are defined over $\mathrm{D}[\Omega]$.

We can now prove the theorem.

Proof of Theorem 6.12. The correctness of the subroutine follows from the preliminaries. The number of arithmetic operation over D is clearly $d^{\mathrm{O}(k)}$. □

6.4. Finding non empty sign conditions

We follow closely [2].

We define the cells of a family of polynomials $\mathcal{P} = \{P_1, \dots, P_s\}$ as the cells of the non empty sign condition realized by $P_1, \dots, P_s$.

We will prove the following theorem.

Theorem 6.16. *Let $\mathcal{P} = \{P_1, \dots, P_s\}$ be a family of polynomials in $k < s$ variables each of degree at most d and each with coefficients in an ordered domain* D *contained in a real closed field* R. *The total numbers of cells of $\mathcal{P}$ is bounded by*

$$\sum_{\ell=1}^{k} \binom{O(s)}{\ell} O(d)^k.$$

There is an algorithm which outputs at least one point in each cell of $\mathcal{P}$ in a real closed field extension K *of* R *and provides the sign vector*

$$(\mathrm{sign}(P_1(x)), \dots, \mathrm{sign}(P_s(x)))$$

of $\mathcal{P}$ at each output point x. The algorithm terminates after at most

$$s \cdot \sum_{\ell=1}^{k} \binom{O(s)}{\ell} d^{O(k)}$$

arithmetic operations in D

6.4.1. Preliminaries The following proposition tells us that in order to find points in every cell of every non empty sign condition on $P_1, \dots, P_s$ it is enough to consider specific algebraic sets.

Proposition 6.17. *Let C be a cell of a non-empty sign condition of the form $P_1 = \cdots = P_\ell = 0, P_{\ell+1} > 0, \dots, P_s > 0$. Then we can find an algebraic set V in $(\mathrm{R}\langle\epsilon\rangle)^k$ defined by equations $P_1 = \cdots = P_\ell = P_{i_1} - \epsilon = \cdots = P_{i_m} - \epsilon = 0$, such that a cell D of V is contained in $C_{\mathrm{R}\langle\epsilon\rangle}$.*

Proof. If C is closed, it is a cell of the algebraic set defined by $P_1 = \cdots = P_\ell = 0$. If not, we consider Γ, the set of all semi-algebraic paths γ in R going from some point $x(\gamma)$ in C to a $y(\gamma)$ in $\bar{C} \setminus C$ such that $\gamma \setminus \{y(\gamma)\}$ is entirely contained in C. For any $\gamma \in \Gamma$, there exists $i > \ell$ such that P_i vanishes at $y(\gamma)$. Then on $\gamma_{\mathrm{R}\langle\epsilon\rangle}$ there exists a point $z(\gamma, \epsilon)$ such that one of the $P_i - \epsilon$ vanishes at $z(\gamma, \epsilon)$ and that on the portion of the path between x and $z(\gamma, \epsilon)$ no such $P_i - \epsilon$ vanishes. We write I_γ for the set of indices between $\ell + 1$ and s such that $i \in I_\gamma$ if and only if $P_i(z(\gamma, \epsilon)) - \epsilon = 0$. Now choose a path $\gamma \in \Gamma$ so that the set $I_\gamma = \{i_1, \dots, i_m\}$ is maximal under set inclusion and let V be defined by $P_1 = \cdots = P_\ell = P_{i_1} - \epsilon = \cdots = P_{i_m} - \epsilon = 0$.

It is clear that at $z(\gamma, \epsilon)$, defined above, we have $P_{\ell+1} > 0, \dots, P_s > 0$ and $P_j - \epsilon > 0$ for every $j \notin I_\gamma$. Let C' be the semi-algebraically connected component of V containing $z(\gamma, \epsilon)$. We shall prove that no polynomial $P_{\ell+1}, \dots, P_s$ vanishes on this cell, and thus that C' is contained in $C_{\mathrm{R}\langle\epsilon\rangle}$.

We suppose on the contrary that some new P_i $(i > \ell, i \notin I_\gamma)$ vanishes on C', say at y_ϵ. We can suppose without loss of generality that y_ϵ is defined over $A[\epsilon]$. Take a semi-algebraic path γ_ϵ defined over $\mathrm{R}[\epsilon]$ connecting $z(\gamma, \epsilon)$ to y_ϵ with $\gamma_\epsilon \subset C'$. Denote by $z(\gamma_\epsilon, \epsilon)$ the first point of γ_ϵ with

$$P_1 = \cdots = P_\ell = P_{i_1} - \epsilon = \cdots = P_{i_m} - \epsilon = P_j - \epsilon = 0$$

for some new j not in I_γ.

For t in R small enough, the set γ_t (obtained by replacing ϵ by t in γ_ϵ) defines a semi-algebraic path from $z(\gamma, t)$ to $z(\gamma_\epsilon, t)$ contained in C. Replacing ϵ by t, in the Puiseux series giving the coordinates of $z(\gamma_\epsilon, \epsilon)$, defines a path γ' containing $z(\gamma_\epsilon, \epsilon)$ from $z(\gamma_\epsilon, t)$ to $y = \mathrm{eval}(z(\gamma_\epsilon, \epsilon))$ (which is a point of $\bar{C} \setminus C$). We consider the new path γ^* consisting of the beginning of γ (up to the point z_t for which $P_{i_1} = \cdots, P_{i_m} = t$), followed by γ_t and then followed by γ'. Now the first point in γ^* such that there exists a new j with $P_j - \epsilon = 0$ is $z(\gamma_\epsilon, \epsilon)$ and thus $\gamma^* \in \Gamma$ with I_{γ^*} strictly larger than I_γ. This is impossible by the maximality of I_γ. □

To compute a point in each cell defined by $\mathcal{P}$ it is thus sufficient to compute a point in each algebraic set defined by a subset of the extended family of $3s$ polynomials $\mathcal{P}_\epsilon = \{P_1, P_1 - \epsilon, P_1 + \epsilon, \dots, P_s, P_s - \epsilon, P_s + \epsilon\}$. In principle, there can be as many as $3 \cdot 2^s$ of these algebraic sets. To limit the number of algebraic sets that our algorithms need to examine, we perturb the given polynomials to bring them into general position. We say that a set of polynomials, $\mathcal{P} = \{P_1, \dots, P_s\}$, in k variables, is in *general position* if no $k+1$ have a common zero in R^k. (This is a weak notion of general position as we assume neither smoothness nor transversality).

Henceforth, for any polynomial $P(X_1, \ldots, X_k)$, we let $P^h(X_0, X_1, \ldots, X_k)$ denote the homogenization of P with respect to an additional variable X_0.

Define

$$H_i = 1 + \sum_{1 \le j \le k} i^j X_j^e.$$

Lemma 6.18. *For any positive integer e, the polynomials*

$$H_i^h = X_0^e + \sum_{1 \le j \le k} i^j X_j^e$$

are in general position in projective (over $\bar{\mathrm{R}} = \mathrm{R}[i]$, the algebraic closure of R*) k-space.*

Proof. Take $H = (X_0{}^e + \sum_{1 \le j \le k} t^j X_j^e)$. If $k+1$ of the H_i^h had a common zero $\bar{x}$, in the projective k-space (over $\bar{\mathrm{R}}$), substituting this root in H would give a non-zero univariate polynomial of degree at most k with $k+1$ distinct roots, which is impossible. □

Since a common zero of a set of H_i would certainly produce a common zero of the corresponding set of H_i^h we have:

Corollary 6.19. *The polynomials H_i are in general position.*

Proposition 6.20. *With $e \ge d$, the polynomials*

$$(1-\delta)P_i - H_{2i-1}, \quad (1-\delta)P_i - \delta H_{2i}$$

are in general position.

Proof. Let $Q_{2i-1,t} = (1-t)P_i - H_{2i-1}$, $Q_{2i,t} = (1-t)P_i - tH_{2i}$. Consider the set T of those t such that the $Q_{j,t}$ are in general position in projective k-space over $\bar{\mathrm{R}}$, which means that no $k+1$ subset of the $Q_{j,t}^h$ has a common zero in projective k-space over $\bar{\mathrm{R}}$. The set T contains 1, according to the preceding lemma. Hence it contains an open interval containing 1, since being in general position in projective k-space over $\bar{\mathrm{R}}$ is a stable condition. It is also Zariski constructible, since it can be defined by a first order formula of the language of algebraically closed fields. So the transcendental element δ belongs to the extension of T to $\mathrm{R}(\delta)$, which proves the lemma. □

The following propositions will guarantee that we recover points in every cell of the original polynomials, after we have computed points in every cell of a perturbed set of polynomials.

Proposition 6.21. *If $S' \subset (\mathrm{R}\langle\epsilon\rangle)^k$ is a semi-algebraic set defined over* $\mathrm{D}[\epsilon]$ *and* $S = \mathrm{eval}_\epsilon(S')$, *then S is a semi-algebraic set. Moreover, if S' is bounded and connected then S is connected.*

Proof. Suppose that $S' \subset (\mathrm{R}\langle\epsilon\rangle)^k$ is described by a quantifier-free formula $\Phi(\epsilon)(X_1, \dots, X_k)$. Introduce a new variable X_{k+1} and denote by

$$\Phi(X_1, \dots, X_k, X_{k+1})$$

the result of substituting X_{k+1} for ϵ in $\Phi(\epsilon)(X_1, \dots, X_k)$.

Embed R^k in R^{k+1} by sending $(X_1, \dots, X_k)$ to $(X_1, \dots, X_k, 0)$ so that S is a subset of $Z(X_{k+1})$. We prove that $S = \overline{T} \cap Z(X_{k+1})$ where

$$\begin{aligned} T = \{ \quad & (x_1, \dots, x_k, x_{k+1}) \in \mathrm{R}^{k+1} | \\ & \Phi((x_1, \dots, x_k, x_{k+1}) \text{ and } x_{k+1} > 0\} \end{aligned}$$

and $\overline{T}$ is the closure of T in the euclidean topology.

If $x \in S$ then there exists $z \in S'$ such that $\mathrm{eval}_\epsilon(z) = x$. Let $B_x(r)$ denote the open ball of radius r centered at x. Since (z, ϵ) belongs to the extension of $B_x(r) \cap T$ to $\mathrm{R}\langle\epsilon\rangle$ it follows that $B_x(r) \cap T$ is non- empty, and hence that $x \in \overline{T}$.

Conversely, let x be in $\overline{T} \cap Z(X_{k+1})$. Using the semi-algebraic curve selection lemma [4], there exists a semi-algebraic function f from $[0, 1]$ to $\overline{T}$ with $f(0) = x$ and $f((0, 1]) \subset T$. This semi-algebraic function defines a point z whose coordinates lie in $\mathrm{R}\langle\epsilon\rangle$ and belongs to S' and moreover $\mathrm{eval}_\epsilon(z) = x$.

If S' is bounded over R by M and connected then there exists a positive t in R such that $T \cap (B_0(M) \times [0, t])$ is semi-algebraically connected. It follows easily that $S = \overline{T} \cap (B_0(M) \times [0, t]) \cap Z(X_{k+1})$ is connected. □

We will also need to consider perturbations with an infintely big element Ω and two additional positive infinitesimals δ, η with the ordering $1/\Omega \gg \delta \gg \eta$.

Proposition 6.22. *Let $\mathcal{P} \subset \mathrm{R}[X_1, \dots, X_k]$ be any set of s polynomials, $P_1, \dots, P_s$, and let C be a cell of the sign condition*

$$\begin{aligned} P_1 = P_2 = \cdots = P_\ell = 0 \\ P_{\ell+1} > 0, \dots, P_s > 0. \end{aligned}$$

Then there exists a cell C' in $\mathrm{R}\langle\delta, \eta\rangle^k$, of the semi-algebraic set defined by the following inequalities:

$$-\eta\delta H_{4i} < (1-\eta)P_i < \eta\delta H_{4i-1}, \ 1 \le i \le \ell,$$

$$(1-\eta)P_i > \delta H_{4i-3}, \ \ell + 1 \le i \le s,$$

$$X_1^2 + \cdots + X_k^2 \le \Omega^2$$

such that $\mathrm{eval}_\eta(C')$ is contained in the extension of C to $\mathrm{R}\langle 1/\Omega, \delta\rangle$.

Proof. If $x \in C$, then x satisfies the following equalities and inequalities

$$-\eta\delta H_{4i} < (1-\eta)P_i < \eta\delta H_{4i-1},\ 1 \le i \le \ell,$$

$$(1-\eta)P_i > \delta H_{4i-3},\ \ell+1 \le i \le s,$$

$$X_1^2 + \cdots + X_k^2 \le \Omega^2$$

in $\mathrm{R}\langle 1/\Omega, \delta, \eta\rangle$. Let C' be the cell of the semi-algebraic set in $(\mathrm{R}\langle 1/\Omega, \delta, \eta\rangle)^k$ defined by the above equalities and inequalities, which contains x.

It is clear that $\mathrm{eval}_\eta(C')$ is contained in the semi- algebraic set defined by the sign condition $P_1 = \cdots = P_\ell = 0, P_{\ell+1} > 0, \ldots, P_s > 0$, in $(\mathrm{R}\langle 1/\Omega, \delta\rangle)^k$ and that it also contains $x \in C$. Since, by Proposition 6.21, $\mathrm{eval}_\delta(C')$ is also connected the statement of the lemma follows. □

6.4.2. Sample Points Subroutine. We are now in a position to outline the subroutine (henceforth, referred to as the Sample Points Subroutine, see [2]) to generate sample points in every cell of $\mathcal{P}$.

The input is a family of s polynomials $\mathcal{P} = \{P_1, \ldots, P_s\} \subset D[X_1, \ldots, X_k]$, each of degree at most d and the output a set of $s^k(O(d))^k$ tuples of the form $(f, g_0, \ldots, g_k)$, where $f, g_0, \ldots, g_k \in D[t]$ are k-representations such that the associated points meet every cell of $\mathcal{P}$.

The algorithm is the following: Replace the set $\mathcal{P}$ by the following set, $\mathcal{P}^*$, of $4s$ polynomials:

$$\mathcal{P}^* = \bigcup_{i=1,\ldots,s} \{(1-\eta)P_i - \delta H_{4i-3}, (1-\eta)P_i + \delta H_{4i-2},$$

$$(1-\eta)P_i - \eta\delta H_{4i-1}, (1-\eta)P_i + \eta\delta H_{4i}\},$$

where e is an even number greater than the degree of any P_i. Enlarge the set of $4s$ polynomials to the following set, $\bar{\mathcal{P}}^*$, of $12s$ polynomials:

$$\bar{\mathcal{P}}^* = \bigcup_{P \in \mathcal{P}^*} \{P, P+\epsilon, P-\epsilon\}$$

where ϵ is a new variable. Note that in the proof of correctness of the algorithm, we make use of the ordering $\delta \gg \eta \gg \epsilon$, and it is useful to keep this ordering in mind while understanding the subroutine, though the subroutine itself makes no use of this ordering.

For every $\ell \le k$-tuple of polynomials $Q_{i_1}, \ldots, Q_{i_\ell}$ consider

$$Q = Q_{i_1}^2 + \cdots + Q_{i_\ell}^2 + (X_1^2 + \cdots + X_k^2 + X_{k+1}^2 - \Omega^2)^2$$

and use the Cell Representatives Subroutine, omitting the first step, with Q as input, and obtain a set of tuples of the form

$$(f(t), g_0(t), \ldots, g_k(t)),$$

where $f(t), g_i(t) \in D[\delta, \eta, \epsilon, \Omega][t]$.

Apply the eval_η map, by letting $\eta, \epsilon, 1/\Omega \to 0$, and obtain polynomials and rational functions defined over $D[\delta]$.

We describe this more precisely. Given a non-zero polynomial $g \in D[\delta, \eta, \epsilon, \Omega, t]$ we write it as $\sum_{i \geq \nu(g)} g^{[i]}(\delta, t)\epsilon^{i_1}\eta^{i_2}\Omega^{i_3}$, where the multi-indices $i = (i_1, i_2, i_3)$ are ordered by the lexicographical ordering and $g^{[i]} \in D[\delta, t]$, $g^{[\nu(g)]} \neq 0$. We call $\nu(g)$ the order of g with respect to (ϵ, η, Ω).

For every tuple $(f, g_0, \ldots, g_k)$ produced in the previous step, we only consider those for which $\nu(f) = \nu(g_0)$, and $\nu(g_i) \geq \nu(g_0)$, and ignore the rest.

For each tuple $(f, g_0, \ldots, g_k)$ retained in the previous step, we replace it by the tuple

$$\left(f^{[\nu(f)]}, g_0^{[\nu(f)]}, \ldots, g_{k+1}^{[\nu(f)]}\right).$$

It follows from Lemma 6.11 that the set of points associated to the tuples obtained as above, includes the image of the set of points bounded over $\mathrm{R}\langle\delta\rangle$, associated to the original set of tuples, under the eval_η map.

The proof of correctness of the above subroutine follows from the preliminaries and the correctness of the Cell Representatives Subroutine.

For the complexity analysis, first note that, since we have introduced only five additional variables, and all the algebraic computations are done in the ordered ring $D[\delta, \eta, \epsilon, \Omega, \zeta]$ using only linear algebra subroutines, the asymptotic complexity is not affected by the introduction of these variables.

The total number of $\ell \leq k$-tuples examined is $\sum_{\ell \leq k} \binom{12s}{\ell}$. Hence, the number of calls to the *Cell Representatives Subroutine* is also bounded by $\sum_{\ell \leq k} \binom{12s}{\ell}$. Each such call costs $d^{O(k)}$ arithmetic operations in D, and produces $d^{O(k)}$ k-representations, whose polynomials have degree bounded by $(O(d))^k$. Thus the number of arithmetic operations performed is bounded by $\sum_{\ell \leq k} \binom{12s}{\ell} d^{O(k)}$ while the number of points output is bounded by $\sum_{\ell \leq k} \binom{12s}{\ell} 0(d)^k$.

In order to ouptut the sign vector

$$(\text{sign}(P_1(x)), \ldots, \text{sign}(P_s(x)))$$

of $\mathcal{P}$ at each output point x, we proceed as follows. Given a univariate representation

$$u(T) = (f(t), g_0(T), \ldots, g_k(T)),$$

and a polynomial $P(X_1, \ldots, X_k) \in \mathcal{P}$, we denote by $P_u(T)$ the polynomial obtained by substituting g_i for X_i in

$$X_0^e P(\frac{X_1}{X_0}, \ldots, \frac{X_k}{X_0}).$$

(where e is the smallest even number greater than the degree of P). We can now use the results of section 5 on univariate sign determinations. The total cost of the algorithm is $s \cdot \sum_{\ell \leq k} \binom{12s}{\ell} d^{O(k)}$ arithmetic operations in D.

This ends the proof of Theorem 6.16.

References

[1] M. E. Alonso, E. Becker, M.-F. Roy, T. Wörmann, Zeroes, Multiplicities and Idempotents for Zerodimensional Systems, MEGA 94.

[2] S. Basu, R. Pollack, M.-F. Roy, A new algorithm to find a point in every cell defined by a family of polynomials, in: Quantifier Elimination and Cylrindical Algebraic Decomposition, Texts Mongr. Symb. Comput. B. Caviness and J. Johnson, eds., Springer-Verlag, Wien–New York, to appear.

[3] M. Ben-Or, D. Kozen, J. Reif, The complexity of elementary algebra and geometry, J. Comput. System Sci. 32 (1986), 251–264.

[4] J. Bochnak, M. Coste, M.-F. Roy, Géométrie algébrique réelle, Ergeb. Math. Grenzgeb. (3) 12, Springer-Verlag, Berlin 1987.

[5] B. Buchberger Gröbner bases: an algorithmic method in polynomial ideal theory, Recent trends in multidimensional systems theory, Reider ed., Bose, 1985.

[6] J. Canny, Some algebraic and geometric computations in PSPACE, in: Proc. Twentieth ACM Symp. on Theory of Computing, 460–467, 1988.

[7] G. Collins, Quantifier elimination for real closed fields by cylindric algebraic decomposition, in: Second GI Conference on Automata Theory and Formal Languages, Lecture Notes in Comput. Sci. 33, pp. 134–183, Springer-Verlag, Berlin 1975.

[8] M. Coste, M.-F. Roy, Thom's lemma, the coding of real algebraic numbers and the topology of semi-algebraic sets, J. Symbolic Comput. 5 (1988), 121–129.

[9] F. Cucker, H. Laneau, B. Mishra, P. Pedersen, M.-F. Roy, NC algorithms for Real algebraic numbers, Appl. Algebra Engrg. Comm. Comput. 3 (1992), 79–98.

[10] G. Frobenius Über das Traegheitsgesetz des quadratischen Formen, Sitzungsber. Preuss. Akad. Wiss. 241–256 (März 1984) und 403–431 (Mai 1884).

[11] Fr. Gantmacher, Théorie des matrices, tome I, Dunod Paris 1966.

[12] L. Gonzalez, H. Lombardi, T. Recio, M.-F. Roy, Spécialisation de la suite de Sturm et sous-résultants I, Informatique théorique et applications 24 (1990), 561–588.

[13] L. Gonzalez, H. Lombardi, T. Recio, M.-F. Roy, Spécialisation de la suite de Sturm, Informatique théorique et applications 28 (1994), 1–24.

[14] L. Gonzalez, H. Lombardi, T. Recio, M.-F. Roy, Sturm–Habicht sequence, determinants and real roots of univariate polynomials, in: Quantifier Elimination and Cylrindical Algebraic Decomposition, Texts Mongr. Symb. Comput. B. Caviness and J. Johnson, eds., Springer-Verlag, Wien–New York 1994, to appear.

[15] W. Habicht, Eine Verallgemeinerung des Sturmschen Wurzelzählverfahrens, Comment. Math. Helv. 21 (1948), 99–116.

[16] C. Hermite, Remarques sur le théorème de Sturm, C. R. Acad. Sci. Paris 36 (1853), 52–54.

[17] Z. Ligatsikas, M.-F. Roy, Séries de Puiseux sur un corps rŕel clos, C. R. Acad. Sci. Paris 311 (1990), 625–628.

[18] R. Loos, Generalized polynomial remainder sequences in: Computer algebra, symbolic and algebraic computation. Springer-Verlag, Berlin 1982.

[19] B. Mishra, Algorithmic Algebra, Texts Monographs Comput. Sci., Springer-Verlag, New York 1993.

[20] P. Pedersen, Counting real zeroes of polynomials, Ph. D. Thesis, Courant Institute, New York University 1991.

[21] P. Pedersen, M.-F. Roy, A. Szpirglas, Counting real zeroes in the multivariate case, Computational algebraic geometry, Eyssette et Galligo, eds. Progr. Math. 109, pp. 203–224, Birkhäuser, Boston 1993.

[22] R. Pollack, M..-F. Roy, On the number of cells defined by a set of polynomials, C. R. Acad. Sci. Paris, 316 (1993), 573–577.

[23] J. Renegar, On the computational complexity and geometry of the first-order theory of the reals, parts I, II and III, J. Symbolic Comput. 13 (1992), 255–352.

[24] F. Rouillier, Formules de Bareiss et réduction de formes quadratiques, C. R. Acad. Sci. Paris 320 (1995), 1273–1278.

[25] M.-F. Roy, A. Szpirglas, Complexity of computations with real algebraic numbers J. Symbolic Comput. 10 (1990), 39–51.

[26] C. Sturm, Mémoire sur la résolution des équations numériques, Inst. France Soc. Math. Phys. 6 (1835).

[27] J. J. Sylvester, On a theory of syzygetic relations of two rational integral functions, comprising an application to the theory of Sturm's function, Trans. Roy. Soc. London (1853).

[28] B. L. Van Der Waerden, Modern Algebra, Volume II, F. Ungar Publishing Co. 1950.

[29] R. J. Walker, Algebraic Curves, Princeton University Press 1950.

IRMAR (URA CNRS 305), Université de Rennes, Campus de Beaulieu,
35042 Rennes Cedex, France
e-mail: `marie-francoise.coste-roy@univ-rennes1.fr`

Nash functions and manifolds

Masahiro Shiota

§1. Introduction

A *semialgebraic set* in $\mathbb{R}^n$ is a finite union of sets of the form

$$\{x \in \mathbb{R}^n : f(x) = 0,\ g_1(x) > 0, \dots, g_k(x) > 0\},$$

where $f, g_1, \dots, g_k$ are polynomials on $\mathbb{R}^n$. An *affine Nash manifold* in $\mathbb{R}^n$ is semialgebraic analytic submanifold of $\mathbb{R}^n$. A *Nash mapping* between affine Nash manifolds is an analytic mapping with semialgebraic graph.

These concepts are equivalently defined as follows. A *semialgebraic set* in $\mathbb{R}^n$ is the image of an algebraic set under the projection $\mathbb{R}^n \times \mathbb{R}^m \to \mathbb{R}^n$ (Tarski–Seidenberg). A C^ω function f on an open semialgebraic set U in $\mathbb{R}^n$ is called *Nash* if it is algebraic, i.e., there is a nonzero polynomial P in $n+1$ variables such that $P(x, f(x)) = 0$ for $x \in U$. A *Nash set* X in U is a common zero set of Nash functions on U, and a *Nash function* f on $X \subset U$ is the restriction of a Nash function on U to X. (We sometimes call X and f simply a Nash set and a Nash function on X, respectively, because their definitions do not depend on U, to be precise, if X is a Nash set in U and if f is a Nash function on $X \subset U$, then X is a Nash set in $\mathbb{R}^n - (\overline{X} - X)$, and f is the restriction of a Nash function on $\mathbb{R}^n - (\overline{X} - X)$ (Theorem 3.12).) A Nash set $X \subset \mathbb{R}^n$ is called an *affine Nash manifold* if one of the following four equivalent conditions is satisfied (Theorem 3.12).

(1) The set X is everywhere C^ω smooth and of constant local dimension.

(2) There exist Nash functions $f_1, \dots, f_k$ on $\mathbb{R}^n - (\overline{X} - X)$ whose common zero set is X and such that for each point $x \in X$, X is C^ω smooth, the local dimension of X at x is constant, and grad $f_1, \dots,$ grad f_k span the normal vector space of X at x in $\mathbb{R}^n$.

(3) Let $N(X)$ denote the ring of Nash functions on X. Then for each $x \in X$, the local ring of $N(X)$ at x is a regular local ring of constant Krull dimension.

(4) Let $\mathcal{N}$ denote the sheaf of Nash function germs on X. Then each stalk $\mathcal{N}_x$ is a regular local ring of constant Krull dimension.

A mapping $f = (f_1, \dots, f_m)$ from a Nash set $X \subset \mathbb{R}^n$ to another $Y \subset \mathbb{R}^m$ is called *Nash* if each f_i is a Nash function. (By a theorem of Malgrange we can replace "C^ω" in the above definitions with "C^∞" (2.27).) Important open problems are whether a semialgebraic set is an affine Nash set if it is the real part of a complex analytic set (Conjecture $\mathrm{II}^{r\prime\prime}$ in §5), and whether an analytic function on a Nash set with semialgebraic graph is a Nash function (Conjecture $\mathrm{III}^{r\prime\prime}$ in §5).

Affine Nash manifolds and mappings form a category. This paper aims to show a theory of this Nash category. The Nash category is not well-known, and may seem to be artificial. But it enjoys many advantages over other categories: the analytic category of real analytic manifolds and analytic mappings, and the algebraic category of affine nonsingular real algebraic sets and smooth rational mappings. The Nash category lies between these categories. The analytic category has too many objects because any C^∞ manifold admits a unique analytic manifold structure, and the algebraic category has too few morphisms to construct a desirable theory (although, for this reason, many interesting singular phenomena happen in polynomial mappings and smooth rational mappings). On the other hand, the Nash category has suitably many objects because any affine Nash manifold is Nash diffeomorphic to an affine nonsingular real algebraic set (Theorem 4.10) and also to the interior of a unique compact Nash manifold possibly with boundary (Theorem 4.6). Moreover, it has enough but not too many morphisms to construct a good theory, which we will elucidate here. In the case of compact affine Nash manifolds we can say that there are as many morphisms as analytic mappings (Theorem 5.6.a). The noncompact case is more interesting. Behavior of Nash mappings at infinity is harmonized. Our main purpose in this article is to show the harmony.

One of the most important properties of algebraic varieties over algebraically closed fields is that the image of an algebraic set under a projection is a finite union of Zariski locally closed sets. But this does not hold for $\mathbb{R}$, and the image is semialgebraic. This is the reason why real algebraic geometry and, of course, Nash geometry are based on semialgebraic sets but not algebraic sets.

It is mainly Lojasiewicz [L] who showed elementary properties of semialgebraic sets. The terminology of an affine Nash manifold comes from a paper of Nash [N], which proves that a compact differentiable manifold is diffeomorphic to a union of connected components of an affine nonsingular real algebraic sets. (Such a union is an affine Nash manifold because a connected component of a semialgebraic set is semialgebraic (2.3).) Artin and Mazur [A-M] paid attention to Nash functions and manifolds, proved a fundamental theorem on them (2.25), and applied it to dynamical systems. The first systematic study of the Nash category is Palais [Pa] (unpublished), and Shiota [S_1] is the second, which contains almost all facts on the Nash category, known at that point of time. We repeat some of the results of [S_1] which are necessary for our theory, without detailed proofs. The reference for their details is [S_1] except for §5.

As preliminaries, we explain in §2 some fundamental facts on semialgebraic sets and Nash functions, and review some theorems of algebraic geometry. In §3 we show the Approximation Theorem (which is the central tool in Nash geometry) and applications (e.g. the Extension Theorem). After that we develop in §4 a theory of affine Nash manifolds, which is the center of this paper. In §5 we clarify relations of the Nash function ring on an affine Nash manifold M, the sheaf of Nash function germs on M, and the germs of all complex Nash sets at M, and we explain four important equivalent conjectures, two of which are Nash versions of Cartan's Theorems A and B. These conjectures holds for compact affine Nash manifolds by Neron's resolution of singularities holds true. §6 is an appendix, and we there treat equivariant Nash geometry.

The main idea to prove theorems is the following. Given a problem on the Nash category, we prove it in the C^r Nash category, $0 < r < \infty$, and then using Approximation Theorem 3.3 we approximate the C^r Nash solution by a Nash solution. Here the C^r Nash category is an extended category of the Nash category, where objects are semialgebraic C^r submanifolds of Euclidean spaces, and morphisms are semialgebraic C^r mappings. This procedure works well for the following two reasons. First we can use a partition of unity in the C^r Nash category and, hence, problems become local. Secondly, the Approximation Theorem holds in relative forms, and a Nash approximation of a C^r Nash solution continues to be a solution.

We consider the Nash category over $\mathbb{R}$. But most of the results hold over any real closed field. We do not treat much C^0 Nash manifolds nor nonaffine Nash manifolds. See [S_1] for them.

The author thanks M. Galbiati and S. Koike for their advices.

§2. Nash functions

Definition 2.1. For $r = 1, 2, \ldots, \infty$ or ω, an *affine C^r Nash manifold* in $\mathbb{R}^n$ is a semialgebraic C^r submanifold of $\mathbb{R}^n$. A *$C^{r'}$ Nash mapping*, $0 \le r' \le r$, from an affine C^r Nash submanifold $M \subset \mathbb{R}^n$ to another $M' \subset \mathbb{R}^{n'}$ is a $C^{r'}$ mapping with semialgebraic graph, and let $N^{r'}(M, M')$ denote the set of all $C^{r'}$ Nash mappings. If $M' = \mathbb{R}$ we write $N^{r'}(M)$. If $r = r' = \omega$ we sometimes call an *affine Nash manifold* and a *Nash mapping*, and write $N(M, M')$ and $N(M)$. An *affine C^r Nash vector bundle* is a vector bundle $\xi = (E, p, B)$ such that the total space E and the base space B are affine C^r Nash manifolds, B is covered by a finite number of semialgebraic coordinate neighborhoods, and all the coordinate functions and the projection p are C^r Nash mappings. An *affine Nash vector bundle* is an affine C^ω Nash vector bundle. In the same way we define an *affine C^r Nash fibre bundle* and an *affine Nash fibre bundle*. A *semialgebraic mapping* between semialgebraic sets is a mapping with semialgebraic graph, which is not necessarily continuous. An *affine C^0 Nash manifold M* in $\mathbb{R}^n$ of dimension m is a semialgebraic set in $\mathbb{R}^n$ such that the pair $(\mathbb{R}^n, M)$ is locally semialgebraically homeomorphic to the pair $(\mathbb{R}^n, \mathbb{R}^m \times 0)$ at each point of M, where $\mathbb{R}^m \times 0 \subset \mathbb{R}^n$. A *$C^0$ Nash mapping* and an *affine C^0 Nash* *vector (fibre) bundle* are similarly defined. A *Nash set* in an affine Nash manifold M is the zero set of a Nash function on M. A *Nash mapping* from a Nash set $X \subset M$ to another $X' \subset M'$ is the restriction to X of a Nash mapping from an open semialgebraic neighborhood of X in M to one of X' in M'.

2.2. For a nonempty semialgebraic set $X \subset \mathbb{R}^n$, the closure $\overline{X}$ is semialgebraic, and $\dim(\overline{X} - X) < \dim X$.

2.3. The number of connected components of a semialgebraic set is finite, and each component is semialgebraic.

2.4. Let X and Y_i, $i = 1, \dots, j$, be semialgebraic sets in $\mathbb{R}^n$. Then X admits a *Nash Whitney stratification* $\{X_k\}$ compatible with $\{Y_i\}$, i.e., each X_k is an affine Nash manifold in $\mathbb{R}^n$, X is the finite disjoint union of X_k, for each X_k, $\overline{X_k} - X_k$ is a union of $X_{k'}$, for each Y_i, $Y_i \cap X$ is a union of X_k, and each pair X_k and $X_{k'}$ satisfies the Whitney condition (see [G-al] for the definition of the Whitney condition).

2.5. Let $X \subset \mathbb{R}^n$, $Y \subset \mathbb{R}^m$ and $Z_i \subset \mathbb{R}^m$, $i = 1, \dots, l$, be semialgebraic sets, and let $f: X \to Y$ be a semialgebraic mapping. Then f admits a *Nash Whitney stratification* $f: \{X_j\} \to \{Y_k\}$ compatible with $\{Z_i\}$, i.e., $\{X_j\}$ is a Nash Whitney stratification of X, $\{Y_k\}$ is a Nash Whitney stratification of Y compatible with $\{Z_i\}$, and for each X_j, $f|_{X_j}$ is a C^ω submersion onto some Y_k.

2.6. The dimension of the Zariski closure of a semialgebraic set in the ambient Euclidean space is the same as the dimension of the semialgebraic set.

2.7. For a nonempty semialgebraic set $X \subset \mathbb{R}^n$, the function

$$\mathbb{R}^n \ni x \longrightarrow \operatorname{dist}(x, X) \in \mathbb{R}$$

is semialgebraic.

2.8 (Lojasiewicz). Let f_1 and f_2 be continuous semialgebraic functions on a locally closed semialgebraic set $X \subset \mathbb{R}^n$ such that

$$f_1^{-1}(0) \subset f_2^{-1}(0).$$

Then there exist a semialgebraic neighborhood U of $f_2^{-1}(0)$ in X and a natural number m such that

$$|f_2(x)|^m \leq |f_1(x)| \quad \text{for } x \in U.$$

2.9. The stereographic projection carries a semialgebraic set, a Nash set and an affine Nash manifold to respective bounded ones included in a sphere.

2.10 (Artin–Mazur, Efroymson, Risler). For a connected affine Nash manifold M and $x \in M$, the set $N(M)$ and the set of Nash function germs at x in M are Noetherian regular rings.

2.11. For a prime ideal $\mathfrak{p}$ of the Nash function germ ring at 0 in $\mathbb{R}^n$, the ideal of the convergent power series ring in n-variables, generated by $\mathfrak{p}$, is prime.

2.12. Partial derivatives of order r' of a C^r Nash function on an open semialgebraic set in $\mathbb{R}^n$ are semialgebraic, $0 < r' \leq r$.

2.13. Let f be a C^r Nash mapping between affine C^r Nash manifolds, $r > 0$. Then the C^1 critical point set of f and the C^1 critical value set are semialgebraic.

2.14. For an affine C^r Nash manifold $M \subset \mathbb{R}^n, r > 0$, the tangent bundle (TM, p, M) and the normal bundle (L, q, M) are affine C^{r-1} Nash vector bundles, where $TM \subset \mathbb{R}^n \times \mathbb{R}^n$ and $L \subset \mathbb{R}^n \times \mathbb{R}^n$. Moreover, there exists a C^{r-1} *Nash tubular neighborhood* U of M in $\mathbb{R}^n$, i.e., U is an affine Nash manifold and the orthogonal projection $\chi: U \to M$ is a C^{r-1} Nash mapping. Here the radius of each fibre $\chi^{-1}(x)$ is not necessarily constant.

2.15. For an affine C^r Nash manifold $M \subset \mathbb{R}^n, r > 0$, let $v_i, i = 1, \ldots, n$, denote the vector fields on M such that for each point $x \in M$, each v_{ix} is the tangent component of $\frac{\partial}{\partial x_i}$ to M at x. Then each v_i is a C^{r-1} Nash cross-section of (TM, p, M). For a C^r Nash function f on M, each $v_i f$ is a C^{r-1} Nash function.

2.16. The universal vector bundle $\xi_{n,m} = (E_{n,m}, \rho, G_{n,m})$ over the Grassmannian manifold $G_{n,m}$ is an affine Nash vector bundle, where $G_{n,m}$ is the set of m-dimensional linear subspaces in $\mathbb{R}^n$, and

$$E_{n,m} = \{(\lambda, x) \in G_{n,m} \times \mathbb{R}^n : x \in \lambda\}.$$

Here $G_{n,m}$ is imbedded in $\mathbb{R}^{n^2}$ by the correspondence

$$G_{n,m} \ni \lambda \longrightarrow \text{the orthogonal projection of } \mathbb{R}^n \text{ onto } \lambda \in \{(n, n)\text{-matrices}\}.$$

2.17. Let $M \subset \mathbb{R}^n$ be an affine C^r Nash manifold of dimension m, $r > 0$. Set $m' = n - m$. Then the following bundle mappings are of class C^{r-1} Nash:

$$\psi_1: (TM, p, M) \to (E_{n,m}, \rho, G_{n,m}),$$

$$\psi_2: (L, q, M) \to (E_{n,m'}, \rho, G_{n,m'}),$$

which are induced by characteristic mappings. Note that for each $x \in M$, $\psi_1|_{p^{-1}(x)}$ and $\psi_2|_{q^{-1}(x)}$ are the identities.

2.18. Let $M \subset \mathbb{R}^n$ be an affine C^r Nash manifold of dimension m, $r \geq 0$. Then there exists a finite system of C^r Nash coordinate neighborhoods $\{(U_i, \psi_i)\}_i$, i.e., each U_i is an open semialgebraic subset of M, and ψ_i is a C^r Nash diffeomorphism (homeomorphism if $r = 0$) from U_i to an open semialgebraic subset of $\mathbb{R}^m$. Here we can choose ψ_i to be the restriction to U_i of some linear projection $\mathbb{R}^n \to \mathbb{R}^m$.

2.19. Let $\xi_1 = (E_1, p_1, B)$, $\xi_2 = (E_2, p_2, B)$ be affine Nash vector bundles. Then the Whitney sum $\xi_1 \oplus \xi_2$, $\mathrm{Hom}(\xi_1, \xi_2)$ and the quotient bundle ξ_1/ξ_2 (if ξ_2 is a Nash subbundle of ξ_1) are affine Nash vector bundles.

2.20 (Mostowski). Given two disjoint closed semialgebraic subsets X_1 and X_2 in $\mathbb{R}^n$, there exists a Nash function f on $\mathbb{R}^n$ such that

$$f > 0 \text{ on } X_1 \quad \text{and} \quad f < 0 \text{ on } X_2.$$

2.21 (Mostowski). (See [Pe$_2$] for an elementary and explicit proof.) For a closed semialgebraic set X in $\mathbb{R}^n$, there exists a C^0 Nash function on $\mathbb{R}^n$ such that $f^{-1}(0) = X$, and the restriction of f to $\mathbb{R}^n - X$ is of class Nash.

2.22 (Corollary of 2.21). An affine C^r Nash manifold, $r \geq 0$, is C^r Nash diffeomorphic to an affine C^r Nash manifold closed in the ambient Euclidean space $\mathbb{R}^n$, and also to one which is bounded and whose boundary as a subset of $\mathbb{R}^n$ consists of at most one point.

2.23 (Corollary of 2.20). Each connected component of a Nash set is a Nash set.

2.24 (Normalization). For an affine algebraic set $X \subset \mathbb{R}^n$ there exists an affine algebraic set $Y \subset \mathbb{R}^n \times \mathbb{R}^m$ (the normalization) with the following properties. Let p denote the restriction to Y of the projection $\mathbb{R}^n \times \mathbb{R}^m \to \mathbb{R}^n$. Then p is a proper finite-to-one mapping onto X. For a nonsingular point a of X, $p^{-1}(a)$ consists of one nonsingular point of Y. For a C^ω immersion (imbedding) f of an analytic manifold M into $\mathbb{R}^n$ with $f(M) \subset X$ and $\dim M = \dim X$, there exist a unique C^ω submanifold M' of Y and a unique surjective C^ω immersion (diffeomorphism, respectively) $g\colon M \to M'$ such that $f = p \circ g$, and M' contains no singular point of Y. Here if M and f are of class Nash, so are M' and g.

2.25 (Artin–Mazur). For an affine Nash manifold $M \subset \mathbb{R}^n$ there exists a Nash mapping $\varphi\colon M \to \mathbb{R}^m$ such that the graph of φ contains no singular points of its Zariski closure in $\mathbb{R}^n \times \mathbb{R}^m$. Moreover, for a Nash mapping $\psi\colon M \to \mathbb{R}^{n'}$ we can choose φ so that the graph of (ψ, φ) contains no singular points of its Zariski closure in $\mathbb{R}^n \times \mathbb{R}^{n'} \times \mathbb{R}^m$.

2.26 (Artin). Let $P(x, y) = (P_1(x, y), \ldots, P_k(x, y))$ be a system of polynomials in variables $(x, y) \in \mathbb{R}^n \times \mathbb{R}^m$, and let $y(x) = (y_1(x), \ldots, y_m(x))$ be a formal power series solution of $P(x, y(x)) = 0$. Then for any natural number l, there exists a Nash mapping germ solution $y'(x)$ such that the partial derivatives of $y(x) - y'(x)$ vanish at 0 up to order l.

2.27 (Malgrange). Let $X \subset \mathbb{R}^n$ be a semialgebraic set and Y be a C^∞ submanifold of $\mathbb{R}^n$ contained in X and of the same dimension as X. Then X is of class C^ω. In particular, an affine C^∞ Nash manifold is an affine Nash manifold, and a C^∞ Nash mapping between affine Nash manifolds is a Nash mapping.

2.28 (Hironaka). Given a real affine algebraic set X, there exist a real affine nonsingular algebraic set Y and a proper smooth rational mapping $f\colon Y \to X$ such that

$$f|_{Y-f^{-1}(\operatorname{Sing} X)}\colon Y - f^{-1}(\operatorname{Sing} X) \longrightarrow X - \operatorname{Sing} X$$

is a diffeomorphism, and its inverse mapping also is smooth rational.

2.29 (Hironaka). Given a real affine nonsingular algebraic set X and an algebraic subset X' of X of smaller dimension, there exist a real affine nonsingular algebraic set Z and a proper surjective smooth regular mapping $g\colon Z \to X$ such that $g^{-1}(X')$ has only normal crossings in Z,

$$g|_{Z-g^{-1}(X')}\colon Z - g^{-1}(X') \longrightarrow X - X'$$

is a diffeomorphism, and its inverse mapping also is smooth regular.

2.30 (Neron's Resolution of Singularities [An], [Sp]). Let $\varphi\colon A \to B$ be a regular homomorphism between Noetherian rings. Then any homomorphism of A-algebras $C \to B$, with C finitely generated over A, factorizes a third finitely generated A-algebra D such that the homomorphism $A \to D$ is regular.

2.31 [C-R-S2]. Assume 2.30. Let $M \subset \mathbb{R}^n$ be a compact affine Nash manifold. Let $P(x, y) = (P_1(x, y), \dots, P_k(x, y))$ be a system of polynomials in variables $(x, y) \in \mathbb{R}^n \times \mathbb{R}^m$, and let $y(x) = (y_1(x), \dots, y_m(x))$ be an analytic mapping solution of $P(x, y(x)) = 0$ on M. Then there exists a Nash mapping solution arbitrarily near $y(x)$ in the C^∞ topology.

Finally we define a Nash structure on a *jet space*. Let r, r', M and M' be the same as above. Let $r' < \infty$. For each $x \in M$ and $y \in M'$, let $J^{r'}_{x,y}(M, M')$ denote the set of all jets of order r' from M to M' with source x and target y, and set

$$J^{r'}(M, M') = \bigcup_{(x,y)\in M\times M'} J^{r'}_{x,y}(M, M').$$

For $f \in N^r(M, M')$ with $f(x) = y$, the class represented by f in $J^{r'}_{x,y}(M, M')$ is called the *r'-jet* of f at x and denoted by $J^{r'} f(x)$. Hence $J^{r'} f$ is a mapping from M to $J^{r'}(M, M')$. We want to imbed $J^{r'}_{x,y}(M, M')$, $(x, y) \in M \times M'$, in a Euclidean space so that $J^{r'}(M, M')$ is an affine $C^{r-r'}$ Nash manifold and $J^{r'}f$ is a $C^{r-r'}$ Nash mapping.

If $M = \mathbb{R}^n$ and $M' = \mathbb{R}$ then for a point $(x_0, y_0) \in \mathbb{R}^n \times \mathbb{R}$, a natural representation of $J^{r'}_{x_0,y_0}(\mathbb{R}^n, \mathbb{R})$ is $x_0 \times y_0 \times P_{x_0,y_0}$, where

$$P_{x_0,y_0} = \Big\{\sum_{\alpha\in A} a_\alpha (x - x_0)^\alpha + y_0 \colon a_\alpha \in \mathbb{R}\Big\},$$

$$A = \{\alpha = (\alpha_1, \dots, \alpha_n) \in \mathbb{N}^n \colon 1 \le |\alpha| \le r'\}.$$

Then we identify $J^{r'}(\mathbb{R}^n, \mathbb{R})$ with $\mathbb{R}^n \times \mathbb{R}^m$, where $m = \sharp A + 1$, by the correspondence

$$J^{r'}(\mathbb{R}^n, \mathbb{R}) \ni (x_0, y_0, \sum_{\alpha\in A} a_\alpha (x - x_0)^\alpha + y_0) \to (x_0, y_0, a_\alpha)_{\alpha\in A} \in \mathbb{R}^n \times \mathbb{R}^m.$$

By this identification, for $f \in N^r(\mathbb{R}^n, \mathbb{R})$ we have

$$J^{r'}f(x) = (x, D^\alpha f(x)/\alpha!)_{\alpha\in\mathbb{N}^n, |\alpha|\le r'},$$

where

$$D^\alpha = \frac{\partial^{|\alpha|}}{\partial x_1^{\alpha_1} \cdots \partial x_n^{\alpha_n}} \quad \text{for } \alpha = (\alpha_1, \dots, \alpha_n) \in \mathbb{N}^n.$$

Hence $J^{r'}f$ is a $C^{r-r'}$ Nash mapping (2.12). Note that P_{x_0,y_0} becomes a commutative $\mathbb{R}$-algebra without unit if we define operations by

$$c\Big(\sum_{\alpha\in A} a_\alpha (x - x_0)^\alpha + y_0\Big) = \sum_{\alpha\in A} c a_\alpha (x - x_0)^\alpha + y_0,$$

$$\sum_{\alpha\in A} a_\alpha (x - x_0)^\alpha + y_0 + \sum_{\alpha\in A} b_\alpha (x - x_0)^\alpha + y_0 = \sum_{\alpha\in A} (a_\alpha + b_\alpha)(x - x_0)^\alpha + y_0,$$

$$\Big(\sum_{\alpha\in A} a_\alpha (x - x_0)^\alpha + y_0\Big)\Big(\sum_{\alpha\in A} b_\alpha (x - x_0)^\alpha + y_0\Big) = \sum_{\alpha+\beta\in A} a_\alpha b_\beta (x - x_0)^{\alpha+\beta} + y_0, \quad \text{for } c, a_\alpha, b_\beta \in \mathbb{R}.$$

Let $M \subset \mathbb{R}^n$ be general and $M' = \mathbb{R}$. In a natural way we give a commutative $\mathbb{R}$-algebra structure also to $J^{r'}_{x_0,y_0}(M, \mathbb{R})$. Then the canonical mapping from P_{x_0,y_0} to $J^{r'}_{x_0,y_0}(M, \mathbb{R})$ is a homomorphism. Let $\mathfrak{m}_{x_0,y_0}$ denote the kernel. Then it consists of $f \in P_{x_0,y_0}$ whose restrictions to M have partial derivatives of order $1, \dots, r'$ which vanish at x_0. Hence by 2.15, $\bigcup_{(x,y)\in M\times\mathbb{R}} x \times y \times \mathfrak{m}_{x,y}$ is semialgebraic in $\mathbb{R}^n \times \mathbb{R}^m$. Moreover, by using a local Nash coordinate system of M we easily see that it is an affine $C^{r-r'}$ Nash manifold in $\mathbb{R}^n \times \mathbb{R}^m$. Since $\mathfrak{m}_{x_0,y_0}$ is a linear subspace of P_{x_0,y_0}, the linear subspace $\mathfrak{m}^\perp_{x_0,y_0}$ of P_{x_0,y_0}, consisting of vectors normal to $\mathfrak{m}_{x_0,y_0}$, is carried bijectively onto $J^{r'}_{x_0,y_0}(M, \mathbb{R})$ by the canonical mapping. Consider $\bigcup_{(x,y)\in M\times\mathbb{R}} x \times y \times \mathfrak{m}^\perp_{x,y}$ and identify it with $J^{r'}(M, \mathbb{R})$. Then $J^{r'}(M, \mathbb{R})$ becomes an affine $C^{r-r'}$ Nash manifold in $\mathbb{R}^n \times \mathbb{R}^m$ such that for $f \in N^r(M)$, $J^{r'}f$ is a $C^{r-r'}$ Nash mapping.

If M is general and $M' = \mathbb{R}^{n'}$, then we can regard $J^{r'}(M, \mathbb{R}^{n'})$ as an affine $C^{r-r'}$ Nash manifold in $\mathbb{R}^n \times \mathbb{R}^{mn'}$ because of

$$J^{r'}(M, \mathbb{R}^{n'}) = \bigcup_{(x,y)=(x,y_1,\dots,y_{n'})\in M\times\mathbb{R}^{n'}} J^{r'}_{x,y_1}(M,\mathbb{R}) \times \cdots \times J^{r'}_{x,y_{n'}}(M,\mathbb{R}).$$

Clearly for $f \in N^r(M, \mathbb{R}^{n'})$, $J^{r'} f$ is a $C^{r-r'}$ Nash mapping.

Let M and M' be general. Let $\rho\colon U \to M'$ be a C^r Nash tubular neighborhood of M' in $\mathbb{R}^{n'}$ (see Corollary 3.8). Then

$$J^{r'}(M, M') = \{f \in J^{r'}_{x,y}(M, \mathbb{R}^{n'})\colon x \in M,\ y \in M,\ (J^{r'}\rho(y)) \circ f = f\}.$$

Hence $J^{r'}(M, M')$ is a semialgebraic subset of $J^{r'}(M, \mathbb{R}^{n'})$. We easily see that it is also an affine $C^{r-r'}$ Nash manifold in $\mathbb{R}^n \times \mathbb{R}^{mn'}$, and for $f \in N^r(M, M')$, $J^{r'} f$ is a $C^{r-r'}$ Nash mapping from M to $J^{r'}(M, M')$. Thus we give an affine $C^{r-r'}$ Nash manifold structure to $J^{r'}(M, M')$.

§3. Approximation Theorem

First we canonically define a topology on $N^r(M, M')$, the set of C^r Nash mappings from an affine C^r Nash manifold $M \subset \mathbb{R}^n$ to another $M' \subset \mathbb{R}^m$, $r \geq 0$, which is very similar to the topology on the space $\mathcal{S}$ of rapidly decreasing C^∞ functions on $\mathbb{R}^n$.

For $r = 0$, a basis of neighborhood system of $f \in N^0(M, M')$ in $N^0(M, N')$ is

$$U_V = \{g \in N^0(M, M')\colon \text{graph}\, g \subset V\}$$

for all semialgebraic neighborhoods V of graph f in $M \times M'$. For general $1 \leq r < \infty$, we define the topology by induction on r so that the mapping

$$N^r(M, M') \ni f \longrightarrow (f, J^1 f) \in N^0(M, M') \times N^{r-1}(M, J^1(M, M'))$$

is a homeomorphism onto the image. If $M = \mathbb{R}^n$ and $M' = \mathbb{R}^m$, this is equivalent to say that the mapping

$$N^r(M, M') \ni f \to \prod_{|\alpha|\leq r} D^\alpha f \in N^0(M, M')^{\sharp\{\alpha:\ |\alpha|\leq r\}}$$

is a homeomorphism onto the image. In the case of $r = \omega$, the topology is the projective limit of the topological spaces $N^r(M, M')$, $0 \leq r < \infty$, and the natural mappings $N^r(M, M') \to N^{r'}(M, M')$ for $r \geq r'$. We call this topology the C^r *topology* or simply the *topology* when $r = \omega$.

By 2.7, 8, 9 it is easy to see the following. If $M = \mathbb{R}^n$ and $M' = \mathbb{R}^m$, then a basis of neighborhood system of $f \in N^r(\mathbb{R}^n, \mathbb{R}^m)$ is

$$U_{r',k,c} = \{g \in N^r(\mathbb{R}^m, \mathbb{R}^m)\colon \sum_{|\alpha| \le r'} |D^\alpha(g - f)(x)| < 1/(c + |x|^k)\},$$

$$r'(\le r),\ k \in \mathbb{N},\ c > 0.$$

Hence $N^r(M, M')$ is metrizable. For noncompact M, the topology is stronger than the compact-open C^r topology and weaker than the C^r Whitney topology. For noncompact M and $M' = \mathbb{R}^m$, $m > 0$, $N^r(M, \mathbb{R}^m)$ is not a linear topological space (indeed, multiplication $\mathbb{R} \times N^r(M, \mathbb{R}^m) \to N^r(M, \mathbb{R}^m)$ is not continuous), which is different from the case of $\mathcal{S}$. We have also:

Lemma 3.1. *The topology does not depend on the ambient Euclidean spaces. To be precise, let $M \to M_1$ and $M' \to M_1'$ be C^r Nash diffeomorphisms (homeomorphisms if $r = 0$) between affine C^r Nash manifolds, $r \ge 0$. Then the bijection*

$$N^r(M, M') \longrightarrow N^r(M_1, M_1'),$$

induced by the diffeomorphisms (homeomorphisms, respectively), is a homeomorphism in the C^r topology.

The C^r topology is adequate to our theory, i.e., it is enough but not too strong. The following lemma shows that the topology is strong enough. The proof is elementary, and we omit it.

Lemma 3.2. *Let M, M' and M'' be affine C^r Nash manifolds, $r \ge 0$, and let $f \in N^r(M, M')$ and $\rho \in N^r(M', M'')$. Then $N^r(M)$ is a topological ring. If $r > 0$, and f is an immersion (a diffeomorphism or a diffeomorphism onto its image), then a strong C^r approximation of f is an immersion (a diffeomorphism or a diffeomorphism onto its image, respectively). If $r > 0$, and f is a diffeomorphism, then*

$$g^{-1} \longrightarrow f^{-1} \quad \textit{as} \quad g \longrightarrow f.$$

The mapping ρ_* from $N^r(M, M')$ to $N^r(M, M'')$, defined by $\rho_*(\varphi) = \rho \circ \varphi$, is continuous.

Now we state the most important tool, which says that the topology is not too strong. The case of $r = 0$ of the theorem was announced by Efroymson and the proof was completed by Pecker.

Approximation Theorem 3.3. *Let M be an affine Nash manifold, and let f be a C^r Nash function on M, $r < \infty$. Then we can approximate f by a Nash function in the C^r topology.*

Sketch of proof. The proof contains two important ideas: reduction to the case of a Euclidean space and a C^r Nash partition of unity.

First we reduce the problem to the case $M = \mathbb{R}^n$. Let $M \subset \mathbb{R}^n$. By 2.2 the boundary $\overline{M} - M$ is semialgebraic and, of course, closed in $\mathbb{R}^n$. By 2.21 we have a C^0 Nash function h on $\mathbb{R}^n$ such that $h^{-1}(0) = \overline{M} - M$, and $h|_{\mathbb{R}^n-(\overline{M}-M)}$ is a Nash function. Consider the graph of $1/h|_M$ in place of M. Then we can suppose that M is closed in $\mathbb{R}^n$. Next we want to extend f to $\mathbb{R}^n$. Let $\rho: U \to M$ be a Nash tubular neighborhood of M in $\mathbb{R}^n$ (2.14). Let γ be the square of the distance function from M in $\mathbb{R}^n$. By 2.7, 8, 9, 13, we have a large number c and a large integer k such that the set

$$U' = \{x \in \mathbb{R}^n: \gamma(x) \leq 1/(c + |x|^{2k})\}$$

is contained in U and γ is of class Nash on U'. Let g be a C^r Nash function on $\mathbb{R}$ such that $g(0) = 1$ and $g(x) = 0$ for $x \geq 1$. Set

$$\zeta(x) = \begin{cases} g(\gamma(x)(c + |x|^{2k})) & \text{for } x \in U' \\ 0 & \text{for } x \notin U'. \end{cases}$$

Then ζ is a C^r Nash function on $\mathbb{R}^n$ with $\zeta = 1$ on M and $\zeta = 0$ outside U'. Extend f to $\mathbb{R}^n$ by setting $f = \zeta \cdot f \circ \rho$ on U and $f = 0$ outside U. Thus we can assume $M = \mathbb{R}^n$.

For the proof we need a C^r Nash partition of unity whose elements satisfy a good estimate and can be approximated by Nash functions. Let $X \subset \mathbb{R}^n$ be an algebraic set, U a semialgebraic neighborhood of X, Y an affine Nash manifold contained in $X-\operatorname{Sing} X$ and of the same dimension as X, V a closed semialgebraic neighborhood of $X - Y$ in $\mathbb{R}^n$, g a C^r Nash function on $\mathbb{R}^n$ *r-flat* at Y (i.e., the partial derivatives of order $\leq r$ vanish at Y), and W a neighborhood of 0 in $N^r(\mathbb{R}^n - V)$ in the C^r topology. Then we have the following by elementary calculations.

Lemma 3.4. *There exists $F \in N^r(\mathbb{R}^n)$ such that $F = 0$ outside U, $F = 1$ in another neighborhood, F can be approximated by a Nash function on $\mathbb{R}^n$ in the C^r topology, and $gF|_{\mathbb{R}^n-V}$ is in W.*

By this lemma we obtain a good C^r Nash partition of unity. Let $\{Y_i\}$ be a finite Nash stratification of $\mathbb{R}^n$, for each i let $g_i \in N^r(\mathbb{R}^n)$ be r-flat at Y_i, and let W be a neighborhood of 0 in $N^r(\mathbb{R}^n)$ in the C^r topology.

Lemma 3.5 (C^r Nash partition of unity). *There exist open semialgebraic neighborhoods $V_i' \subset V_i$ of Y_i and $H_i \in N^r(\mathbb{R}^n)$ such that*

(i) $$\overline{V_i'} \subset V_i \cup \overline{Y_i},$$

(ii) *if $V_i \cap V_j \neq \emptyset$ and $i \neq j$ then $Y_i \subset \overline{Y_j}$ or $Y_j \subset \overline{Y_i}$,*

(iii) $$H_i = \begin{cases} 1 & \text{on } V_i' - Z_i \\ 0 & \text{outside } V_i - Z_i', \end{cases}$$

where

$$Z_i = \bigcup_{Y_j \subset \overline{Y_i} - Y_i} V_j \quad \text{and} \quad Z'_i = \bigcup_{Y_j \subset \overline{Y_i} - Y_i} V'_j,$$

(iv) *H_i can be approximated by Nash functions on $\mathbb{R}^n$ in the C^r topology,*

(v) $$\Sigma H_i = 1,$$

(vi) $$g_i H_i \in W.$$

For any $f \in N^r(\mathbb{R}^n)$ we easily obtain a finite Nash stratification $\{Y_i\}$ of $\mathbb{R}^n$ such that for each partial derivative D^α of order $\leq r$ and for each i, $(D^\alpha f)|_{Y_i}$ is of class Nash. Moreover, we have:

Lemma 3.6. *There exist a finite Nash stratification $\{Y_i\}$ of $\mathbb{R}^n$ and $f_i \in N^r(\mathbb{R}^n)$ such that each f_i can be approximated by a Nash function on $\mathbb{R}^n$, and $f - f_i$ is r-flat at Y_i.*

Now we can prove the theorem. Let W be a neighborhood of 0 in $N^r(\mathbb{R}^n)$ in the C^r topology. Let $\{Y_i\}_{i=1,\dots,k}$ and f_i be the results in Lemma 3.6. Apply Lemma 3.5 to $\{Y_i\}_{i=1,\dots,k}$, $\{f - f_i\}_{i=1,\dots,k}$ and W. Then we obtain $H_i \in N^r(\mathbb{R}^n)$, $i = 1, \dots, k$, which satisfy (iv) and (v) and such that $(f - f_i)H_i \in W$. By (iv) in Lemma 3.5 and Lemma 3.6 we have $\tilde{H}_i, \tilde{f}_i \in N(\mathbb{R}^n)$ such that $f_i H_i - \tilde{f}_i \tilde{H}_i \in W$. Then we have

$$f - \sum_{i=1}^{k} \tilde{f}_i \tilde{H}_i = \sum_{i=1}^{k} (f H_i - f_i H_i) + \sum_{i=1}^{k} (f_i H_i - \tilde{f}_i \tilde{H}_i) \in 2kW,$$

which proves the theorem. □

Remark 3.7. Using a Nash tubular neighborhood and Lemma 3.2 we can generalize Theorem 3.3 as follows. Let M and M' be affine Nash manifolds, and let f be a C^r Nash mapping from M to M', $r < \infty$. Then we can approximate f by a Nash mapping. Consequently, by Lemma 3.2 if two affine Nash manifolds are C^r Nash diffeomorphic, $r > 0$, then they are Nash diffeomorphic.

Another immediate generalization is a relative version. Let M and f be the same as in the theorem. Assume that f is of the form $\sum_{i=0}^{k} f_i g_i$, where f_i are C^r Nash functions and g_i are Nash functions. Then we can approximate f by a Nash function of the form $\sum_{i=1}^{k} f'_i g_i$, where f'_i are Nash functions. This is clear because $N(M)$ is a topological ring (Lemma 3.2).

Easy corollaries are:

Corollary 3.8. *An affine C^r Nash manifold in $\mathbb{R}^n$, $r > 0$, has a C^r Nash tubular neighborhood. (The projection mapping is not necessarily the orthogonal projection.)*

Here the condition $r > 0$ is not necessary [S_1].

Proof. Let $M \subset \mathbb{R}^n$ be an affine C^r Nash manifolds of codimension m. Let $\psi: M \to G_{n,m}$ denote the characteristic mapping of the normal bundle of M in $\mathbb{R}^n$, which is of class C^{r-1} Nash (2.17). Extend ψ to an open semialgebraic neighborhood of M in $\mathbb{R}^n$, approximate the extension by a Nash mapping, and restrict it to M. Then we have a C^r Nash approximation ψ' of ψ. Using ψ' we can construct a C^r Nash tubular neighborhood of M. □

Corollary 3.9. *An affine C^r Nash manifold, $r > 0$, is C^r Nash diffeomorphic to an affine Nash manifold.*

Proof. Let M, ψ and ψ' be the same as in the above proof, and let U be a C^r Nash tubular neighborhood of M in $\mathbb{R}^n$ defined by ψ'. Then there exists a C^r Nash mapping Ψ from U to $E_{n,m}$ such that Ψ is transversal to $G_{n,m}$, and $\Psi^{-1}(G_{n,m}) = M$. Approximate Ψ by a Nash mapping Ψ'. Then $(\Psi')^{-1}(G_{n,m})$ is what we want. □

The first half of the following theorem, in the case where φ is defined on M, is due to Efroymson and Pecker, and the general case and the latter half are due to [S_2].

Extension Theorem 3.10. *Let $M \subset \mathbb{R}^n$ be an affine Nash manifold, let φ and g be Nash functions on an open semialgebraic set U in M, and let S be a closed semialgebraic set in M with $U \subset S$. Assume that $\varphi^{-1}(0)$ is closed in M. Then there exists a Nash function h on M such that $g - h|_U$ is the product of φ and a Nash function on U.*

Assume, moreover, that φ is extended to a C^0 function $\tilde{\varphi}$ on S so that $\varphi^{-1}(0) = \tilde{\varphi}^{-1}(0)$. Then there exists a positive C^0 function f on S such that $f\tilde{\varphi}$ is extendible to a Nash function on M.

The next relative approximation theorem immediately follows from Approximation Theorem 3.3 and the last theorem.

Relative Approximation Theorem 3.11. *Let M and f be the same as in Approximation Theorem 3.3. Let φ be a Nash function on an open semialgebraic set U in M. Assume that f is of class Nash on U, and $\varphi^{-1}(0)$ is closed in M. Then we can approximate f by a Nash function g such that $(f - g)|_U$ is the product of φ and a Nash function on U.*

By using a Nash tubular neighborhood, the last two theorems imply:

Theorem 3.12. *Let $M_1 \subset M$ be affine Nash manifolds such that M_1 is closed in M.*

(1) *A Nash set in an open semialgebraic set $U \subset M$ is a Nash set in M if it is closed in M. In particular, M_1 is a Nash set in M.*

Moreover,

(2) *there exist Nash functions $f_1, \dots, f_k$ on M whose common zero set is M_1, and whose gradients span the normal bundle of M_1 in M;*

(3) *a Nash function on M_1 is extendible to M;*

(4) *let f be a C^r Nash function on M, $r \geq 0$, whose restriction to M_1 is of class Nash. Then fixing it on M_1 we can approximate f by a Nash function on M.*

§4. Nash manifolds

In singularity theory, the Transversality Theorem [Th$_4$] and the First and Second Isotopy Lemmas [Th$_{2,3}$] play the most important role. Nash versions of the Transversality Theorem and the First Isotopy Lemma hold, though the other is still open. The following theorem also says that the topology on $N(M)$ is suited for our theory.

Transversality Theorem 4.1 [S$_3$]. *Let M and M' be affine Nash manifolds, and let $\{X_i\}$ be a finite Nash Whitney stratification of a closed semialgebraic set in $J^r(M, M')$, $0 < r < \infty$. Then the subset of $N(M, M')$ consisting of f such that $J^r f$ are transversal to each X_i is open and dense in $N(M, M')$.*

Sketch of proof. Openness: For simplicity of notation we assume $\dim X_1 < \dim X_2 < \cdots$. Let $f \in N(M, M')$ be such that $J^r f$ is transversal to each X_i. By elementary calculations we find an open neighborhood U_1 of f in $N(M, M')$ so that for $f' \in U_1$, $J^r f'$ is transversal to X_1. By the Whitney condition, there is an open semialgebraic neighborhood O_1 of X_1 in $J^r(M, M')$ such that for each $f' \in U_1$, $J^r f'$ is transversal to each $O_1 \cap X_i$. Find smaller $U_2 \subset U_1$ so that for $f' \in U_2$, $J^r f'$ is transversal to $X_2 - O_1$, and so on. Thus we see openness.

Density: The problem is local in the C^∞ case [Th$_{2,3}$], and global in our Nash case. But the proofs are similar. By the above arguments we can assume $\{X_i\} = \{X_1\}$. Moreover, using a Nash tubular neighborhood of M' we can suppose that $M' = \mathbb{R}^m$, and M is open in $\mathbb{R}^n$. Let $f = (f_1, \dots, f_m) \in N(M, \mathbb{R}^m)$, and let ε be a small positive Nash function on M. Set

$$\{\alpha^1, \dots, \alpha^{m'}\} = \{\alpha \in \mathbb{N}^n : |\alpha| \leq r\}.$$

Define Nash mappings $\theta: M \times I^n \to \mathbb{R}^n$, $\lambda_j: M \to \mathbb{R}$ $(j = 1, \dots, m')$, $\tilde{f}: M \times I^{mm'} \to \mathbb{R}^m$ and $\hat{f}: M \times I^n \times I^{mm'} \to \mathbb{R}^m$, where $I = [-1, 1]$, by

$$\theta(x, a) = (x_1 + a_1\varepsilon(x), \dots, x_n + a_n\varepsilon(x)) \quad \text{for} \quad (x, a) = (x_1, \dots, x_n, a_1, \dots, a_n) \in M \times I^n,$$

$$\lambda_j(x) = x^{\alpha^j}\varepsilon(x) \quad \text{for} \quad x \in M,$$

$$\tilde{f}(x,\beta) = (f_1(x) + \sum_{j=1}^{m'} \beta_{1j}\lambda_j(x), \ldots, f_m(x) + \sum_{j=1}^{m'} \beta_{mj}\lambda_j(x))$$

$$\text{for} \quad x \in M \text{ and } \beta = (\beta_{ij}) \in I^{mm'},$$

$$\hat{f}(x,a,\beta) = \tilde{f}(\theta_a^{-1}(x),\beta) \quad \text{for} \quad (x,a,\beta) \in M \times I^n \times I^{mm'},$$

where θ_a is the map from M to $\mathbb{R}^n$ defined by $\theta_a(\cdot) = \theta(\cdot, a)$. Choose ε so small that for any $a \in I^n$ and $\beta \in I^{mm'}$, $\theta(\cdot, a)$ is a diffeomorphism of M, and $\hat{f}(\cdot, a, \beta)$ is a strong approximation of f. (Here we construct ε as follows. By 2.22 we can assume that M is bounded, and $\overline{M} - M$ consists of the origin. (We forget the assumption that M is open in $\mathbb{R}^n$.) Then $\varepsilon(x) = c|x|^{2k}$ for a sufficiently small positive number c and a sufficiently large integer k.) If the following statement (1) holds then there exist a and β such that the mapping $J^r\hat{f}(\cdot, a, \beta) : M \to J^r(M, M')$ is transversal to X_1, i.e., $\hat{f}(\cdot, a, \beta)$ is what we want.

(1) Each point of M has a compact neighborhood V such that the set

$$\{(a,\beta) \in I^n \times I^{mm'} \colon J^r\hat{f}(\cdot, a, \beta) \text{ is transversal to } X_1 \text{ at each point of } V\}$$

is open and dense in $I^n \times I^{mm'}$.

Here the set is always open. Hence we require only density. (1) follows from the following easier statement (2). Regard $\hat{f}$ as an x-mapping and set

$$F(x,a,\beta) = (J^r\hat{f})(\theta(x,a),a,\beta) \quad \text{for} \quad (x,a,\beta) \in M \times I^n \times I^{mm'}.$$

Then F is a Nash mapping from $X \times I^n \times I^{mm'}$ to $J^r(M, \mathbb{R}^m)$ such that

$$F(x,0,0) = J^r\hat{f}(x,0,0) = J^r f(x).$$

(2) For each point $(x_0, a_0, \beta_0) \in M \times I^n \times I^{mm'}$, x_0 has a compact neighborhood V such that (a_0, β_0) is adherent to the set

$$\{(a,\beta) \in I^n \times I^{mm'} \colon F(\cdot, a, \beta) \text{ is transversal to } X_1 \text{ at each point of } V\}.$$

To prove (2) we use only the property of F that for each $(x_0, a_0, \beta_0) \in M \times I^n \times I^{mm'}$, $F|_{x_0 \times I^n \times I^{mm'}}$ is locally a diffeomorphism at (x_0, a_0, β_0). Hence what we need to prove is:

(3) Let $g\colon \mathbb{R}^n \times \mathbb{R}^{m''} \to \mathbb{R}^{m''}$, $m'' = n + mm'$, be a C^∞ mapping such that $g(0) = 0$, and $g|_{0 \times \mathbb{R}^{m''}}$ is a local diffeomorphism at 0×0. Let $X_1 \subset \mathbb{R}^{m''}$ be a C^∞ submanifold such that $0 \in X_1$. Then there exist an open neighborhood V of 0 in $\mathbb{R}^n$ and a subset Z of $\mathbb{R}^{m''}$ such that $0 \in \overline{Z}$, and for each $a \in Z$, $g|_{V \times a}$ is transversal to X_1.

Since the problem is local at 0×0 in $\mathbb{R}^n \times \mathbb{R}^{m''}$ and 0 in $\mathbb{R}^{m''}$, we can reduce (3) to the case of $X_1 = \{0\}$. In this case we prove easily (3) by calculation of first partial derivatives and by Sard's theorem. □

A proper C^∞ submersion onto $\mathbb{R}^n$ is C^∞ trivial, which is easily proved by the integrations of vector fields. But the Nash version is not easy to prove because the integration of a Nash vector field is not necessarily of class Nash.

Triviality Theorem 4.2 [C-S$_{1,2}$]. *Let M be an affine Nash manifold, and let $f\colon M \to \mathbb{R}^n$ be a proper Nash submersion. Then f is Nash trivial, i.e., there exists a Nash diffeomorphism π from M to $f^{-1}(0) \times \mathbb{R}^n$ of the form $\pi = (\pi', f)$.*

Sketch of proof. We proceed through two stages. First we prove that there exists a finite Nash stratification $\{U_i\}$ of $\mathbb{R}^n$ such that f is Nash trivial over each U_i, i.e., the restriction

$$f|_{f^{-1}(U_i)}\colon f^{-1}(U_i) \longrightarrow U_i$$

is Nash trivial. Next we prove the theorem in this situation. The first stage is generalized for an induction procedure as follows.

Statement. Let X, P and S be semialgebraic sets in $\mathbb{R}^n$, let U be an affine Nash manifold in $\mathbb{R}^{n'}$ of dimension m, let $f\colon \mathbb{R}^n \to U$ be a Nash mapping, and set

$$f_X = f|_X, \quad f_P = f|_P, \quad f_S = f|_S, \quad f_{X-S} = f|_{X-S},$$
$$X_y = f_X^{-1}(y), \quad P_y = f_P^{-1}(y), \quad S_y = f_S^{-1}(y) \quad \text{for each } y \in U.$$

Assume that S is a subset of X and of P, $X - S$ is a C^ω smooth and of dimension $k+m$, P is C^ω smooth and of dimension $k+m+1$, P includes a neighborhood of S in X, f_X is proper, f_{X-S} is a surjective C^ω submersion, for each $y \in U$, P_y is C^ω smooth, the dimension of S_y is 0, and there exists a Nash function φ on P such that $\varphi^{-1}(0) = X \cap P$, and for each $y \in U$, $\varphi|_{P_y}$ takes the Morse type at each point of S_y. Shrink P. Then there exists a finite Nash stratification $\{U_i\}_i$ of U and for each U_i and $y_i \in U_i$, a Nash imbedding

$$\pi_i\colon f_X^{-1}(U_i) \cup f_P^{-1}(U_i) \longrightarrow (X_{y_i} \cup P_{y_i}) \times U_i$$

of the form $\pi_i = (\pi_i', f)$ such that

$$\pi_i(f_X^{-1}(U_i)) = X_{y_i} \times U_i, \quad \pi_i(f_P^{-1}(U_i)) \subset P_{y_i} \times U_i,$$
$$\varphi \circ \pi_i' = \varphi \quad \text{on } f_P^{-1}(U_i).$$

If this statement holds, we apply it to $X = M$ and $P = S = \emptyset$. Then the first stage of the theorem follows. Hence it suffices to prove the statement. Since the case of $k = 0$ is trivial, we prove the statement by induction on k. For simplicity of notation we assume $0 \in U$. By stratifying U we can suppose that S is C^ω smooth and f_S is a proper local homeomorphism. Shrink P. Then f_P is a C^ω submersion. Hence, shrinking P once more, we can define a Nash imbedding $\pi\colon P \to P_0 \times U$ of the form $\pi = (\pi', f)$ such that

$$\pi(P \cap X) = (P_0 \cap X_0) \times U, \quad \varphi \circ \pi' = \varphi.$$

It suffices to stratify U and to extend π to X. Hence we reduce the problem to the case of $P = S = \emptyset$. (Here f_X is no longer proper, and we need some modification of the induction procedure. Moreover, the extension can not be of class Nash. It is a C^r Nash diffeomorphism, $0 < r < \infty$. Hence we use Approximation Theorem 3.3 to obtain a Nash diffeomorphism.) For a moment, assume that there exists a Nash function ψ on X such that for each $y \in U$, $\psi|_{X_y}$ takes the Morse type at each critical point. For each $y \in U$, let Z_y denote the critical point set of $\psi|_{X_y}$. Set $Z = \bigcup_{y \in U} Z_y$. Then we can assume that ψ is constant on each connected component of Z. Apply the induction hypothesis to $\psi^{-1}(\psi(Z))$, X, Z, f and ψ. Then we obtain a Nash trivialization of the restriction of f to a semialgebraic neighborhood V of $\psi^{-1}(\psi(Z))$ in X. Next we can extend the trivialization to $X - V$ when we once more apply the induction hypothesis to

$$(f, \psi)|_{X-V}: X - V \longrightarrow \mathbb{R}^{m+1}.$$

(Here also we need to apply the Approximation Theorem.)

It remains to construct ψ. Let a subset Q of $J^2(X, \mathbb{R})$ be given so that

$$Q \cap J^2_{x,t}(X, \mathbb{R}) = \{J^2\lambda(x): \lambda \in N(X, \mathbb{R}),$$
$$x \text{ is a degenerate critical point of } \lambda|_{X_{f(x)}}\} \quad \text{for } (x, t) \in X \times \mathbb{R},$$

and let $\{X_i\}$ be a finite Nash Whitney stratification of Q. Then by Transversality Theorem 4.1 there exists ψ such that $\dim (J^2\psi)^{-1}Q < \dim U$. Hence stratifying U we suppose that $(J^2\psi)^{-1}Q$ is empty, which implies that for each $y \in U$, the restriction $\psi|_{X_y}$ takes the Morse type at each critical point. Thus the first stage is proved.

We prove the second stage as follows. Since f is proper, we can extend Nash triviality of f over each U_i to an open semialgebraic neighborhood of U_i, using a Nash tubular neighborhood of $f^{-1}(U_i)$ in M. Hence there is a finite open semialgebraic covering $\{V_i\}$ of $\mathbb{R}^n$ such that f is Nash trivial over each V_i, i.e., f is an affine Nash fibre bundle. Thus what we prove is:

An affine Nash fibre bundle $f: M \to \mathbb{R}^n$ is Nash trivial.

Let $\{V_i\}$ be the same as above. Let σ_0 be an n-simplex. Then we can construct a Nash diffeomorphism τ from $\mathbb{R}^n$ to $\operatorname{Int}\sigma_0$ and a fine simplicial decomposition K of σ_0 such that for each $\sigma \in K$, an open semialgebraic neighborhood $W(\sigma)$ of $\sigma \cap \operatorname{Int}\sigma_0$ in $\operatorname{Int}\sigma_0$ is carried by τ^{-1} into some V_i, and K collapses on some vertex v of K. Replace $\mathbb{R}^n$ with $\operatorname{Int}\sigma_0$ through τ. Let $K = K_1 \overset{e}{\searrow} K_2 \overset{e}{\searrow} \cdots \overset{e}{\searrow} K_l = v$ be a sequence of elementary collapses. (See [R-S] for the notation of a collapsing and an elementary collapse.) Set

$$W_j = \bigcup_{\sigma \in K_j} W(\sigma), \qquad j = 1, \dots, l.$$

Then each $W_j - W_{j+1}$ is included in some V_i. Hence f is Nash trivial over each of W_l, $W_{l-1} - W_l, \dots, W_1 - W_2$. Moreover, we can choose $W(\sigma)$'s so that if $K_{j+1} \cup \{\sigma\} = K_j$ then $W_{j+1} \cap W(\sigma)$ is Nash diffeomorphic to $\mathbb{R}^n$. Then we can extend C^r Nash triviality of f on W_l to W_{l-1} and so on until W_1, $0 < r < \infty$, which completes, together with Approximation Theorem 3.3, the proof. □

Remark 4.3 [C-S$_2$]. The above theorem holds for an affine Nash manifold with boundary. To be precise, let M be an affine Nash manifold with boundary, and let $f: M \to \mathbb{R}^n$ be a proper Nash submersion such that $f|_{\partial M}$ also is a proper Nash submersion. Then we can choose a Nash diffeomorphism π from M to $f^{-1}(0) \times \mathbb{R}^n$ of the form $\pi = (\pi', f)$ such that

$$\pi(\partial M) = (f^{-1}(0) \cap \partial M) \times \mathbb{R}^n.$$

Here if π is previously defined on ∂M then we can extend it to M.

Moreover, we can generalize this as follows. Let M be as above, let M' be another affine Nash manifold, and let $f = (f_1, f_2) : M \to M' \times \mathbb{R}^n$ be a proper Nash submersion such that $f|_{\partial M}$ also is a proper Nash submersion. Then there exists a Nash diffeomorphism π from M to $f_2^{-1}(0) \times \mathbb{R}^n$ of the form (π', f_2) such that

$$\pi(\partial M) = (f_2^{-1}(0) \cap \partial M) \times \mathbb{R}^n,$$

$$f_1 \circ \pi' = f_1 \quad \text{on } M.$$

In this case also we can extend π from ∂M to M.

In the above proof we prove the fact that an affine Nash fibre bundle over $\mathbb{R}^n$ is Nash trivial also in the case with boundary.

A corollary of the last theorem is:

Corollary 4.4 [C-S$_1$]. *Let R be a real closed field. An affine Nash manifold in R^n is Nash diffeomorphic to an affine Nash manifold in some $R^{n'}$ defined by polynomials with coefficients in the field $\mathbb{R}_{\mathrm{alg}}$ of real algebraic numbers.*

First Isotopy Lemma 4.5 [C-S$_2$]. *Let X be a semialgebraic set in $\mathbb{R}^n$, let $\{X_i\}$ be a Nash Whitney stratification of X, and let $f: X \to \mathbb{R}^m$ be a proper C^0 mapping such that the restriction of f to each stratum is a Nash submersion, and f is extendible to $\mathbb{R}^n$ as a C^1 mapping. Then f is semialgebraically trivial, i.e., there exists a semialgebraic homeomorphism π from X to $f^{-1}(0) \times \mathbb{R}^m$ of the form $\pi = (\pi', f)$ such that for each i, $\pi|_{X_i}$ is a Nash diffeomorphism onto $(f^{-1}(0) \cap X_i) \times \mathbb{R}^m$.*

Sketch of proof. In the C^∞ case, the unique known method of proof is to integrate vector fields. We can not use this.

For simplicity of notation we assume $\dim X_1 < \dim X_2 < \cdots$. There exists a C^2 Nash tube system $\{T_i = (|T_i|, \tau_i, \rho_i)\}$ of $\{X_i\}$, where each $\tau_i: |T_i| \to X_i$ is a C^2 Nash tubular neighborhood of X_i in $\mathbb{R}^n$, and ρ_i is a nonnegative C^2 Nash function on $|T_i|$ such that $\rho_i^{-1}(0) = X_i$, each point x of X_i is the unique and nondegenerate critical point of the restriction of ρ_i to $\tau_i^{-1}(x)$,

$$f \circ \tau_i = f \quad \text{on } |T_i|,$$

and for each pair i and j with $i < j$,

$$\tau_i \circ \tau_j = \tau_i \quad \text{and} \quad \rho_i \circ \tau_j = \rho_i \quad \text{on } |T_i| \cap |T_j|.$$

By induction on i we construct a C^2 Nash diffeomorphism π_i from X_i to $(f^{-1}(0) \cap X_i) \times \mathbb{R}^m$ of the form $\pi_i = (\pi_i', f)$ such that for each $j < i$,

$$\pi_i'(|T_j| \cap X_i) \subset |T_j|,$$
$$\tau_j \circ \pi_i' = \pi_j' \circ \tau_j \quad \text{and} \quad \rho_j \circ \pi_i' = \rho_j \quad \text{on } |T_j| \cap X_i,$$

and the mapping, defined to be π_j on each X_j, $j \leq i$, is continuous. (Here we shrink $|T_j|$'s if necessary.) By Triviality Theorem 4.2 this is clear for $i = 1$. Hence assume that we have $\pi_1, \dots, \pi_{i-1}$. To construct π_i we use a downward induction. Let $1 \leq k \leq i$. Let U_k be the intersection of X_i and a small open semialgebraic neighborhood of $X_k \cup \cdots \cup X_{i-1}$ in $\mathbb{R}^n$. Then we want a C^2 Nash imbedding $\kappa_k \colon U_k \to (f^{-1}(0) \cap U_k) \times \mathbb{R}^m$ of the form $\kappa_k = (\kappa_k', f)$ such that for each $k \leq j < i$,

$$\kappa_k'(|T_j| \cap U_k) \subset |T_j|,$$
$$\pi_j' \circ \tau_j = \tau_j \circ \kappa_k' \quad \text{and} \quad \rho_j \circ \kappa_k' = \rho_j \quad \text{on } |T_j| \cap U_k.$$

If $k = i$ then $U_k = \emptyset$. Hence assuming κ_{k+1} we construct κ_k. Shrink U_{k+1} and modify T_k. Then the set $V_k = |T_k| \cap X_i - U_{k+1}$ is an affine C^2 Nash manifold with boundary, and the mapping $(\tau_k, \rho_k) \colon V_k \to X_k \times]0, 1[$ is a proper C^2 Nash submersion whose restriction to ∂V_k also is so. Since we have shrunk U_{k+1} we can assume that κ_{k+1} is defined on ∂V_k. Then the equality $\rho_k \circ \kappa_{k+1}' = \rho_k$ on ∂V_k follows from the above equalities. Apply Triviality Theorem 4.2 (Remark 4.3) to the mapping

$$(\pi_k' \circ \tau_k, f, \rho_k) \colon V_k \longrightarrow (f^{-1}(0) \cap X_k) \times \mathbb{R}^m \times]0, 1[.$$

Then we can extend κ_{k+1} to $\kappa_k \colon U_{k+1} \cup V_k \to (f^{-1}(0) \cap (U_{k+1} \cup V_k)) \times \mathbb{R}^m$, which completes κ_1.

It remains to extend κ_1 to X_i. Shrink U_1 and $|T_j|$, $j = 1, \dots, i-1$, and modify κ_1 so that κ_1 is a diffeomorphism onto $(f^{-1}(0) \cap U_1) \times \mathbb{R}^m$. Then by Triviality Theorem 4.2 we can extend κ_1 to the required diffeomorphism π_i as in the proof of 4.3. □

One of the most important open problems on semialgebraic geometry is the second isotopy lemma. If this holds true then we can replace the term "C^0 equivalence" on polynomial functions, mappings, and germs with the term "semialgebraic C^0 equivalence". These two equivalence classes are distinct. Indeed, there are two polynomial mapping germs which are right-left C^0 equivalent but not right-left semialgebraically C^0 equivalent [C-S3], and there are two smooth rational functions on an affine nonsingular real algebraic set, which are C^0 equivalent but not semialgebraically C^0 equivalent [S3].

Compactification Theorem 4.6. *Given a noncompact affine Nash manifold M, there exists a unique compact affine Nash manifold with boundary whose interior is Nash* *diffeomorphic to M. Here uniqueness means that if we have two Nash compactifications, then they are Nash diffeomorphic.*

In the C^∞ case, this does not hold. Indeed a C^∞ manifold is not necessarily C^∞ compactifiable (hence it does not always admit an affine Nash manifold structure), and even if it is so, its C^∞ compactification is not necessarily unique. On the other hand, a compact C^∞ manifold possibly with boundary admits a unique affine Nash manifold structure. (We give a structure to compact M by taking the double DM of M in some $\mathbb{R}^n$, and approximating the characteristic mapping of the normal bundle of DM in $\mathbb{R}^n$ by a smooth rational mapping as in 3.8. The uniqueness is shown in the next theorem.)

Proof. Let $M \subset \mathbb{R}^n$. Then by 2.7 and Approximation Theorem 3.3 we have a proper positive Nash function f on M. By 2.13 the critical value set of f is finite. Hence for some $c > 0$, f is C^1 regular on $f^{-1}(]c, \infty[)$. Then by Triviality Theorem 4.2 f is Nash trivial over $]c, \infty[$. Clearly $f^{-1}([0, 2c])$ is a compact affine Nash manifold with boundary, and by the Approximation Theorem its interior is Nash diffeomorphic to M.

See the next theorem for the uniqueness. □

Theorem 4.7. *Let L_1 and L_2 be compact affine Nash manifolds possibly with boundary, and let M_1 and M_2 denote their interiors. Then the following conditions are equivalent.*

(i) *L_1 and L_2 are C^1 diffeomorphic.*

(ii) *L_1 and L_2 are Nash diffeomorphic.*

(iii) *M_1 and M_2 are Nash diffeomorphic.*

Sketch of proof. The implication (ii) $\Rightarrow$ (iii) is trivial.

(iii) $\Rightarrow$ (i): We can assume that L_1 and L_2 have boundary. Let $\tau: M_1 \to M_2$ be a Nash diffeomorphism. We have nonnegative Nash functions h_1 on L_1 and h_2 on L_2 such that

$$h_i^{-1}(0) = \partial L_i, \qquad i = 1, 2,$$

and h_1 and h_2 are C^1 regular at the respective boundaries. Then L_1 and L_2 are Nash diffeomorphic to

$$L_{1\varepsilon} = \{x \in M_1 \colon h_1(x) \geq \varepsilon\} \quad \text{and} \quad L_{2\varepsilon} = \{x \in M_2 \colon h_2(x) \geq \varepsilon\},$$

respectively, for small $\varepsilon > 0$. Compare h_1 and $h_2 \circ \tau$ on M_1. Then by the following lemma, $L_{1\varepsilon}$ and $L_{2\varepsilon}$ are Nash diffeomorphic.

(i) $\Rightarrow$ (ii): Let $\tau: L_1 \to L_2$ be a C^1 diffeomorphism. Let $L_2 \subset \mathbb{R}^{n_2}$, and let $\pi: U \to \partial L_2$ be a Nash tubular neighborhood of ∂L_2 in $\mathbb{R}^{n_2}$. Using π and a polynomial approximation, we can modify τ so that the germ of τ at ∂L_1 is a Nash diffeomorphism germ onto the germ of L_2 at ∂L_2. Next by Extension Theorem 3.10

we extend $\tau|_{\partial L_1}$ to a Nash mapping $\tau'\colon L_1 \to \mathbb{R}^{n_2}$. Lastly as in 3.7 we approximate $\tau - \tau'$ by a Nash mapping $\tau''\colon L_1 \to \mathbb{R}^{n_2}$ such that $\tau'' = 0$ on ∂L_1. Then $\pi' \circ (\tau' + \tau'')$ is what we want, where π' is the projection of a Nash tubular neighborhood of L_2 in $\mathbb{R}^{n_2}$. □

Lemma 4.8. *Let f_1 and f_2 be proper positive Nash functions on an affine Nash manifold M. Then there exist open semialgebraic subsets U_1 and U_2 of M and a Nash diffeomorphism $\pi\colon U_1 \to U_2$ such that $M - U_1$ and $M - U_2$ are compact, and $f_1 \circ \pi = f_2$ on U_1.*

Sketch of proof. For each $i = 1, 2$, let v_i denote the vector field of the gradient of f_i. Then v_1 and v_2 are nonsingular and do not point in the opposite directions at each point outside a compact subset C of M. Hence the vector field v, defined by

$$v_x = v_{1x}/|v_{1x}| + v_{2x}/|v_{2x}| \quad \text{for } x \in M - C,$$

is nonsingular outside C, and f_1 and f_2 are monotone increasing on each integral curve of v. Define π on each curve so that $f_1 \circ \pi = f_2$. Then π is of class C^∞. We can approximate π by a Nash one in the same way as the above proof of (i) $\Rightarrow$ (ii). □

A compactification of a Nash manifold with a Nash mapping is:

Theorem 4.9. *Let M_1 be a noncompact affine Nash manifold, let L_2 be a compact affine Nash manifold possibly with corners, and let*

$$f\colon M_1 \longrightarrow M_2 = \operatorname{Int} L_2$$

be a Nash mapping. Then there exist a compact affine Nash manifold with corners L_1 and a Nash diffeomorphism π from Int L_1 *to M_1 such that $f \circ \pi$ is extendible to a Nash mapping from L_1 to L_2.*

Sketch of proof. To treat only compact manifolds, we reduce M_1 and M_2 to the case of

$$M_1 \subset S^{n-1} \subset \mathbb{R}^n, \quad M_2 = S^{m-1} \subset \mathbb{R}^m, \quad \overline{M}_1 - M_1 = \{\text{a point } a\}.$$

Let $M \subset \mathbb{R}^n \times \mathbb{R}^m$ denote the graph of f, and let

$$p_1\colon \mathbb{R}^n \times \mathbb{R}^m \longrightarrow \mathbb{R}^n, \quad p_2\colon \mathbb{R}^n \times \mathbb{R}^m \longrightarrow \mathbb{R}^m$$

denote the projections. By 2.24, 25 there exists a compact affine algebraic set X of dimension $= \dim M$, a Nash submanifold M' of X, and a smooth regular mapping q from X to $\mathbb{R}^n \times \mathbb{R}^m$ such that M' contains no singular points of X, and $q|_{M'}$ is a Nash diffeomorphism onto M. Apply 2.28, 29 to X and $q^{-1}(a \times \mathbb{R}^m)$. Then we have a compact affine nonsingular algebraic set Y, an algebraic subset Y_1 of Y, a union M'' of connected components of $Y - Y_1$, and a smooth rational mapping $r\colon Y \to \mathbb{R}^n \times \mathbb{R}^m$ such that Y_1 has only normal crossings in Y, and $r|_{M''}$ is a Nash diffeomorphism onto

M. Set

$$L_1 = \overline{M''} \quad \text{and} \quad \pi = p_1 \circ r|_{M''}.$$

If $Y_1 \cap L_1$ is not the boundary of L_1, we clip L_1 along $Y_1 \cap L_1$. Then L_1 is an affine Nash manifold with corners, and $f \circ \pi$ is extendible to L_1 because

$$f \circ \pi = p_2 \circ r|_{M''}. \qquad \square$$

Algebraic Model Theorem 4.10. *An affine Nash manifold is Nash diffeomorphic to an affine nonsingular algebraic set.*

See [S$_1$] for its proof.

§5. Sheaf theory of Nash function germs

The contents of this section is due to [C-R-S$_1$] if not otherwise specified. Throughout this section, let $M \subset \mathbb{R}^n$ denote an affine Nash manifold, $\mathcal{N}$ the sheaf of Nash function germs on M (we write $\mathcal{N}_M$ if we need to emphasize M), and $\mathcal{N}_x$ the stalk of $\mathcal{N}$ at a point x of M. First we state two fundamental facts from [S$_1$].

5.1. For an ideal $\mathfrak{p} \subset N(M)$, we have

$$\mathfrak{p} = H^0(M, \mathfrak{p}\mathcal{N}),$$

where $\mathfrak{p}\mathcal{N}$ denotes the sheaf of $\mathcal{N}$-ideals generated by $\mathfrak{p}$.

5.2 (Strong Coherence). The sheaf $\mathcal{N}$ is coherent as a sheaf of $\mathcal{N}$-modules. Moreover, for each $f = (f_1, \dots, f_k) \in N(M, \mathbb{R}^k)$, the *relation* $\mathfrak{R}$ of f is finitely generated by its global cross-sections. Here the relation means the sheaf of submodules of $\mathcal{N}^k$ defined by

$$\mathfrak{R}_x = \Big\{(\varphi_1, \dots, \varphi_k) \in \mathcal{N}_x^k : \sum_{i=1}^k \varphi_i f_{ix} = 0\Big\}, \quad x \in M.$$

Since the concept of a coherent sheaf of $\mathcal{N}$-ideals is too wide (e.g. the sheaf defined by $\mathbb{Z}$ in the case of $M = \mathbb{R}$), we introduce the following two concepts. A coherent sheaf of $\mathcal{N}$-ideals $\mathcal{I}$ is called *finite* if $\mathcal{I}$ does not admit a nontrivial infinite decomposition $\mathcal{I} = \bigcap \mathcal{I}_i$ into coherent sheaves of $\mathcal{N}$-ideals $\mathcal{I}_i$. Here a decomposition $\mathcal{I} = \bigcap \mathcal{I}_i$ is called *nontrivial* if for each i, $\mathcal{I} \neq \bigcap_{j \neq i} \mathcal{I}_j$. In the following arguments, decompositions are always assumed to be nontrivial. The coherent sheaf $\mathcal{I}$ is called *strongly finite* if there exists a finite open semialgebraic covering $\{U_i\}$ of M such that for each i, $\mathcal{I}|_{U_i}$ is generated by Nash functions on U_i. It is clear that if M is compact then these three concepts coincide with each other. Moreover:

5.3. A coherent sheaf of $\mathcal{N}$-ideals $\mathcal{I}$ admits a finite or countable decomposition into irreducible coherent sheaves of $\mathcal{N}$-ideals. If $\mathcal{I}$ is not reduced then the decomposition is not necessarily unique, but if $\mathcal{I}$ is reduced then it is unique.

5.4. A strongly finite sheaf of $\mathcal{N}$-ideals is finite. If a finite sheaf of $\mathcal{N}$-ideals is reduced, or its stalks are all principal, then it is strongly finite. But it seems plausible that in general, a finite sheaf of $\mathcal{N}$-ideals is strongly finite.

Let $\mathcal{O}$ denote the sheaf of analytic function germs on M, and let $\mathcal{O}(M)$ denote the ring of analytic functions on M. Now we have four important conjectures.

Conjecture I. *For a prime ideal* $\mathfrak{p} \subset N(M)$, $\mathfrak{p}\mathcal{O}(M)$ *is prime.*

Conjecture II. *A finite sheaf of* $\mathcal{N}$*-ideals* $\mathcal{I}$ *is generated by global Nash functions.*

Conjecture III. *For the same* $\mathcal{I}$ *as above, the natural homomorphism*

$$H^0(M, \mathcal{N}) \longrightarrow H^0(M, \mathcal{N}/\mathcal{I})$$

is surjective.

Conjecture IV. *Given a Nash function* f *on* M *and an analytic factorization* $f = f_1 f_2$, *there exist Nash functions* g_1 *and* g_2 *on* M *and positive analytic functions* φ_1 *and* φ_2 *such that*

$$\varphi_1\varphi_2 = 1, \quad f_1 = \varphi_1 g_1 \quad \text{and} \quad f_2 = \varphi_2 g_2.$$

When we treat only $\mathfrak{p}$ of height 1 in Conjecture I, we call it **Conjecture $\mathbf{I_1}$**. **Conjecture $\mathbf{II_1}$** and **$\mathbf{III_1}$** are the restrictions of Conjecture II and III to the case where all the ideals $\mathcal{I}_x$, $x \in M$, are principal, **Conjecture $\mathbf{II^r}$** and **$\mathbf{III^r}$** are the restrictions to the case where $\mathcal{I}$ is reduced, and **Conjecture $\mathbf{II^s}$** and **$\mathbf{III^s}$** are defined for strongly finite $\mathcal{I}$, respectively. Similarly we also consider **Conjecture $\mathbf{II^r_1}$** and **$\mathbf{III^r_1}$**.

Note that even if M is compact, Conjectures II nor III do not hold for general coherent sheaves of $\mathcal{N}$-modules [S_1, IV.2.10] and that Conjectures I and IV locally hold (see 2.11, 26).

Theorem 5.5. *Conjectures* I, II^r, III^r, II^s *and* III^s *are equivalent.*
Conjectures I_1, II_1, II^r_1, III_1, III^r_1 *and* IV *are equivalent.*

Later we give a sketch of the proof. For an ideal $\mathfrak{p}$ of $N(M)$, let $\mathfrak{p}^{-1}(0)$ denote the common zero set of the elements of $\mathfrak{p}$. We call $\mathfrak{p}$ *real* if it is the set of Nash functions vanishing on $\mathfrak{p}^{-1}(0)$. For a sheaf of $\mathcal{N}$-ideals $\mathcal{I}$, $\operatorname{supp}\mathcal{I}$ denotes the support.

Theorem 5.6. *The above conjectures hold in the following cases:*

(a) *The manifold M is compact* [C-R-S_2].

(b) *In* I, $\dim \mathfrak{p}^{-1}(0) = 0$.

(b)* *In* II *and* III, $\dim\ \operatorname{supp}\mathcal{I} = 0$.

(c) *In* I, *the ideal $\mathfrak{p}$ is real, and $\mathfrak{p}^{-1}(0)$ has only isolated Nash singularities, where Nash singularity is defined by Nash functions on M in the same way as algebraic singularity.*

(d) *In* I, coheight $\mathfrak{p} \le 1$.

(d)* *In* II *and* III, *the coheight of the sheaf $\mathcal{I}$ is everywhere* ≤ 1.

(e) *In* II, *each stalk $\mathcal{I}_x$ is real, and the intersection of M with the Zariski closure of* $\operatorname{supp}\mathcal{I}$ *in $\mathbb{R}^n$ has only isolated Nash singularities.*

(f) *In* IV, *the set $f^{-1}(0) \cap \operatorname{Sing} f$ is discrete* [S_2].

(g) *In* I, *the ring $N(M)/\mathfrak{p}$ is normal.*

(g)* *In* II *and* III, *all the rings $\mathcal{N}_x/\mathcal{I}_x$, $x \in M$, are normal.*

(h) $\dim M \le 2$.

The most useful tool for our theory is complexification. Let us explain it. Let $M^{\mathbb{C}} \subset \mathbb{C}^n$ denote an analytic complexification of $M \subset \mathbb{R}^n$, which is semialgebraic when we regard $\mathbb{C}^n$ as $\mathbb{R}^{2n}$. For an ideal $\mathfrak{p}$ of $N(M)$, the *complex Nash set germ* of $\mathfrak{p}$ at M is the germ at M of the common zero set of $f_1^{\mathbb{C}}, \dots, f_k^{\mathbb{C}}$, which are complexifications of generators $f_1, \dots, f_k$ of $\mathfrak{p}$, defined on an open semialgebraic neighborhood of M in $M^{\mathbb{C}}$. We denote it by $X(\mathfrak{p})$. (We omit the term "at M" when no confusion is possible.) For a complex Nash set germ X at M, let $\mathfrak{p}(X)$ denote the ideal of $N(M)$ consisting of functions whose complexifications have germs at M which vanish at X. We define the *invariant complex analytic set germ* $X(\mathfrak{p})$ of an ideal $\mathfrak{p}$ of $\mathcal{O}(M)$ in the same way. Here an invariant set (germ) means a subset (germ) of $\mathbb{C}^n$ invariant under the complex conjugation. Then we have:

5.7. For an ideal $\mathfrak{p}$ of $N(M)$, $\mathfrak{p}(X(\mathfrak{p})) = \sqrt{\mathfrak{p}}$ (see [S_1]), and a similar equality holds also for an ideal of $\mathcal{O}(M)$. Hence the following correspondences are bijective:

$$\{\text{reduced ideals of } N(M)\} \ni \mathfrak{p} \longrightarrow X(\mathfrak{p}) \in \{\text{complex Nash set germs at } M\},$$
$$\{\text{prime ideals of } N(M)\} \ni \mathfrak{p} \longrightarrow X(\mathfrak{p}) \in \{\text{irreducible complex Nash set germs}\},$$
$$\{\text{reduced ideals of } \mathcal{O}(M)\} \ni \mathfrak{p} \longrightarrow X(\mathfrak{p}) \in \{\text{invariant complex analytic set germs}\},$$
$$\{\text{prime ideals of } \mathcal{O}(M)\} \ni \mathfrak{p} \longrightarrow X(\mathfrak{p}) \in \{\text{irreducible invariant complex analytic set germs}\}.$$

5.8. A complex Nash set germ X admits a unique finite decomposition $X = \bigcup X_i$ into irreducible complex Nash set germs and also into irreducible invariant complex analytic set germs.

By 5.7, 8, Conjecture I is equivalent to

Conjecture I″. *An irreducible complex Nash set germ at M is irreducible as an invariant complex analytic set germ at M.*

For a coherent sheaf of $\mathcal{O}$-ideals $\mathcal{I}$ on M, we define the *invariant complex analytic set germ* $X(\mathcal{I})$ of $\mathcal{I}$ at M as above. Conversely, for an invariant complex analytic set germ X at M we define the reduced coherent sheaf of $\mathcal{O}$-ideals $\mathcal{I}(X)$ of X on M.

5.9. For a coherent sheaf of $\mathcal{O}$-ideals $\mathcal{I}$ on M, $\mathcal{I}(X(\mathcal{I})) = \sqrt{\mathcal{I}}$. Hence we have the following bijective correspondences:

$$\{\text{reduced coherent sheaves of } \mathcal{O}\text{-ideals on } M\} \ni \mathcal{I} \longrightarrow X(\mathcal{I}) \in \{\text{invariant complex analytic set germs at } M\},$$
$$\{\text{irreducible reduced coherent sheaves of } \mathcal{O}\text{-ideals}\} \ni \mathcal{I} \longrightarrow X(\mathcal{I}) \in \{\text{irreducible invariant complex analytic set germs}\}.$$

An invariant complex analytic set germ X is called *finite* if it admits a finite decomposition into irreducible ones. We call X *semialgebraic* if it is the germ of a semialgebraic set.

5.10. A finite invariant complex analytic set germ is semialgebraic if it is locally semialgebraic at each point of M. This implies that each component of the decomposition of a complex Nash set germ into irreducible invariant complex analytic set germs is semialgebraic, and that the following correspondences are bijective.

$$\{\text{reduced finite sheaves of } \mathcal{N}\text{-ideals on } M\} \ni \mathcal{I} \longrightarrow X(\mathcal{I}) \in \{\text{semialgebraic invariant complex analytic set germs at } M\},$$
$$\{\text{irreducible reduced finite sheaves of } \mathcal{N}\text{-ideals}\} \ni \mathcal{I} \longrightarrow X(\mathcal{I}) \in \{\text{semialgebraic irreducible invariant complex analytic set germs}\}.$$

Hence Conjecture II^r is equivalent to

Conjecture II^{r}″. *A semialgebraic invariant complex analytic set germ at M is a complex Nash set germ at M.*

5.11. For a semialgebraic irreducible invariant complex analytic set germ X at M, there exists a unique irreducible complex Nash set germ Y at M such that X is one component of the decomposition of Y into irreducible invariant complex analytic set germs.

Let $\mathcal{I}$ be a finite reduced sheaf of $\mathcal{N}$-ideals. Set $X = X(\mathcal{I})$ and $M_1 = M \times \mathbb{R}$. A *semialgebraic invariant complex analytic set germ* over X is the intersection of the germ of $X \times \mathbb{C}$ at M_1 and the germ of the graph of $\Phi^{\mathbb{C}}$ at M_1 if it is semialgebraic, where $\Phi^{\mathbb{C}}$ is a complexification of an analytic function Φ on M. Let φ be a cross-section of $\mathcal{N}/\mathcal{I}$. Then φ defines a semialgebraic invariant complex analytic set germ X_1 over X because by Cartan's Theorem B φ is the restriction to X of an analytic function Φ on M.

5.12. The following correspondence is bijective:

$$H^0(M, \mathcal{N}/\mathcal{I}) \ni \varphi \longrightarrow$$
$$X_1 \in \{\text{semialgebraic invariant complex analytic set germs over } X\}.$$

Hence Conjecture III^r is equivalent to:

Conjecture III$^{r''}$. *Let X be a semialgebraic invariant complex analytic set germ at M, and let X_1 be a semialgebraic invariant complex analytic set germ over X. Then there exists a Nash function f on M such that X_1 is the intersection of the germ of $X \times \mathbb{C}$ at $M \times \mathbb{R}$ and the germ at $M \times \mathbb{R}$ of the graph of a complexification of f.*

Let $\mathcal{I}$ be a coherent sheaf of $\mathcal{N}$-ideals on M. For each integer $1 \le i \le \dim M$, the *associated sheaf* $\mathcal{A}_i(\mathcal{I})$ is defined so that $\mathcal{A}_i(\mathcal{I})_x$, $x \in M$, is the intersection of the associated prime ideals of $\mathcal{I}_x$ of height i.

5.13. Each $\mathcal{A}_i(\mathcal{I})$ is reduced and coherent. Let $\mathcal{I} = \bigcap \mathcal{I}_i$ be a decomposition into irreducible coherent sheaves of $\mathcal{N}$-ideals. Then for each i, $\mathcal{A}_j(\mathcal{I}_i) \neq \mathcal{N}$ for one and only one j, and such an $\mathcal{A}_j(\mathcal{I}_i)$ is an irreducible component of $\mathcal{A}_j(\mathcal{I})$. For each irreducible component $\mathcal{J}$ of $\mathcal{A}_j(\mathcal{I})$, the number of $\mathcal{I}_i$ such that $\mathcal{A}_j(\mathcal{I}_i) = \mathcal{J}$ is finite. Consequently, $\mathcal{I}$ is finite if and only if all $\mathcal{A}_j(\mathcal{I})$'s are finite.

Let k be a nonnegative integer, and let $\mathcal{I}$ be a coherent sheaf of $\mathcal{N}$-submodules of $\mathcal{N}^k$ on M. For each $1 \le i \le k$, let p_i denote the projection of $\mathcal{N}^k$ onto the i-th factor. The *factor sheaf* $\mathfrak{B}_i(\mathcal{I})$ of $\mathcal{I}$ is defined to be

$$p_i(\mathcal{I} \cap (\overbrace{0 \times \cdots \times 0 \times \mathcal{N}}^{i} \times \cdots \times \mathcal{N})).$$

Clearly $\mathfrak{B}_i(\mathcal{I})$ is a coherent sheaf of $\mathcal{N}$-ideals. We call $\mathcal{I}$ *finite* if so is each $\mathfrak{B}_i(\mathcal{I})$. We call $\mathcal{I}$ *strongly finite* if there exists a finite open semialgebraic covering $\{U_i\}$ of M such that for each i, $\mathcal{I}|_{U_i}$ is generated by global cross-sections. Note that $\mathcal{I}$ is strongly finite if and only if each factor sheaf of $\mathcal{I}$ is strongly finite. We generalize Conjecture II and III.

Conjecture II′. *For any k, a finite sheaf of $\mathcal{N}$-submodules $\mathcal{I}$ of $\mathcal{N}^k$ is generated by global cross-sections.*

Conjecture III′. *For the same $\mathcal{I}$ as above, the natural homomorphism*

$$H^0(M, \mathcal{N}^k) \longrightarrow H^0(M, \mathcal{N}^k/\mathcal{I})$$

is surjective.

We generalize also the first statement of Theorem 5.5 and Theorem 5.6 as follows:

Theorem 5.5′. *Conjectures* I, II′^{r}, II′^{s}, III′^{r} *and* III′^{s} *are equivalent.*

Theorem 5.6′. *In* 5.6 *we can add the following cases*:

(b)′ *In* II′ *and* III′, *for each i,* $\dim \operatorname{supp} \mathfrak{B}_i(\mathcal{I}) = 0$.

(d)′ *In* II′ *and* III′, *the coheight of each factor sheaf* $\mathfrak{B}_i(\mathcal{I})$ *is everywhere* ≤ 1.

(g)′ *In* II′ *and* III′, *all the rings* $\mathcal{N}_x/\mathfrak{B}_i(\mathcal{I})_x$, $x \in M$, *are normal.*

(h)′ *In* II′ *and* III′, $\dim M \leq 2$.

Proof that Conjectures I *and* II^r *are equivalent.* Clear because by 5.11, Conjecture I″ and II$^{r\prime\prime}$ are equivalent. □

Sketch of proof that if Conjecture II^r *holds then Conjecture* III^r *does.* Assuming Conjecture II$^{r\prime\prime}$ we prove Conjecture III$^{r\prime\prime}$. Let M, X and X_1 be given as in Conjecture III$^{r\prime\prime}$. Set $M_1 = M \times \mathbb{R}$, and let $p\colon M_1 \to M$ denote the projection. Then X_1 is a complex Nash set germ at M_1. (Here we use the hypothesis.) Let $\mathcal{I}$ and $\mathcal{I}_1$ denote the sheaves of $\mathcal{N}_M$-ideals and $\mathcal{N}_{M_1}$-ideals generated by $\mathfrak{p} = \mathfrak{p}(X) \subset N(M)$ and $\mathfrak{p}_1 = \mathfrak{p}(X_1) \subset N(M_1)$, respectively. Let φ be the cross-section of $\mathcal{N}_M/\mathcal{I}$ which defines X_1. For each point $x \in M$, let Φ_x be an element of $\mathcal{N}_{Mx}$ whose image in $\mathcal{N}_{Mx}/\mathcal{I}_x$ is φ_x.

Let $f_1, \ldots, f_k$ be generators of $\mathfrak{p}_1$. Then for every $(x_0, t_0) \in X_1 \cap M_1$ we can write

$$g_i f_{i(x_0,t_0)} = t - \Phi_{x_0} \mod \mathcal{I}_{x_0}\mathcal{N}_{M_1(x_0,t_0)}$$

for some i and $g_i \in \mathcal{N}_{M_1(x_0,t_0)}$. It follows that the sets

$$\begin{aligned} U_i &= \{(x,t) \in M_1 \colon \mathcal{I}_{1(x,t)} = (f_{i(x,t)}) + \mathcal{I}_x\mathcal{N}_{M_1(x,t)}\} \\ &= \{(x,t) \in M_1 \colon t - \Phi_x \in (f_{i(x,t)}) + \mathcal{I}_x\mathcal{N}_{M_1(x,t)}\}, \qquad i = 1, \ldots, k, \end{aligned}$$

cover $X_1 \cap M_1$. Clearly each U_i is open in M_1. Moreover, it is semialgebraic because f_i is C^1 regular with respect to t at every point of $U_i \cap X_1$. Shrink each U_i so that f_i is C^1 regular with respect to t at every point of U_i. Then $f_i^{-1}(0) \cap U_i$ is the graph of a Nash function F_i on $V_i = p(U_i)$. Then $F_i = \varphi$ on $V_i \cap X$, and the germ at V_i of a Nash complexification of F_i equals φ on the germ of X at V_i.

Next, using a partition of unity we paste the F_i's. Then we have a C^1 Nash extension of φ to M. Lastly apply the relative approximation theorem. (Here we use the property that X_1 is a Nash set.) Then we obtain a Nash extension of φ to M. □

Sketch of proof that if Conjecture III^r *holds then Conjecture* I *does.* Let $\mathfrak{p}$ be a reduced ideal of $N(M)$ such that $\mathfrak{p}\mathcal{O}(M)$ is not prime. We want to see that $\mathfrak{p}$ is not prime. There are finite reduced sheaves of $\mathcal{N}_M$-ideals $\mathcal{I}_1$ and $\mathcal{I}_2$ such that

$$\mathcal{I} = \mathcal{I}_1 \cap \mathcal{I}_2 = \mathfrak{p}\mathcal{N} \quad \text{and} \quad \dim X(\mathcal{I}_1) = \dim X(\mathcal{I}_2) > \dim X(\sqrt{\mathcal{I}_1 + \mathcal{I}_2}).$$

Considering the Zariski closure of $X(\sqrt{\mathcal{I}_1 + \mathcal{I}_2})$ we obtain a Nash function f on M whose complexification has a germ at M which vanishes on $X(\sqrt{\mathcal{I}_1 + \mathcal{I}_2})$ but not on $X(\mathcal{I}_1)$. Since Conjecture III^r holds, there exists a Nash function F on M whose image in $H^0(M, \mathcal{N}/\mathcal{I})$ coincides with the image of f in $H^0(M, (\mathcal{I}_1 + \mathcal{I}_2)/\mathcal{I}_1) = H^0(M, \mathcal{I}_2/\mathcal{I}) \subset H^0(M, \mathcal{N}/\mathcal{I})$. Then the germ at M of a complexification of F vanishes on $X(\mathcal{I}_2)$ and does not vanish on $X(\mathcal{I}_1)$. In the same way we obtain a Nash function G on M whose complexification has germ at M which vanishes on $X(\mathcal{I}_1)$ and does not vanish on $X(\mathcal{I}_2)$. Then neither F nor G is an element of $\mathfrak{p}$, while FG is so, which implies that $\mathfrak{p}$ is not prime. □

Sketch of Proof that if Conjectures II^r *and* III^r *hold then Conjecture* III^s *does.* It suffices to prove the following statement because Conjecture III is trivial for $\mathcal{I} = \mathcal{N}$, and an increasing sequence of finite reduced sheaves of $\mathcal{N}$-ideals is stationary.

(1) Let $\mathcal{I}$ be a strongly finite sheaf of $\mathcal{N}$-ideals. Assume that Conjecture III holds for any strongly finite sheaf of $\mathcal{N}$-ideals $\mathcal{I}'$ with $\mathcal{I}' \supset \sqrt{\mathcal{I}}$. Then Conjecture III holds for $\mathcal{I}$.

Let $\mathcal{J}$ be a strongly finite sheaf of $\mathcal{N}$-ideals with $\sqrt{\mathcal{J}} \supset \sqrt{\mathcal{I}}$. We introduce an invariant $\alpha_{\mathcal{J}}$ of $\mathcal{J}$ as follows. Let $f_1, \ldots, f_k$ generate $\sqrt{\mathcal{I}}$. For each $1 \le i \le k$, let α_i denote the minimal number α such that $f_i^{\alpha+1} \in H^0(M, \sqrt{\mathcal{I}} \cap \mathcal{J})$. Then $\alpha_{\mathcal{J}}$ is the minimal of all the sums $\sum_{i=1}^k \alpha_i$ as $\{f_1, \ldots, f_k\}$ runs through the families of generators of $\sqrt{\mathcal{I}}$. If $\alpha_{\mathcal{J}} = 0$, then $\mathcal{J} \supset \sqrt{\mathcal{I}}$ and, hence, by assumption in (1) Conjecture III holds for $\mathcal{J}$. Hence, if the following statement holds then Conjecture III holds for $\mathcal{J}$ and, in particular, for $\mathcal{I}$.

(2) Assume that Conjecture III holds for any strongly finite sheaf $\mathcal{I}'$ with $\sqrt{\mathcal{I}'} \supset \sqrt{\mathcal{I}}$ and $\alpha_{\mathcal{I}'} < \alpha_{\mathcal{J}}$. Then Conjecture III holds for $\mathcal{J}$.

Let us prove (2). We replace Conjecture III for $\mathcal{J}$ with the equivalent Conjecture III$'$ for $\tilde{\mathcal{J}}$, where $\tilde{\mathcal{J}}$ is defined as follows. Let $f_1, \ldots, f_k$ be generators of $\sqrt{\mathcal{I}}$ which define $\alpha_{\mathcal{J}}$. Let $f_0, \ldots, f_{-l}$, $l > 0$, be generators of $\sqrt{\mathcal{J}}$. Define a homomorphism $\pi \colon \mathcal{N}^{k+l+1} \to \mathcal{N}$ by

$$\mathcal{N}_x^{k+l+1} \ni (\varphi_{-l}, \ldots, \varphi_k) \longrightarrow \sum_{i=-l}^{k} \varphi_i f_{ix} \in \mathcal{N}_x, \qquad x \in M.$$

Set $\tilde{\mathcal{J}} = \pi^{-1}(\mathcal{J})$. Then since Im $\pi = \sqrt{\mathcal{J}}$, it suffices to show Conjecture III′ for $\tilde{\mathcal{J}}$.

For each $1 \le i \le k+l+1$, set

$$\tilde{\mathcal{J}}_i = \tilde{\mathcal{J}} \cap (\overbrace{0 \times \cdots 0 \times \mathcal{N}}^{i} \times \mathcal{N} \cdots \times \mathcal{N}),$$

$$\tilde{\mathcal{K}}_i = \tilde{\mathcal{J}}_i + \overbrace{0 \times \cdots \times 0}^{i} \times \mathcal{N} \times \cdots \times \mathcal{N}.$$

Then

$$\tilde{\mathcal{J}}_1 = \tilde{\mathcal{J}}, \quad \tilde{\mathcal{J}}_{k+l+1} = 0,$$

and the following sequence is exact:

$$0 \longrightarrow H^0(M, \tilde{\mathcal{K}}_i/\tilde{\mathcal{J}}_i) \longrightarrow H^0(M, \overbrace{0 \times \cdots \times \mathcal{N}}^{i} \times \mathcal{N} \times \cdots \times \mathcal{N}/\tilde{\mathcal{J}}_i)$$
$$\longrightarrow H^0(M, \overbrace{0 \times \cdots \times \mathcal{N}}^{i} \times \mathcal{N} \times \cdots \times \mathcal{N}/\tilde{\mathcal{K}}_i).$$

Let $\mathcal{K}_i$ denote the image of $\tilde{\mathcal{K}}_i$ under the projection of $\mathcal{N}^{k+l+1}$ onto the i-th factor. Then we can regard

$$\tilde{\mathcal{K}}_i/\tilde{\mathcal{J}}_i \quad \text{and} \quad \overbrace{0 \times \cdots \times 0 \times \mathcal{N}}^{i} \times \mathcal{N} \times \cdots \times \mathcal{N}/\tilde{\mathcal{K}}_i$$

as

$$\overbrace{0 \times \cdots \times 0}^{i} \times \mathcal{N} \times \cdots \times \mathcal{N}/\tilde{\mathcal{J}}_{i+1} \quad \text{and} \quad \mathcal{N}/\mathcal{K}_i,$$

respectively. Hence we can reduce Conjecture III′ for $\tilde{\mathcal{J}}$ to Conjecture III for $\mathcal{K}_i$, $i = 1, \dots, k+l+1$. By the above choice of $f_1, \dots, f_k$, each $\mathcal{K}_i$ is a strongly finite sheaf with $\sqrt{\mathcal{K}_i} \supset \sqrt{\mathcal{I}}$ and $\alpha_{\mathcal{K}_i} < \alpha_{\mathcal{J}}$. Hence by the assumption in (2), Conjecture III holds for all $\mathcal{K}_i$. □

Sketch of proof that if Conjecture III^s *holds then Conjecture* II^s *does.* For the same reason as in the above proof it suffices to prove the following.

As above let $\mathcal{I}$ and $\mathcal{J}$ be strongly finite sheaves of $\mathcal{N}$-ideals with $\sqrt{\mathcal{J}} \supset \sqrt{\mathcal{I}}$. Assume that any strongly finite sheaf $\mathcal{I}'$, with $\sqrt{\mathcal{I}'} \supset \sqrt{\mathcal{I}}$ and $\alpha_{\mathcal{I}'} < \alpha_{\mathcal{J}}$, is generated by global Nash functions. Then $\mathcal{J}$ is generated by global Nash functions.

Let π, $\tilde{\mathcal{J}}$, $\tilde{\mathcal{J}}_i$, $\tilde{\mathcal{K}}_i$ and $\mathcal{K}_i$ be defined as above. Then we need only show that $\tilde{\mathcal{J}}$ is generated by global cross-sections. Consider the sequence $\tilde{\mathcal{J}}/\tilde{\mathcal{J}}_2 = \tilde{\mathcal{J}}_1/\tilde{\mathcal{J}}_2, \dots,$ $\tilde{\mathcal{J}}_{k+l+1}/\tilde{\mathcal{J}}_{k+l+2} = \tilde{\mathcal{J}}_{k+l+1}$, where $\tilde{\mathcal{J}}_{k+l+2} = 0$. By hypothesis we can prove that for each i, $\tilde{\mathcal{J}}_i$ mod $\tilde{\mathcal{J}}_{i+1}$ is generated by global cross-sections. Then the generators for all i generate $\tilde{\mathcal{J}}$. □

Sketch of proof that if Conjectures II *and* III *hold (for strongly finite sheaves) then Conjecture* II′ *and* III′ *do (for strongly finite sheaves, respectively)* [S_1]. As in the last two sketches, using factor sheaves we reduce Conjectures II′ and III′ to Conjectures II and III. □

Proof that if Conjectures II^s *or* III^s *holds then Conjectures* II^r *or* III^r *does, respectively.* Clear because a reduced finite sheaf is strongly finite. □

Sketch of proof that Conjectures I_1, II_1, III_1, II_1^r *and* III_1^r *are equivalent.* In the same way as above we can prove that I_1, II_1^r and III_1^r are equivalent. Hence it suffices to show that if Conjecture II_1^r holds then do Conjectures II_1 and III_1. Let $\mathcal{I}$ be a finite sheaf of $\mathcal{N}$-ideals such that all the ideals $\mathcal{I}_x$, $x \in M$, are principal. Then we have a unique factorization $\mathcal{I} = \prod_{i=1}^k \mathcal{I}_i^{\alpha_i}$ into irreducible coherent sheaves of $\mathcal{N}$-ideals. By hypothesis, each $\mathcal{I}_i$ is generated by global Nash functions. Hence $\mathcal{I}$ also is generated by global ones. Thus Conjecture II_1 holds for $\mathcal{I}$.

We prove Conjecture III_1 by induction on $\alpha = \sum_{i=1}^k \alpha_i$. A key of proof is Thom's Realization Theorem [Th$_1$]. If $\alpha = 1$ then Conjecture III_1 follows. Hence assume that Conjecture III_1 holds for smaller α. We can assume $\alpha_1 > 1$. Let $\varphi \in H^0(M, \mathcal{N}/\mathcal{I})$. We prove that φ is the image of some element of $H^0(M, \mathcal{N})$. For a moment we suppose that Conjecture III_1 holds for $\mathcal{I}_1^2$. Then, since the sequence

$$0 \longrightarrow H^0(M, \mathcal{I}_1^2/\mathcal{I}) \longrightarrow H^0(M, \mathcal{N}/\mathcal{I}) \longrightarrow H^0(M, \mathcal{N}/\mathcal{I}_1^2)$$

is exact, we can reduce φ to the case where $\varphi \in H^0(M, \mathcal{I}_1^2/\mathcal{I})$. Let f_1 be a generator of $\mathcal{I}_1^2$, e.g. the sum of the squares of a finite number of generators of $\mathcal{I}_1$. Define an isomorphism from $\mathcal{N}$ to $\mathcal{I}_1^2$ by

$$\mathcal{N}_x \ni \rho \longrightarrow \rho f_{1x} \in \mathcal{I}_{1x}^2, \quad x \in M.$$

Then it induces a commutative diagram

$$\begin{array}{ccc} H^0(M, \mathcal{N}) & \overset{\cong}{\longrightarrow} & H^0(M, \mathcal{I}_1^2) \\ \downarrow & & \downarrow \\ H^0(M, \mathcal{N}/\mathcal{I}_1^{\alpha_1 - 2} \prod_{i \neq 1} \mathcal{I}_i^{\alpha_i}) & \overset{\cong}{\longrightarrow} & H^0(M, \mathcal{I}_1^2/\mathcal{I}). \end{array}$$

Hence we can replace φ with an element of $H^0(M, \mathcal{N}/\mathcal{I}_1^{\alpha_1-2} \prod_{i\neq 1} \mathcal{I}_i^{\alpha_i})$. Then by induction hypothesis, φ is the image of an element of $H^0(M, \mathcal{N})$. Thus Conjecture III_1 is proved for $\mathcal{I}$.

We need only prove the case of $\mathcal{I} = \mathcal{I}_1^2$. For the same reason as above we can assume that φ is an element of $H^0(M, \mathcal{I}_1/\mathcal{I}_1^2)$.

By Compactification Theorem 4.6, the manifold M can be the interior of a compact affine Nash manifold possibly with boundary M'. Let $\dim M = n$. Let $[\operatorname{supp} \mathcal{I}_1]$ denote the fundamental class of the support in $H_{n-1}(\operatorname{supp} \mathcal{I}_1 \cup \partial M', \partial M'; \mathbb{Z}_2)$. Let $[\operatorname{supp} \mathcal{I}_1]_*$ denote its image in $H_{n-1}(M', \partial M'; \mathbb{Z}_2)$. There are two cases: $[\operatorname{supp} \mathcal{I}_1]_* = 0$ or $\neq 0$.

Consider the first case. The ideal $H^0(M, \mathcal{I}_1)$ is principal for the following reason. Let M'' be a union of some connected components of $M - \operatorname{supp}\mathcal{I}_1$ such that the fundamental class of the closure $\overline{M''}$ in $H_n(M', \operatorname{supp}\mathcal{I}_1 \cup \partial M'; \mathbb{Z}_2)$ is carried to $[\operatorname{supp}\mathcal{I}_1]$ by the boundary homomorphism. Define a function g_1 on M so that

$$g_1^2 = f_1 \quad \text{on} \quad M \quad \text{and} \quad g_1 \begin{cases} > 0 & \text{on } M'' - \operatorname{supp}\mathcal{I}_1 \\ < 0 & \text{on } M - M'' - \operatorname{supp}\mathcal{I}_1. \end{cases}$$

Then g_1 is a generator of $H^0(M, \mathcal{I}_1)$. Hence $H^0(M, \mathcal{I}_1)$ is principal.

Using g_1 as above, we obtain a commutative diagram

$$\begin{array}{ccc} H^0(M, \mathcal{N}) & \xrightarrow{\cong} & H^0(M, \mathcal{I}_1) \\ \downarrow & & \downarrow \\ H^0(M, \mathcal{N}/\mathcal{I}_1) & \xrightarrow{\cong} & H^0(M, \mathcal{I}_1/\mathcal{I}_1^2). \end{array}$$

Hence we can reduce φ to the case where it is an element of $H^0(M, \mathcal{N}/\mathcal{I}_1)$ as above. Then, since $\mathcal{I}_1$ is reduced, by assumption, φ is the image of an element of $H^0(M, \mathcal{N})$.

Let us consider the case of $[\operatorname{supp}\mathcal{I}_1]_* \neq 0$. By [Th$_1$] there exists a compact C^∞ submanifold L of M' possibly with boundary and of dimension $n-1$ such that $\partial L \subset \partial M'$, L is transversal to $\partial M'$ in M', and $[\operatorname{Int} L]_* = [\operatorname{supp}\mathcal{I}_1]_*$. We can assume that L is of class Nash and $\dim(L \cap \operatorname{supp}\mathcal{I}_1) < n-1$. Let $\mathcal{J}$ denote the sheaf of $\mathcal{N}$-ideals on M defined by $L - \partial L$. Set $\mathcal{K} = \mathcal{I}_1\mathcal{J}$. Then $\mathcal{K}$ is reduced, and we have $[\operatorname{supp}\mathcal{K}]_* = 0$. As above we have a commutative diagram

$$\begin{array}{ccc} H^0(M, \mathcal{N}) & \xrightarrow{\cong} & H^0(M, \mathcal{J}^2) \\ \downarrow & & \downarrow \\ H^0(M, \mathcal{N}/\mathcal{I}_1^2) & \xrightarrow{\cong} & H^0(M, \mathcal{J}^2/\mathcal{K}^2). \end{array}$$

This reduces the problem to the case where φ is an element of $H^0(M, \mathcal{J}^2/\mathcal{K}^2)$. Since $[\operatorname{supp}\mathcal{K}]_* = 0$, we can find an element of $H^0(M, \mathcal{J}^2)$ whose image is φ in the same way as above. □

Proof that if Conjecture II$_1$ *holds then Conjecture* IV *does.* Let $f = \prod_{i=1}^k f_i^{\alpha_i}$ be a factorization of a Nash function on M into irreducible analytic functions. Then it suffices to find Nash functions g_i, $i = 1, \dots, k$, on M and nowhere vanishing analytic functions φ_i such that $f_i = \varphi_i g_i$. For each i let $\mathcal{I}_i$ denote the sheaf of $\mathcal{O}$-ideals generated by f_i. Then $\prod_{i=1}^k \mathcal{I}_i^{\alpha_i}$ is generated by f, and the sheaf of $\mathcal{N}$-ideals $\mathcal{J}_i = \mathcal{I}_i \cap \mathcal{N}$ is finite. By hypothesis $\mathcal{J}_i$ is generated by global cross-sections. Then as in the last proof we have a nonnegative function generator F_i of $\mathcal{J}_i^2$. By definition f_i^2 is a generator of $\mathcal{I}_i^2$. Hence there exists a positive analytic function ψ_i on M such that $F_i = \psi_i f_i^2$. Then $g_i = \psi_i^{1/2} f_i$ and $\varphi_i = \psi_i^{1/2}$ fulfill the requirements.

Proof that if Conjecture IV *holds then Conjecture* II$_1^r$ *does.* Let $\mathcal{I}$ be a finite reduced sheaf of $\mathcal{N}$-ideals such that all the ideals $\mathcal{I}_x$, $x \in M$, are principal. We need to prove that $X(\mathcal{I})$ is a complex Nash set germ. By 5.11 there exists a sheaf of $\mathcal{N}$-ideals $\mathcal{J}$ with the same properties as $\mathcal{I}$ such that $\mathcal{I} \cap \mathcal{J}$ is generated by global Nash functions, say, $f_1, \dots, f_k$, and $\dim X(\sqrt{\mathcal{I}+\mathcal{J}}) < \dim M - 1$. Let f denote the sum $f_1^2 + \cdots + f_k^2$, which is a generator of $(\mathcal{I} \cap \mathcal{J})^2$. By Cartan's Theorem A we have analytic functions $g_1, g_2, \dots$ on M, which generate $\mathcal{I}\mathcal{O}$, and whose complexifications are defined on a common Stein open neighborhood of M in $M^{\mathbb{C}}$. Let $c_1, c_2, \dots$ be sufficiently small positive numbers. Then $g = \sum_{i=1}^{\infty} c_i g_i^2$ is an analytic function on M, and it is a generator of $\mathcal{I}^2\mathcal{O}$. Let h be a nonnegative function generator of $\mathcal{J}^2\mathcal{O}$. Then $f = ghf'$ for some positive analytic function f'. Since we assume that Conjecture IV holds, we can replace g with a Nash function, whose complexification has a zero set germ at M that coincides with $X(\mathcal{I})$. Hence $X(\mathcal{I})$ is a complex Nash set germ. □

We omit the proof of Theorem 5.6.

§6. Nash groups

If a semialgebraic set or an affine Nash manifold admits a group structure then it should be expected to have many interesting properties. But at present we know only a few properties. This section explains elementary facts and three results of Hrushovski–Pillay [H-P], Madden–Stanton [M-S] and Pillay [Pi]. Other references are [N-P], [N-P-R], [Raz] and [St], which treat groups model-theoretically and, hence, work in the semialgebraic category.

An *affine Nash group* is an affine Nash manifold with group operations

$$G \times G \ni (x, y) \longrightarrow xy \in G \quad \text{and} \quad G \ni x \longrightarrow x^{-1} \in G$$

that are Nash mappings. We similarly define a *semialgebraic group* to be a semialgebraic set with semialgebraic group operations. Here we do not require the group operations to be continuous.

For example, $G_0 = [0, 1[$ equipped with noncontinuous group operation

$$a \cdot b = \begin{cases} a+b & \text{if } a+b < 1 \\ a+b-1 & \text{otherwise} \end{cases}$$

is a semialgebraic group. But keeping its group and semialgebraic structures we can modify G_0 so that the group operation will be continuous as follows. Let π be a continuous semialgebraic bijection from G_0 to $S^1 \subset \mathbb{R}^2$, and transfer group structure from G_0 to S^1 through π. Then S^1 is a semialgebraic group with continuous group operation. By Theorem 6.1 below, such a modification is possible for any semialgebraic group.

It is natural to ask, moreover, if there exists a semialgebraic group isomorphism from G_0 to some affine Nash group. But this is impossible as shown below.

Assume that there exists a semialgebraic group isomorphism π' from G_0 to an affine Nash group G'. Then G' would be homeomorphic to S^1 because otherwise it would be nonconnected or C^0 group isomorphic to $\mathbb{R}$. On the other hand, for any open semialgebraic subset U of G', $\pi^{-1}(U)$ contains inner points and, hence, there exists a positive integer k such that

$$k(\pi^{-1}(U)) = G_0 \qquad (\text{i.e. } kU = G'),$$

which is impossible. By Theorem 6.3 below, G' should be Nash group isomorphic to $(\mathbb{R}, \sin)/\pi\mathbb{Z}$ or $(\mathbb{R}, \wp_\alpha)/\mathbb{Z}$, where $(\mathbb{R}, \sin)/\pi\mathbb{Z}$ is $\mathbb{R}/\pi\mathbb{Z}$ as a topological group and is given a Nash structure so that the function sin: $\mathbb{R}/\pi\mathbb{Z} \to \mathbb{R}$ is a Nash mapping (or equivalently $(\mathbb{R}, \sin)/\pi\mathbb{Z}$ is the circle $\{z \in \mathbb{C} : |z| = 1\}$ in $\mathbb{C} = \mathbb{R}^2$ with the group structure of $\mathbb{C}$), and $(\mathbb{R}, \wp_\alpha)/\mathbb{Z}$ is similarly defined by a Weierstrass $\wp$-function $\wp_\alpha$ (see the details below). Consider the case of $(\mathbb{R}, \sin)/\pi\mathbb{Z}$ (the other case is treated in the same way). On the other hand, by 2.5, π' is continuous at some point of G_0 because it is semialgebraic. Hence π' is everywhere continuous because it is a group homomorphism. In conclusion, if we let p: $\mathbb{R} \to (\mathbb{R}, \sin)/\pi\mathbb{Z}$ denote the canonical covering then $(p|_{[0,\pi[})^{-1} \circ \pi'$ or $(p|_{]-\pi,0]})^{-1} \circ \pi'$ is a homeomorphism from $[0, 1[$ to $[0, \pi[$ or $]-\pi, 0]$, respectively, which is of the form $t \to \pi t$ or $t \to -\pi t$, respectively, by the group structure of $(\mathbb{R}, \sin)/\pi\mathbb{Z}$. But by the affine Nash structure of $(\mathbb{R}, \sin)/\pi\mathbb{Z}$, $\sin \circ (p|_{[0,\pi[})^{-1} \circ \pi'$ or $\sin \circ (p|_{]-\pi,0]})^{-1} \circ \pi'$ is semialgebraic. This means that $\sin \pi t$ or $\sin -\pi t$ is semialgebraic, which is impossible. Thus there does not exist a semialgebraic group isomorphism from G_0 to an affine Nash group.

In spite of the above example we want to give a good structure to a semialgebraic group. For that we extend the definitions of an affine Nash manifold and group as follows. A *Nash manifold* of dimension n is an analytic manifold with a finite system of coordinate neighborhoods $\{(U_i, \psi_i)\}_i$ such that each $\psi_i(U_i)$ is an open semialgebraic subset of $\mathbb{R}^n$ and for each pair of (U_i, ψ_i) and $(U_{i'}, \psi_{i'})$, the mapping

$$\psi_i \circ \psi_{i'}^{-1}: \psi_{i'}(U_i \cap U_{i'}) \longrightarrow \psi_i(U_i \cap U_{i'})$$

is a Nash diffeomorphism. If $(M, \{U_i, \psi_i\})$ and $(M', \{U'_{i'}, \psi'_{i'}\})$ are Nash manifolds then a *Nash mapping* f: $M \to M'$ is a C^0 mapping such that for any i and i'

$$\psi'_{i'} \circ f \circ \psi_i^{-1}: \psi_i(f^{-1}(U'_{i'}) \cap U_i) \longrightarrow \psi'_{i'}(U'_{i'})$$

is a Nash mapping between affine Nash manifolds.

Note that if a Nash manifold is Nash imbedded into a Euclidean space so that the image is a C^0 manifold then the image is an affine Nash manifold (we call a Nash manifold *imbeddable* if there exists such a Nash imbedding) and that an affine Nash manifold is a Nash manifold (2.18). The simplest example of a nonimbeddable Nash manifold is $\mathbb{R}/\mathbb{Z}$ whose Nash structure is given by it of $\mathbb{R}$ through the canonical covering p: $\mathbb{R} \to \mathbb{R}/\mathbb{Z}$ (Mazur). (If there were a Nash imbedding τ: $\mathbb{R}/\mathbb{Z} \to \mathbb{R}^n$ then $\tau \circ p$: $\mathbb{R} \to \mathbb{R}^n$ would be a Nash mapping such that $(\tau \circ p)^{-1}(\tau \circ p(0))$ is infinite, which is impossible.) See [S_1] for more results on nonimbeddable Nash manifold. Here we note only that the composite of two Nash mappings is a Nash mapping.

A *Nash group* is a Nash manifold with Nash group operations. Clearly $\mathbb{R}/\mathbb{Z}$ is a Nash group, and G_0 is semialgebraically group isomorphic to $\mathbb{R}/\mathbb{Z}$. This is possible for any semialgebraic group as follows.

Theorem 6.1 [Pi]. *A semialgebraic group is semialgebraically group isomorphic to a Nash group.*

Proof. Let $G \subset \mathbb{R}^n$ be a semialgebraic group of dimension m. By 2.5 there is an open affine Nash submanifold G_0 of G such that $\dim(G - G_0) < m$, the inversion $G_0 \ni x \to x^{-1} \in G$ is a Nash diffeomorphism onto its image, and the image G_0^{-1} is open in G. Note that $\dim(G - G_0^{-1}) < m$. Replace G_0 with $G_0 \cap G_0^{-1}$. Then we can assume $G_0^{-1} = G_0$. It follows $G_0 G_0 = G$ because for each $x \in G$, $\dim xG_0 = m$.

Next by 2.5 there is an open semialgebraic subset X of G_0G_0 such that $\dim(G_0G_0 - X) < 2m$, and the multiplication is a Nash mapping from X to G_0. Shrink G_0 so that for each $y \in G_0$

$$\text{(i)} \qquad \dim\{x \in G_0 \colon (x, y) \notin X\} < m.$$

This is possible because the subset of G_0, consisting of points y such that

$$\dim\{x \in G_0 \colon (x, y) \notin X\} = m$$

is of dimension $< m$. Moreover, considering in the same way the image of $G_0 \times y$ under the mapping

$$G_0G_0 \ni (x, y) \longrightarrow (x^{-1}, xy) \in G_0G,$$

we assume that for each $y \in G_0$

$$\text{(ii)} \qquad \dim\{x \in G_0 \colon (x^{-1}, xy) \notin X\} < m.$$

It follows from (i) and (ii) that

(iii) for any $a, b \in G$ the set

$$G_1 = \{x \in G_0 \colon axb \in G_0\}$$

is open in G_0, and the mapping

$$G_1 \ni x \longrightarrow axb \in aG_1b$$

is a Nash diffeomorphism for the following reason.

Let $x_0 \in G_1$. We show that x_0 has an open semialgebraic neighborhood G_2 in G_1 such that the mapping $G_2 \ni x \to axb \in G_0$ is a Nash imbedding. First write $b = b_1b_2$ with $b_1, b_2 \in G_0$. Then we have $c \in G_0$ such that

$$(ca, x_0),\ (cax_0, b_1),\ (cax_0b_1, b_2),\ (c^{-1}, caxb) \in X,$$

because by (i) and (ii)

$$\dim\{c \in G_0 \colon (ca, x_0) \notin X\} < m, \ldots, \dim\{c \in G_0 \colon (c^{-1}, caxb) \notin X\} < m.$$

Hence for some G_2, the mappings

$$G_2 \ni x \longrightarrow cax \in G_0, \qquad caG_2 \ni cax \longrightarrow caxb_1 \in G_0,$$
$$caG_2b_1 \ni caxb_1 \longrightarrow caxb \in G_0, \qquad caG_0b \ni caxb \longrightarrow axb \in G_0$$

are Nash imbeddings, whose composition is the mapping $G_2 \ni x \to axb \in G_0$.

In the same way we can prove that

(iv) for any $a, b \in G$ the set

$$Y = \{(x, y) \in G_0G_0\colon axby \in G_0\}$$

is open in G_0G_0, and the corresponding mapping $Y \to G_0$ is a Nash mapping.

Now we define a semialgebraic bijection π from G to a Nash manifold M. As shown below there are points $a_1, \dots, a_k \in G$ such that $\bigcup_{i=1}^k a_iG_0 = G$. Let $G_0^1, \dots, G_0^k$ denote copies of G_0. Define semialgebraic bijections $\pi^i\colon a_iG_0 \to G_0^i$, $i = 1, \dots, k$, by $\pi^i(a_ix) = x$. For each pair i and i', $\pi^i(a_iG_0 \cap a_{i'}G_0)$ and $\pi^{i'}(a_iG_0 \cap a_{i'}G_0)$ are open and semialgebraic in G_0^i and $G_0^{i'}$, respectively, and we can identify them through the mapping

$$\pi^i(a_iG_0 \cap a_{i'}G_0) \ni x \longrightarrow \pi^{i'} \circ (\pi^i)^{-1}(x) \in \pi^{i'}(a_iG_0 \cap a_{i'}G_0),$$

which is a Nash diffeomorphism, because by (iii) the mapping

$$\{x \in G_0\colon a_{i'}a_i^{-1}x \in G_0\} \ni x \longrightarrow a_{i'}a_i^{-1}x \in G_0$$

is a Nash imbedding. Hence $M = \bigcup_{i=1}^k G_0^i$ is a Nash manifold.

We need to find $a_1, \dots, a_k$. For that it suffices to prove the following statement.

Let H be a semialgebraic subset of G of dimension m' ($< m$). Then there exist points $b_0, \dots, b_l$ of G such that $\dim b_0H \cap \dots \cap b_lH < m'$.

Let $\overline{H_0}$ denote the closure in $\mathbb{R}^n \times \mathbb{R}^n$ of the set

$$H_0 = \{(x, xy) \in G \times G\colon y \in H\},$$

and let $p\colon \overline{H_0} \to \mathbb{R}^n$ denote the restriction of the projection from $\mathbb{R}^n \times \mathbb{R}^n$ onto the first factor. Then by 2.5 there exists an open semialgebraic subset G_3 of G such that $\dim(G - G_3) < m$, and for each $x \in G_3$, $\dim p^{-1}(x) = m'$. Moreover, we have a nonempty open semialgebraic subset G_4 of G_3 and a Nash stratification $\{H_i\}_{i=1,\dots,l}$ of $p^{-1}(G_4)$ compatible with H_0 such that $p|_{p^{-1}(G_4)}\colon \{H_i\}_{i=1,\dots,l} \to G_4$ is a stratification. Shrink G_4 and substratify $\{H_i\}_{i=1,\dots,l}$ if necessary. Then we can assume that for each H_i of dimension $m+m'$, $q(H_i)$ is an open subset of G_0, and $q|_{H_i}\colon H_i \to G_0$ is a Nash submersion, where q is the projection from $\mathbb{R}^n \times \mathbb{R}^n$ onto the second factor, because $\dim G_4H = m$. This implies that for each $b_0 \in G_4$ and H_i of dimension $m + m'$ there exists $x \in G_4$ such that $\dim q(H_i \cap p^{-1}(b_0)) \cap q(H_i \cap p^{-1}(x)) < m'$ because $q(H_i \cap p^{-1}(b_0))$ and $q(H_i \cap p^{-1}(x))$ are analytic manifolds. Moreover, the set of such x is semialgebraic, and open and dense in G_4. Hence so is also the following set:

$$\bigcap_{i=1}^{l}\{x \in G_4\colon \dim q(H_i \cap p^{-1}(b_0)) \cap q(H_i \cap p^{-1}(x)) < m'\}.$$

Let b_1 be a point of this set. Note that this does not necessarily imply that $\dim q(p^{-1}(b_0)) \cap q(p^{-1}(b_1)) < m'$. Indeed it may happen that

$$\dim q(H_i \cap p^{-1}(b_0)) \cap q(H_{i'} \cap p^{-1}(b_1)) = m'$$

for some i and i'. Next choose $b_2, \dots, b_l$ so that for each i and $0 \le j < j' \le l$,

$$\dim q(H_i \cap p^{-1}(b_j)) \cap q(H_i \cap p^{-1}(b_{j'})) < m'.$$

Then for any $1 \le i_0, \dots, i_l \le l$,

$$\dim \bigcap_{j=0}^{l} q(H_{i_j} \cap p^{-1}(b_j)) < m',$$

because $H_{i_j} = H_{i_{j'}}$ for some $j \ne j'$. Hence

$$\dim \bigcap_{j=0}^{l} q(p^{-1}(b_j)) < m',$$

which means that $\dim b_0 H \cap \cdots \cap b_l H < m'$.

Lastly we give a group structure to M by

$$xy = \pi(\pi^{-1}(x)\pi^{-1}(y)) \quad \text{for } (x, y) \in M \times M.$$

Then it remains to prove that the operations of inversion and multiplication in M are Nash mappings. But we omit the details because by (iii) and (iv) we can prove this in the same way as in the above proof that M is an manifold (see [Pi]). □

Remark 6.2. A C^r, $0 \le r < \infty$, Nash group, which we define in a natural way, is imbeddable. Hence a semialgebraic group is semialgebraically group isomorphic to an affine C^r Nash group.

Sketch of proof. It suffices to prove that a C^r Nash manifold is imbeddable [S_1]. Let M be a C^r Nash manifold, and let $\{\psi_i\colon U_i \to \mathbb{R}^n\}_{i=1,\dots,k}$ be a finite system of C^r Nash coordinate neighborhoods of M. Assume $n > 0$. Each $\psi_i(U_i)$ is an open semialgebraic subset of $\mathbb{R}^n$. Hence by 2.22 we have a Nash imbedding $\psi_i'\colon \psi_i(U_i) \to \mathbb{R}^{n'}$ for some n' such that the image is an affine bounded Nash manifold in $\mathbb{R}^{n'}$, and the boundary

$$\overline{\psi_i' \circ \psi_i(U_i)} - \psi_i' \circ \psi_i(U_i)$$

consists of a point, say 0. Then we can extend $\psi_i' \circ \psi_i$ to M as a C^0 Nash mapping by setting $= 0$ on U_i^c. We want to modify $\psi_i' \circ \psi_i$ so that the extension is of class C^r.

Set

$$\eta(x) = |x|^{2l}x \quad \text{for} \quad x \in \mathbb{R}^{n'} \quad \text{and} \quad g_i = \eta \circ \psi_i' \circ \psi_i \quad \text{on} \quad U_i$$

for a sufficiently large integer l. Then g_i is a C^r Nash imbedding of U_i, and by 2.8 it is extendible to M as a C^r Nash mapping. Let $\tilde{g}_i$ denote the extension. Then the following mapping is the required one:

$$\prod_{i=1}^{k} \tilde{g}_i \colon M \longrightarrow \mathbb{R}^{kn'}.$$

□

For a classification of all (affine) Nash groups, the category of (affine) Nash groups and Nash group homomorphisms is not large enough. For example, when $S^1 = \{x \in \mathbb{C} : |x| = 1\}$ is given an affine Nash structure by $\mathbb{R}^2 = \mathbb{C}$ and a group structure by complex multiplication, the universal covering space $\tilde{S}^1$ of S^1 has naturally a group structure and locally a Nash structure, but $\tilde{S}^1$ is not a Nash group. (If $\tilde{S}^1$ were so, then $p^{-1}(x)$ would be a finite set, where $p\colon \tilde{S}^1 \to S^1$ is the covering mapping, and x is a point of S^1.) This $\tilde{S}^1$ requires an infinite system of Nash coordinate neighborhoods. A *locally Nash manifold* is an analytic manifold with a (possibly infinite) system of Nash coordinate neighborhoods. We call a mapping f from a locally Nash manifold M_1 to another M_2 *locally Nash* if each point x of M_1 and $f(x)$ have Nash neighborhoods U_1 in M_1 and U_2 in M_2, respectively, such that $f(U_1) \subset U_2$, and $f|_{U_1}\colon U_1 \to U_2$ is a Nash mapping. In a natural way we define a *locally Nash group* and a *locally Nash group homomorphism* between locally Nash groups.

Note the following facts.

(i) The composite of two locally Nash mappings is a locally Nash mapping, which is clear. Let $f\colon M_1 \to M_2$ be a locally Nash mapping between Nash manifolds. Then f is not necessarily a Nash mapping even if f is a diffeomorphism.

For example,

$$M_1 = M_2 = \mathbb{R} \times (\mathbb{R}/\mathbb{Z}),$$
$$f(x, y) = (x, y + p(x)) \quad \text{for} \quad (x, y) \in \mathbb{R} \times (\mathbb{R}/\mathbb{Z}),$$

where $p\colon \mathbb{R} \to \mathbb{R}/\mathbb{Z}$ is the covering mapping (indeed, $\mathbb{R} \times 0 \cap p^{-1}(\mathbb{R} \times 0) = \mathbb{Z} \times 0$).

Let us consider the reason why such a phenomenon happens. Let $U_1 \subset M_1$ and $U_2 \subset M_2$ be open Nash submanifolds, and let $\psi_1\colon U_1 \to \mathbb{R}^{n_1}$ and $\psi_2\colon U_2 \to \mathbb{R}^{n_2}$ be Nash imbeddings whose images are manifolds. Then:

(ii) f is a Nash mapping if $\psi_1(U_1 \cap f^{-1}(U_2))$ is semialgebraic in $\mathbb{R}^{n_1}$ for every U_1 and U_2 of given systems of Nash coordinate neighborhoods of M_1 and M_2, respectively, because $\psi_1(U_1 \cap f^{-1}(U_2))$ is an affine Nash submanifold of $\psi_1(U_1)$ in this case, and

(iii) a locally Nash mapping between affine Nash manifolds is a Nash mapping.

(iv) $\psi_1(U_1 \cap f^{-1}(U_2))$ is semialgebraic if and only if $U_1 \cap f^{-1}(U_2)$ has only a finite number of connected components, because:

(v) in general, each connected component of $\psi_1(U_1 \cap f^{-1}(U_2))$ is semialgebraic.

It follows from (iii) and (v) that

(vi) if M_2 is affine then f is a Nash mapping.

Proof of (iii). Assume that M_1 and M_2 are affine Nash manifolds, which are closed in the ambient Euclidean spaces $\mathbb{R}^{n_1}$ and $\mathbb{R}^{n_2}$, respectively (2.22). Then graph f is closed in $\mathbb{R}^{n_1} \times \mathbb{R}^{n_2}$, locally semialgebraic at any of its points, and an analytic submanifold of $\mathbb{R}^{n_1} \times \mathbb{R}^{n_2}$, and it has only a finite number of connected components. Then by 2.25 graph f is the image under a Nash mapping of some union of connected components on an affine algebraic set. Hence graph f is semialgebraic, which implies that f is a Nash mapping. □

Proof of (v). We can assume that $M_1 = U_1 \subset \mathbb{R}^{n_1}$ and $\psi_1 = \mathrm{id}$. Then each connected component of graph $\psi_2 \circ f|_{f^{-1}(U_2)}$ is closed in $U_1 \times \psi_2(U_2)$, locally semialgebraic at any of its points, and an analytic submanifold of $U_1 \times \psi_2(U_2)$. Hence, for the same reason as in the proof of (iii), the component is semialgebraic in $\mathbb{R}^{n_1} \times \mathbb{R}^{n_2}$. Therefore, each connected component of $U_1 \cap f^{-1}(U_2)$ is semialgebraic. □

A locally Nash manifold is not necessarily a Nash manifold even if it is affine (i.e., it is a locally Nash submanifold of a Euclidean space). An example is $\mathbb{Z}$ ($\subset \mathbb{R}$) (which is also an affine locally Nash group), and another is an open subset of $\mathbb{R}^n$, $n > 1$. But:

(vii) if a connected locally Nash group is affine then it is an affine Nash group.

Proof of (vii). Let $G \subset \mathbb{R}^n$ be an affine connected locally Nash group of dimension m. It suffices to prove that G is semialgebraic. Note that G is an analytic submanifold of $\mathbb{R}^n$ and that the dimension of the Zariski closure of G in $\mathbb{R}^n$ is m. Hence by 2.22, 25, 28 we can assume that G is an open subset of a compact nonsingular algebraic set M in $\mathbb{R}^n$.

Let H denote the intersection of G and the sphere with center at the unit of G and with small radius in $\mathbb{R}^n$. Set

$$N = \{(x, y) \in G \times G\colon y \in xH\} \subset M \times M.$$

Let $\widehat{N}$ denote the Zariski closure of N in $\mathbb{R}^n \times \mathbb{R}^n$, and let $p_1, p_2\colon M \times M \to M$ denote the projections onto the first and second factors, respectively. By 2.5 we have a Nash stratification $p_1\colon \{X_i\}_i \to \{Y_j\}_j$ compatible with $\widehat{N}$. Then we show that $\{Y_j\}$ is compatible with G (hence G is semialgebraic). If it were not so, then some Y_j would contain points x_0 of G and x_1 of $\overline{G} - G$ which are joined by a path C in $G \cap Y_j$. Since $N \cap p_1^{-1}(x_0)$ is an affine Nash submanifold of the algebraic set $\widehat{N} \cap p_1^{-1}(x_0)$ of the same dimension, $N \cap p_1^{-1}(x_0)$ is compatible with $\{X_i \cap p_1^{-1}(x_0)\}_i$. Let $N \cap p_1^{-1}(x_0)$

be the union of $X_i \cap p_1^{-1}(x_0)$, $i \in J$. Then for each $x \in C - x_1$

$$N \cap p_1^{-1}(x) = \bigcup_{i \in J} X_i \cap p_1^{-1}(x),$$

and $\overline{N} \cap p_1^{-1}(x_1)$ would be homeomorphic to $N \cap p_1^{-1}(x_0)$. On the other hand,

$$p_2(\overline{N} \cap p_1^{-1}(x_1)) = \lim_{C - x_1 \ni x \to x_1} xH.$$

(We denote this set by $x_1 H$.) Hence $x_1 H$ would be homeomorphic to $x_0 H$. Note $x_1 H \subset M$.

Let H_1 denote the interior of H in G. Then the above arguments say that $x_1 H_1$ $(= \lim_{C - x_1 \ni x \to x_1} x H_1)$ would be homeomorphic to $x_0 H_1$. Clearly $x_1 H_1 \subset M$. Since $\dim x_0 H_1 = m$, $x_1 H_1$ would have an inner point in M, which contradicts the definition of $x_1 H_1$. Thus G is semialgebraic. □

A connected Lie group of dimension 1 is $\mathbb{R}$ or $\mathbb{R}/\mathbb{Z}$, and Nash group structures on $\mathbb{R}/\mathbb{Z}$ are decided by locally Nash group structures on the covering space $\mathbb{R}$. Hence let us consider locally Nash group structures on $\mathbb{R}$. Given a structure, there is a local Nash coordinate neighborhood (U, ψ) at 0. We can assume $U =]-\varepsilon, \varepsilon[$, $\varepsilon > 0$. Then ψ is an analytic imbedding of U into $\mathbb{R}$,

(1) the mapping

$$\psi(U/2) \times \psi(U/2) \ni (x, y) \longrightarrow \psi(\psi^{-1}(x) + \psi^{-1}(y)) \in \psi(U)$$

is semialgebraic, where $U/2 =]-\varepsilon/2, \varepsilon/2[$, and

(2) the mapping

$$\psi(U/2) \ni x \longrightarrow \psi(-\psi^{-1}(x)) \in \psi(U/2)$$

is also semialgebraic.

Conversely, given a local analytic coordinate neighborhood $(U =]-\varepsilon, \varepsilon[, \psi)$ at 0 with the property (1), then a locally Nash group structure of $\mathbb{R}$ is defined as follows. For each $j \in \mathbb{Z}$, set

$$U_j =]j\varepsilon/2, j\varepsilon/2 + \varepsilon[,$$
$$\psi_j(x) = \psi(x - (j+1)\varepsilon/2) \quad \text{for} \quad x \in U_j.$$

Then $\{(U_j, \psi_j)\}_{j \in \mathbb{Z}}$ is a system of analytic coordinate neighborhoods of $\mathbb{R}$. By (1), the transformations of local coordinates are all semialgebraic. Hence $\mathbb{R}$ is given a locally Nash structure. Moreover, (2) follows from (1), and hence the group operations of $\mathbb{R}$ are of class locally Nash in this structure. We denote $\mathbb{R}$ with this locally Nash group structure by $(\mathbb{R}, \psi)$.

Note that for (U, ψ) and (V, φ), $(\mathbb{R}, \psi)$ and $(\mathbb{R}, \varphi)$ are locally Nash group isomorphic if and only if the mapping

$$\psi(U \cap V/\alpha) \ni x \longrightarrow \varphi(\alpha\psi^{-1}(x)) \in \varphi(\alpha U \cap V)$$

is semialgebraic for some $\alpha \in \mathbb{R} - 0$. In the above arguments, U is a neighborhood of 0. But this is not essential. For example, let $U =]0, \varepsilon[$, and let $\psi: U \to \mathbb{R}$ be an analytic imbedding such that (1) holds. Then

$$U' =]-\varepsilon/2, \varepsilon/2[\quad \text{and} \quad \psi'(x) = \psi(x + \varepsilon/2)$$

satisfy (1), and (U, ψ) is a local Nash coordinate neighborhood of $(\mathbb{R}, \psi')$. Hence in such a case also we write $(\mathbb{R}, \psi)$.

A complete classification of (affine) (locally) Nash groups of dimension 1 is:

Theorem 6.3 [M-S]. (i) (a) *Every locally Nash group homeomorphic to* $\mathbb{R}$ *is locally Nash group isomorphic to one of* $(\mathbb{R}, \mathrm{id})$ *(=* $\mathbb{R}$ *with addition),* $(\mathbb{R}, \exp)$ *(= positive reals with multiplication),* $(\mathbb{R}, \sin)$ *(= circle) or* $(\mathbb{R}, \mathfrak{p}_\alpha)$ *(= connected nonsingular projective cubics), where* $\mathfrak{p}_\alpha$ *is the restriction to* $]0, 1[$ *of the Weierstrass* $\mathfrak{p}$*-function with lattice of periods* $\mathbb{Z} + i\alpha\mathbb{Z}$, $\alpha > 0$.

(b) *These four groups are not locally Nash group isomorphic to one another.*

(c) $(\mathbb{R}, \mathfrak{p}_\alpha)$ *and* $(\mathbb{R}, \mathfrak{p}_{\alpha'})$ *are locally Nash group isomorphic if and only if* α/α' *is rational.*

(d) *The first two groups are imbeddable.*

(e) *The remaining two are not imbeddable.*

(ii) (a) *Every Nash group homeomorphic to* S^1 *is Nash group isomorphic to one of* $(\mathbb{R}, \mathrm{id})/\mathbb{Z}$, $(\mathbb{R}, \exp)/\beta\mathbb{Z}$, $(\mathbb{R}, \sin)/\beta\mathbb{Z}$ *or* $(\mathbb{R}, \mathfrak{p}_\alpha)/\beta\mathbb{Z}$, $\alpha, \beta > 0$.

(b) *These are not Nash group isomorphic to one another.*

(c) *If* $\varphi = \exp$ *or* $\sin$, *then the groups* $(\mathbb{R}, \varphi)/\beta\mathbb{Z}$ *and* $(\mathbb{R}, \varphi)/\beta'\mathbb{Z}$ *are Nash group isomorphic if and only if* β/β' *is rational.*

(d) $(\mathbb{R}, \mathfrak{p}_\alpha)/\beta\mathbb{Z}$ *and* $(\mathbb{R}, \mathfrak{p}_{\alpha'})/\beta'\mathbb{Z}$ *are so if and only if* α/α' *and* β/β' *are both rational.*

(e) *Neither* $(\mathbb{R}, \mathrm{id})/\mathbb{Z}$ *nor* $(\mathbb{R}, \exp)/\beta\mathbb{Z}$ *is imbeddable.*

(f) $(\mathbb{R}, \sin)/\beta\mathbb{Z}$ *and* $(\mathbb{R}, \mathfrak{p}_\alpha)/\beta\mathbb{Z}$ *are imbeddable if and only if* β *is a rational multiple of* π *and rational, respectively.*

Proof. (i): Let $(\mathbb{R}, \psi)$ be a locally Nash group. Let $\psi^{\mathbb{C}}$ denote a complexification of ψ. By the above property (1) there exists a nonzero polynomial Ψ on $\mathbb{C}^3$ such that

$$\Psi(\psi^{\mathbb{C}}(x), \psi^{\mathbb{C}}(y), \psi^{\mathbb{C}}(x + y)) = 0$$

on a neighborhood of 0 in $\mathbb{C}^2$, i.e., $\psi^{\mathbb{C}}$ satisfies an *algebraic addition theorem*. Then by Weierstrass' Theorem $\psi^{\mathbb{C}}$ is algebraic over $\mathbb{C}[z]$, $\mathbb{C}[\exp \alpha z]$, $\alpha \in \mathbb{C}$, or $\mathbb{C}[\mathfrak{p}(z)]$, where $\mathfrak{p}$ is a Weierstrass $\mathfrak{p}$-function.

If $\psi^{\mathbb{C}}$ is algebraic over $\mathbb{C}[z]$ then we can replace ψ with id because ψ is semialgebraic. In the other cases, since ψ is real valued, the set Λ of all periods of $\exp \alpha z$ or $\mathfrak{p}$ is symmetric with respect to the real axis. Hence, if $\psi^{\mathbb{C}}$ is algebraic over $\mathbb{C}[\exp \alpha z]$, then α is real or purely imaginary. In the first case, i.e. where α is real, the mapping

$$(\mathbb{R}, \psi) = (\mathbb{R}, \exp \alpha x) \ni x \longrightarrow \alpha x \in (\mathbb{R}, \exp)$$

is a locally Nash group isomorphism. In the second case, $(\mathbb{R}, \psi)$ is locally Nash group isomorphic to $(\mathbb{R}, \sin)$. If $\psi^{\mathbb{C}}$ is algebraic over $\mathbb{C}[\mathfrak{p}(z)]$ then

$$\Lambda = \beta\mathbb{Z} + i\alpha\mathbb{Z} \quad \text{or} \quad \Lambda = \beta\mathbb{Z} + (\beta/2 + i\alpha)\mathbb{Z}, \qquad \alpha, \beta > 0.$$

Here we can assume $\beta = 1$ because the mapping

$$(\mathbb{R}, \psi(x/\beta)) \ni x \longrightarrow x/\beta \in (\mathbb{R}, \psi)$$

is a locally Nash group isomorphism. Moreover, we suppose $\Lambda = \mathbb{Z} + i\alpha\mathbb{Z}$ because $\mathbb{Z} + (1/2 + i\alpha)\mathbb{Z}$ is a sublattice of $(1/2)\mathbb{Z} + i\alpha\mathbb{Z}$ and, hence, by the addition theorem the Weierstrass $\mathfrak{p}$-function corresponding to $\mathbb{Z} + (1/2 + i\alpha)\mathbb{Z}$ satisfies an algebraic relation with the Weierstrass $\mathfrak{p}$-function corresponding to $(1/2)\mathbb{Z}+i\alpha\mathbb{Z}$. Hence $(\mathbb{R}, \psi)$ is locally Nash group isomorphic to some $(\mathbb{R}, \mathfrak{p}_\alpha)$. Thus (a) is proved.

(b) Clearly $(\mathbb{R}, \mathrm{id})$, $(\mathbb{R}, \exp)$ and $(\mathbb{R}, \sin)$ are not locally Nash group isomorphic to one another.

(c) The groups $(\mathbb{R}, \mathfrak{p}_\alpha)$ and $(\mathbb{R}, \mathfrak{p}_{\alpha'})$ are locally Nash group isomorphic if α/α' is rational (c), because $\mathbb{Z} + i\alpha\mathbb{Z}$ and $\mathbb{Z} + i\alpha'\mathbb{Z}$ are sublattices of some lattice and, hence, by the addition theorem, the functions $\mathfrak{p}_\alpha$ and $\mathfrak{p}_{\alpha'}$ have an algebraic relation. The converse is proved as follows. Assume that $(\mathbb{R}, \mathfrak{p}_\alpha)$ and $(\mathbb{R}, \mathfrak{p}_{\alpha'})$ are locally Nash group isomorphic. Then there are $\rho \neq 0 \in \mathbb{R}$ and a nonzero polynomial P such that

$$P(\mathfrak{p}_\alpha(x), \mathfrak{p}_{\alpha'}(\rho x)) = 0 \quad \text{for} \quad x \in]0, \min(1, 1/\rho)[.$$

Let $P^{\mathbb{C}}$, $\mathfrak{p}^{\mathbb{C}}_\alpha$ and $\mathfrak{p}^{\mathbb{C}}_{\alpha'}$ denote the complexifications of P, $\mathfrak{p}_\alpha$ and $\mathfrak{p}_{\alpha'}$, respectively. Then

$$P^{\mathbb{C}}(\mathfrak{p}^{\mathbb{C}}_\alpha(z), \mathfrak{p}^{\mathbb{C}}_{\alpha'}(\rho z)) = 0 \quad \text{for} \quad z \in \mathbb{C} - (\mathbb{Z} + i\alpha\mathbb{Z}) \cup ((1/\rho)\mathbb{Z} + (i\alpha'/\rho)\mathbb{Z}).$$

Fix such a point z_0. Then

$$P^{\mathbb{C}}(\mathfrak{p}^{\mathbb{C}}_\alpha(z_0), \mathfrak{p}^{\mathbb{C}}_{\alpha'}(\rho z_0 + \rho m + i\alpha\rho n)) = 0 \quad \text{for all } m, n \in \mathbb{Z}.$$

Hence, since $P^{\mathbb{C}}(a, z) = 0$ has only a finite number of solutions for general $a \in \mathbb{C}$, $\rho\mathbb{Z} + i\rho\alpha\mathbb{Z}$ and $\mathbb{Z} + i\alpha'\mathbb{Z}$ are sublattices of some lattice. This means that ρ, $\alpha'/\alpha\rho$ and, hence, α/α' are rational.

The above arguments show also that $(\mathbb{R}, \mathrm{id})$, $(\mathbb{R}, \exp)$ or $(\mathbb{R}, \sin)$ is not locally Nash group isomorphic to $(\mathbb{R}, \mathfrak{p}_\alpha)$, completing the proof of (b).

(d) Since id and exp are global imbeddings of $\mathbb{R}$ into $\mathbb{R}$, the groups $(\mathbb{R}, \mathrm{id})$ and $(\mathbb{R}, \exp)$ are imbeddable.

By the above arguments on periods, any analytic imbedding of $\mathbb{R}$ or an open half line into $\mathbb{R}$, that is algebraic over sin or $\mathfrak{p}_\alpha$, is not a finite-to-one mapping. Hence such a function cannot be a local coordinate neighborhood of $(\mathbb{R}, \sin)$ nor $(\mathbb{R}, \mathfrak{p}_\alpha)$, i.e., $(\mathbb{R}, \sin)$ and $(\mathbb{R}, \mathfrak{p}_\alpha)$ require infinite number of local Nash coordinate neighborhoods. Therefore, they are not Nash groups, showing (e).

(ii): (a) follows from (a) of (i) because $(\mathbb{R}, \mathrm{id})/\mathbb{Z}$ is Nash group isomorphic to $(\mathbb{R}, \mathrm{id})/\beta\mathbb{Z}$ for any $\beta > 0$. (b) is clear by (b) of (i). If $\varphi, \varphi' \neq \mathrm{id}$ then the groups $(\mathbb{R}, \varphi)\beta\mathbb{Z}$ and $(\mathbb{R}, \varphi')\beta'\mathbb{Z}$ are Nash group isomorphic if and only if the function $x \rightarrow \varphi'(\pm\varphi^{-1}(x)\beta'/\beta)$ is semialgebraic. Hence we easily see (c) and, by the arguments on periods in (i), (d). (e) is clear when we consider the canonical covering groups. The group $(\mathbb{R}, \sin)/2\pi\mathbb{Z}$ is imbeddable because the mapping $x \to (\sin x, \cos x)$ is a Nash imbedding (f). The group $(\mathbb{R}, \mathfrak{p}_\alpha)/\mathbb{Z}$ is imbeddable in $\mathbb{R}^4 = \mathbb{C}^2$ by the mapping

$$x \longrightarrow (\mathfrak{p}_\alpha^{\mathbb{C}}(x+a), \mathfrak{p}_\alpha^{\mathbb{C}}(x+b)) \in \mathbb{C}^2$$

for some $a, b \in \mathbb{C}$ (f). The rest of (f) is clear by the arguments on periods. □

An equivariant version of Algebraic Model Theorem 4.10 is the following.

Theorem 6.4 [H-P]. *Let G be a connected affine Nash group. Then there exist a complex algebraic group H, defined over $\mathbb{R}$, and a Nash isogeny (i.e. a finite covering group homomorphism) from G onto the connected component of the $\mathbb{R}$-rational point set of H containing the unit.*

See [B0] for terminology. The proof requires many facts on algebraic groups, and we omit it. Note only that H is quasi-projective, the $\mathbb{R}$-rational point set is Nash imbeddable, and that we can not expect the isogeny to be injective.

References

[An] M. André, Cinq exposés sur la désingularization, preprint, 1991.

[A] M. Artin, Algebraic approximation of structures over complete local rings, Publ. Math. IHES 36 (1969), 23–57.

[A-M] M. Artin, B. Mazur, On periodic points, Ann. of Math. 81 (1965), 82–99.

[B-C-R] J. Bochnak, M. Coste, M. F. Roy, Géométrie algébrique réelle, Ergeb. Math. Grenzgeb. 12, Springer-Verlag 1987.

[Bo] A. Borel, Linear algebraic groups, Benjamin, New York 1969.

[C-R-S$_1$] M. Coste, J. M. Ruiz, M. Shiota, Separation, factorization and finite sheaves on Nash manifolds, preprint.

[C-R-S$_2$] M. Coste, J. M. Ruiz, M. Shiota, Approximation in compact Nash manifolds, Amer. J. Math. 117 (1995), 905–927.

[C-S$_1$] M. Coste, M. Shiota, Nash triviality in families of Nash manifolds, Inv. Math. 108 (1992), 349–368.

[C-S$_2$] M. Coste, M. Shiota, Thom's first isotopy lemma: semialgebraic version, with uniform bound, in: Real ananlytic and algebraic geometry, Walter de Gruyter, Berlin 1995, 83–101.

[C-S$_3$] M. Coste, M. Shiota, to appear.

[E$_1$] G. Efroymson, A nullstellensatz for Nash rings, Pacific J. Math. 54 (1974), 101–112.

[E$_2$] G. Efroymson, The extension theorem for Nash functions, Lecture Notes in Math. 959, Springer-Verlag 1982, 343–357.

[G-al] C. G. Gibson et al., Topological stability of smooth mappings, Lecture Notes in Math. 552, Springer-Verlag 1976.

[Hi] H. Hironaka, Resolution of singularities of an algebraic variety over a field of characteristic zero, I, II, Ann. of Math. 79 (1964), 109–326.

[H-P] E. Hrushovski, A. Pillay, Groups definable in local fields and pseudo-finite fields, preprint.

[L] S. Lojasiewicz, Ensembles semi-analytiques, IHES, 1965.

[M-S] J. J. Madden, C. M. Stanton, One-dimensional Nash groups, Pacific J. Math. 154 (1992), 331–344.

[Ma] B. Malgrange, Ideals of differentiable functions, Oxford Univ. Press, 1966.

[Mo] T. Mostowski, Some properties of the ring of Nash functions, Ann. Scu. Norm. Sup. Pisa III 2 (1976), 245–266.

[N] J. F. Nash, Real algebraic manifolds, Ann. of Math. 56 (1952), 405–421.

[N-P] A. Nesin, A. Pillay, Some model theory of compact Lie groups, Trans. Amer. Math. Soc. 326 (1991), 453–463.

[N-P-R] A. Nesin, A. Pillay, V. Razenj, Groups of dimension two and three over 0-minimal structures, Ann. Pure Appl. Logic 53 (1991), 279–296.

[Pa] R. Palais, Equivalent real algebraic differential topology, Part 1, smoothness categories and Nash manifolds, Notes, Brandeis Univ. 1972.

[Pe_1] D. Pecker, On Efroymson's extension theorem for Nash functions, J. Pure Appl. Alg. 37 (1985), 193–203.

[Pe_2] D. Pecker, On the elimination of algebraic inequalities, Pacific J. Math. 146 (1990), 305–314.

[Pi] A. Pillay, On groups and fields definable in 0-minimal structures, J. Pure Appl. Alg. 53 (1988), 239–255.

[Ra] M. Raimondo, On the global ring of Nash functions, Boll. Un. Mat. Ital. 18.A (1981), 317–321.

[Raz] V. Razenj, One-dimensional groups over an 0-minimal structure, Ann. Pure Appl. Logic 53 (1991), 269–277.

[Ri] J. J. Risler, Sur l'anneau des fonctions de Nash globales, Ann. Sci. Ecole Norm. Sup. 8 (1975), 365–378.

[S_1] M. Shiota, Nash manifolds, Lecture Notes in Math. 1269, Springer-Verlag 1987.

[S_2] M. Shiota, Extension et factorisation de fonctions de Nash C^ω, C. R. Acad. Sci. Paris 308 (1989), 253–256.

[S_3] M. Shiota, Geometry of subanalytic and semialgebraic sets, to appear.

[Sp] M. Spivakovski, Smoothing of ring homomorphisms, approximation theorems and the Bass-Quillen conjecture, preprint.

[St] A. W. Strzebonski, Introduction to semialgebraic and definable groups, thesis, State Univ. of New York, 1993.

[Th_1] R. Thom, Quelques propriétés globales des variétés différentiables, Comm. Math. Helv. 28 (1954), 17–81.

[Th_2] R. Thom, La stabilité topologique des applications polynomiales, L'Enseignement Math. 8 (1962), 24–33.

[Th_3] R. Thom, Ensembles et Morphismes stratifiés, Bull. Amer. Math. Soc. 75 (1969), 240–284.

[Th_4] R. Thom, Singularities of differentiable mappings, Lecture Notes in Math. 192, Springer-Verlag 1971, 1–89.

[T-T] A. Tancredi, A. Tognoli, On the extension of Nash functions, Math. Ann. 288 (1990), 595–604.

Graduate School of Polymathematics, Nagoya University, Chkusa-Ku,
Nagoya, 464-01, Japan
e-mail: d42806@nucc.cc.nagoya-u.ac.jp

Approximation theorems in real analytic and algebraic geometry

*A. Tognoli**

Introduction

This article is a global survey of approximation theorems; more precisely it contains approximation theorems for sections of sheaves $\mathcal{F} \otimes \mathcal{E}_X$, where $\mathcal{F}$ is an analytic (or algebraic) coherent sheaf defined over a coherent real analytic space X (or a real affine variety X) and $\mathcal{E}_X$ is the sheaf of germs of C^∞ functions on X.

The paper is divided in three parts: in the first one, the analytic case, we prove that the set of sections of $\mathcal{F}$ is dense in the strong (or Whitney) topology in the space of sections of $\mathcal{F} \otimes \mathcal{E}_X$.

In §1 we define the strong topology on the set $\Gamma(X, \mathcal{F})$ of sections of a coherent sheaf $\mathcal{F}$ over a coherent real analytic space X. This topology is defined using a family of seminorms and in a similar way we construct the strong topology on the set $\Gamma(X, \mathcal{F}^\infty)$ of sections of the sheaf $\mathcal{F} \otimes \mathcal{E}_X$.

Essential tools to prove that $\Gamma(X, \mathcal{F})$ is dense in $\Gamma(X, \mathcal{F}^\infty)$ (Theorem 3.1), are:

(1) a generalization of Whitney approximation theorem for functions (here we do not suppose that X can be embedded in some euclidean space);

(2) Theorems A and B (proved by H. Cartan in the real analytic case);

(3) a result of B. Malgrange about the ideal sheaf of C^∞ functions vanishing on a coherent real analytic set.

The second part of this paper deals with the algebraic case. In this context we shall use the classical Weierstrass theorem and we shall obtain approximation in the weak (or compact open) topology.

In the real algebraic context, Theorems A and B are not true, so we shall study two subcategories of coherent sheaves: A-coherent and B-coherent sheaves. They are studied in §6 and §7; we shall prove approximation theorems for sections of $\mathcal{F} \otimes \mathcal{E}_X$ by means of sections of $\mathcal{F}$.

---

* Member of G.N.S.A.G.A. of C.N.R. The author is partially supported by M.U.R.S.T.

As an application of the results just sketched we shall prove that if we have an analytic linear system

$$\sum_k a_{hk}(x)\, y_k \;=\; g_h(x), \quad h = 1, \dots, q, \tag{1}$$

where the $a_{hk}(x)$, $g_h(x)$ are analytic functions, defined over an open set of $\mathbb{R}^n$, then any C^∞ solution of (1) can be approximated by an analytic solution, in the strong topology. A similar result holds in the algebraic case for the weak topology.

The third part of this article is devoted to fiber bundles.

Approximation theorems were known also for sections of fiber bundles. In §8, via a duality theory, we find new approximation results for sections of fiber bundles of a more general class, which are not, in particular, locally trivial.

Locally trivial algebraic vector bundles do not have the usual good behaviour. For example, the fiber is not in general generated by global sections. Bundles having the latter property are called strongly algebraic and are studied in §9.

We prove that an algebraic vector bundle is strongly algebraic if, and only if, its total space is an affine variety.

By applying the approximation results to the sheaf $\mathrm{Hom}\,(\mathcal{F}, \mathcal{G})$ we can extend to general vector bundles the results about the equivalence of the analytic and C^∞ classification.

For strongly algebraic vector bundles, if X is compact, we have similar results about C^∞ and algebraic classification.

We wish to remark that the material contained in Sections 1, 2, 3, 4, 7 and 10 reproduces the article [ABT], while the material in the other sections is scattered throughout the literature and not always easily found: the purpose of this paper is to give a comprehensive and self-contained treatment of the subject.

Finally the author wishes to thank Francesca Acquistapace who read the first version of this article and suggested many valuable changes.

I. The analytic case

1. The Whitney topology for sections of a sheaf

Let X be a paracompact, locally compact space and $\mathcal{F} = \{\mathcal{F}_U, r^U_V\}$ be a sheaf of real vector spaces; so, for any $A \subset X$, the set $\Gamma(A, \mathcal{F})$ has the structure of a real vector space and the restriction maps are linear.

Definition 1.1. A *local system of seminorms* $\mathcal{L}$ in $\mathcal{F}$ is given by the following data:

(1) A locally finite open covering $\mathfrak{A} = \{U_\lambda\}_{\lambda \in \Lambda}$ of X by relatively compact open sets.

(2) For any compact set $K \subset U_\lambda$, for any open neighbourhood U of K and any natural number p, a seminorm $\| \ \|^p_{K,\lambda}$ (depending on λ) defined on $\Gamma(U, \mathfrak{F})$ with the following properties:

a) If $\gamma_1 \in \Gamma(U_1, \mathfrak{F})$, $\gamma_2 \in \Gamma(U_2, \mathfrak{F})$ and $r^{U_1}_U \gamma_1 = r^{U_2}_U \gamma_2$ for an open neighbourhood U of K, then for any p

$$\|\gamma_1\|^p_{K,\lambda} = \|\gamma_2\|^p_{K,\lambda} = \|r^{U_1}_U \gamma_1\|^p_{K,\lambda}.$$

b) If $K \subset U_\lambda \cap U_{\lambda'}$, for each integer p there exist two positive numbers α and β such that for each $\gamma \in \Gamma(U, \mathfrak{F})$

$$\alpha\|\gamma\|^p_{K,\lambda} \leq \|\gamma\|^p_{K,\lambda'} \leq \beta\|\gamma\|^p_{K,\lambda}.$$

c) If $K = \bigcup_{i=1}^{n} K_i$ is a decomposition of K as finite union of compact sets, then for each $\gamma \in \Gamma(U, \mathfrak{F})$ and each p

$$\|\gamma\|^p_{K,\lambda} = \sup_{i=1,\ldots,n} \|\gamma\|^p_{K_i,\lambda}.$$

In particular if $K \subset K' \subset U_\lambda$ then $\|\gamma\|^p_{K,\lambda} \leq \|\gamma\|^p_{K',\lambda}$.

d) If $U \supset \overline{U}_\lambda$ then $\sup\limits_{\substack{K \subset U_\lambda \\ K \text{compact}}} \|\gamma\|^p_{K,\lambda} < \infty$, for any $\gamma \in \Gamma(U, \mathfrak{F})$.

Let now K be any compact set in X and suppose $K \cap U_\lambda = \emptyset$ if λ is different from $\lambda_1, \ldots, \lambda_q$.

Definition 1.2. For any $\gamma \in \Gamma(U, \mathfrak{F})$ with $K \subset U$ we define

$$\|\gamma\|^p_K = \sup_{i=1,\ldots,q} \sup_{\substack{H \subset U_{\lambda_i} \cap K \\ H \text{compact}}} \|\gamma\|^p_{H,\lambda_i}.$$

Property d) of Definition 1.1 ensures that $\|\gamma\|^p_K < \infty$.

Definition 1.3. Let $U \subset X$ be an open set. The *weak topology defined by the local system of seminorms* $\mathcal{L}$ for $\Gamma(U, \mathfrak{F})$ is the topology having the family

$$\mathfrak{U}^{\mathcal{L}}_{K,p,\varepsilon} = \{\gamma \in \Gamma(U, \mathfrak{F}) \mid \|\gamma\|^p_K < \varepsilon, \quad K \text{ compact set, } K \subset U\}$$

as a fundamental system of neighbourhoods of 0.

Remark 1.4. The restriction maps r^U_V are continuous with respect to the weak topology.

Now we are ready to define the Whitney topology, as usual, as a limit of the weak topology.

Consider a local system $\mathcal{L}$ of seminorms on the sheaf $\mathfrak{F}$. Let $U \subset X$ be an open set.

Take:

(1) an exhaustive sequence of compact sets

$$\mathcal{K} = \{K_i\}_{i \in \mathbb{N}} \quad K_i \subset \mathring{K}_{i+1} \quad \bigcup K_i = U,$$

(2) a sequence $\mathcal{M} = \{m_i\}_{i \in \mathbb{N}}$ of natural numbers,

(3) a sequence $\mathcal{E} = \{\varepsilon_i\}_{i \in \mathbb{N}}$ of positive numbers.

Then:

Definition 1.5. A fundamental system of neighbourhoods of $0 \in \Gamma(U, \mathfrak{F})$ for the Whitney topology on $\Gamma(U, \mathfrak{F})$ is given by the sets

$$\mathfrak{A}^{\mathcal{L}}_{\mathcal{K},\mathcal{M},\mathcal{E}} = \{\gamma \in \Gamma(U, \mathfrak{F}) \mid \forall n \sup_{p \le m_n} \|\gamma\|^p_{K_n - \mathring{K}_{n-1}} < \varepsilon_n\}.$$

Remarks 1.6.

1. The weak topology can be given by a countable family of seminorms, namely $\| \ \|^{p_n}_{K_n}$ for an exhaustive sequence of compact sets. Hence the weak topology is induced by a metric. This is not true for the Whitney topology because the family $\mathfrak{A}^{\mathcal{L}}_{\mathcal{K},\mathcal{M},\mathcal{E}}$ is not countable and does not have any countable cofinal subfamily.
2. If X is compact then the weak and the Whitney topologies coincide.

Definition 1.7. Two local system of seminorms over X

$$\mathcal{L} = \left\{\{U_\lambda\}, \| \ \|^p_{K,\lambda}\right\} \quad \text{and} \quad \mathcal{L}' = \left\{\{U'_{\lambda'}\}, \| \ \|^p_{K,\lambda'}\right\}$$

are said to be *equivalent* if for each compact $K \subset X$ and any p there exist two positive numbers α, β such that

$$\alpha \left(\|\gamma\|^p_K\right)_{\mathcal{L}} \le \left(\|\gamma\|^p_K\right)_{\mathcal{L}'} \le \beta \left(\|\gamma\|^p_K\right)_{\mathcal{L}}$$

for each $\gamma \in \Gamma(U, \mathfrak{F})$, with $K \subset U$.

Lemma 1.8. *If $\mathcal{L}$ and $\mathcal{L}'$ are equivalent, they induce the same weak topology and the same Whitney topology on $\Gamma(U, \mathfrak{F})$, for any open set $U \subset X$.*

Proof. This is clear by the definitions. □

In the following we shall omit the restriction maps when there is no risk of confusion.

Examples.

(1) If $U \subset \mathbb{R}^n$ is an open set, we have the classical seminorms for functions in $C^\infty(U)$ or in $C^\omega(U)$

$$\|f(x_1,\dots,x_n)\|_K^p = \sup\left[\sup_K |f(x)|, \sup_{\substack{K \\ j_1+\dots+j_n=j\le p}} \left|\frac{\partial^j f(x_1,\dots,x_n)}{\partial x_1^{j_1},\dots,\partial x_n^{j_n}}\right|\right]$$

which give to $C^\infty(U)$ and $C^\omega(U)$ the usual compact open topology (or weak topology) and Whitney (or strong) topology.

(2) Let $(X, \mathcal{O}_X)$ be a reduced real coherent analytic space. We can find a locally finite open covering $\{U_\lambda\}$ of X such that for each λ there exists an isomorphism $j_\lambda : U_\lambda \to X_\lambda$, where X_λ is a closed real analytic subset of an open set Ω_λ in $\mathbb{R}^{n_\lambda}$. The isomorphism j_λ induces a surjective map $\pi_\lambda : C^\omega(\Omega_\lambda) \to \Gamma(U_\lambda, \mathcal{O}_X)$ which is the composition of j_λ^{-1} with the quotient map. So for each $K \subset U_\lambda$ and each $f \in \Gamma(U_\lambda, \mathcal{O}_X)$ we can define

$$\|f\|_K^p = \inf_{g\in\pi_\lambda^{-1}(f)} \|g\|_{j_\lambda(K)}^p.$$

By this local system of seminorms we can define the weak and the strong topology on $\Gamma(U, \mathcal{O}_X)$ for any open set $U \subset X$.

If X is not coherent we can extend any analytic function on a local model to a C^∞ function on Ω_λ and then use C^∞ seminorms.

(3) Let $(X, \mathcal{O}_X)$ be a reduced complex analytic space and $\mathcal{F}$ be a coherent sheaf of $\mathcal{O}_X$-modules. Then we can find an open covering $\{U_\lambda\}$ of X by holomorphically convex open sets and for each λ a resolution of $\mathcal{F}$ on U_λ

$$\mathcal{O}^p_{U_\lambda} \longrightarrow \mathcal{O}^q_{U_\lambda} \longrightarrow \mathcal{F}_{U_\lambda} \longrightarrow 0.$$

This induces a surjective map

$$\beta : \Gamma(U, \mathcal{O}^q) \longrightarrow \Gamma(U, \mathcal{F}) \longrightarrow 0$$

for each open Stein set $U \subset U_\lambda$. Hence for any compact set $K \subset U$ we can define

$$\|\gamma\|_K = \inf_{\substack{\overline{\gamma}=(\gamma_1,\dots,\gamma_q) \\ \beta(\overline{\gamma})=\gamma}} \left\{\sup_K \left(|\gamma_1| + \dots + |\gamma_q|\right)\right\}.$$

With this local system of seminorms the weak topology gives to $\Gamma(U, \mathcal{F})$ the structure of a Frechét space (see [G.R] Chap. VII).

(4) Let now $(X, \mathcal{O}_X)$ be a reduced coherent real analytic space and $\mathcal{F}$ be a coherent sheaf of $\mathcal{O}_X$-modules. We can take the same definition as before; namely, if $\{U_\lambda\}$ is an open covering of X such that on each U_λ we have a resolution of $\mathcal{F}$, we can

define for $K \subset U_\lambda$ compact, $p \in \mathbb{N}$ and $\gamma \in \Gamma(K, \mathcal{F})$

$$\|\gamma\|_K^p = \inf_{\substack{\overline{\gamma}=(\gamma_1,\dots,\gamma_q) \\ \beta(\overline{\gamma})=\gamma}} \left(\|\gamma_1\|_K^p + \cdots + \|\gamma_q\|_K^p\right).$$

(5) In the same situation as (4) we can define

$$\mathcal{F}^\infty = \mathcal{F} \otimes_{\mathcal{O}_X} \mathcal{E}_X,$$

where $\mathcal{E}_X$ is the sheaf of germs of C^∞-functions on X.[1] Since $\mathcal{F}$ is coherent, the stalk $\mathcal{F}_x$ is generated by a finite number of global sections (Theorem A). Hence we can construct an open covering $\mathfrak{A} = \{U_\lambda\}$ and for each λ we can find $f_1^\lambda, \dots, f_{q(\lambda)}^\lambda$ in $\Gamma(X, \mathcal{F})$ such that they generate $\mathcal{F}_x$ as $\mathcal{O}_{X,x}$-module for each $x \in U_\lambda$. Let γ be an element in $\Gamma(U_\lambda, \mathcal{F}^\infty)$. Then we can write (not in a unique way) $\gamma = \sum_{i=1}^{q(\lambda)} \alpha_i f_i^\lambda$ with $\alpha_i \in C^\infty(U_\lambda)$. In fact this can be done locally by definition of $\mathcal{F}^\infty$ and then can be globalized by using a C^∞ partition of unity. For $K \subset U_\lambda$ and $p \in \mathbb{N}$ we can define

$$\|\gamma\|_{K,\lambda}^p = \inf_{\alpha_1,\dots,\alpha_{q(\lambda)}} \left(\sum_{i=1}^{q(\lambda)} \|\alpha_i\|_K^p\right).$$

(The inf is taken on all the system of coefficients $\alpha_1, \dots, \alpha_{q(\lambda)}$ describing γ with respect to the chosen generators $f_1, \dots, f_{q(\lambda)}$). This is a local system of seminorms that we shall always use to define the weak and the Whitney topology on $\Gamma(U, \mathcal{F}^\infty)$ if we do not specify otherwise.

Remark 1.9. The morphisms between coherent sheaves of $\mathcal{O}_X$-modules induce continuous maps between the spaces of sections (see [GR] for the complex case: the same proof works in the real one).

2. A Whitney approximation theorem

This section is devoted to the proof of a Whitney-like approximation theorem for smooth functions defined on a real analytic space X.

1 A map $\varphi: X \to \mathbb{R}$ is C^∞ if for any $x \in X$ there exist a neighbourhood W_x of x, an embedding $W_x \to \mathbb{R}^n$ as a locally closed analytic set and $\varphi_{|W_x}$ extends to a smooth function on some neighbourhood of W_x in $\mathbb{R}^n$. If $\mathcal{E}_\varphi$ is the set of such extensions we can define, for a compact set $K \subset W_x$,

$$\|\varphi\|_K^p = \inf_{\overline{\varphi} \in \mathcal{E}_\varphi} \|\overline{\varphi}\|_K^p.$$

If X is coherent we shall get in the next section a similar result for sections of any coherent sheaf of $\mathcal{O}_X$-modules.

Our proof is similar to the classical one found in [W], [N], [T6] , [T8]. Under the hypothesis that X is an analytic submanifold of $\mathbb{R}^n$ and $\mathcal{F}$ a subsheaf of $\mathcal{O}$, Theorem 2.9 is proved in [BKS], where the Whitney topology is called the *very strong topology*.

We shall use the following standard notations for $x \in \mathbb{R}^n$ (or $\mathbb{C}^n$), φ a C^∞ function on $\mathbb{R}^n$ (or $\mathbb{C}^n$), $\alpha = (\alpha_1, \dots, \alpha_n)$, $\beta = (\beta_1, \dots, \beta_n)$ in $\mathbb{N}^n$.

$$|\alpha| = \alpha_1 + \cdots + \alpha_n, \qquad \alpha! = \alpha_1! \dots \alpha_n!,$$

$$\binom{\alpha}{\beta} = \frac{\alpha!}{\beta!(\alpha-\beta)!}, \qquad (\text{if } \beta_j \le \alpha_j \text{ for } j = 1, \dots, n)$$

$$|x| = \max_j |x_j|, \qquad \|x\| = \Big(\sum_j |x_j|^2\Big)^{\frac{1}{2}},$$

$$x^\alpha = x_1^{\alpha_1} \dots x_n^{\alpha_n} \qquad D^\alpha \varphi = \frac{\partial^{\alpha_1 + \cdots + \alpha_n} \varphi}{\partial x_1^{\alpha_1} \dots \partial x_n^{\alpha_n}}.$$

Let $(X, \mathcal{O}_X)$ be a real analytic space, not necessarily coherent; we suppose that $(X, \mathcal{O}_X)$ is the real part of a reduced complex analytic space $(\widetilde{X}, \mathcal{O}_{\widetilde{X}})$.

This means that there exists a complex analytic space $(\widetilde{X}, \mathcal{O}_{\widetilde{X}})$ which is defined over $\mathbb{R}$, and an antiinvolution $\sigma\colon \widetilde{X} \to \widetilde{X}$ such that $(X, \mathcal{O}_X)$ is isomorphic to the real analytic space $X' = \{x \in \widetilde{X} : \sigma(x) = x\}$ endowed with the structure sheaf $\mathcal{O}_X$ consisting of all σ-invariant germs.

In this situation X has in $\widetilde{X}$ an invariant neighbourhood $U = \sigma(U)$ which is a Stein space. So, in the following, we shall assume that $(\widetilde{X}, \mathcal{O}_{\widetilde{X}})$ is a reduced Stein space defined over $\mathbb{R}$. In the case when the real part is coherent we can assume that $(\widetilde{X}, \mathcal{O}_{\widetilde{X}})$ is its complexification (see [T10] and [T11]).

Consider three compact sets

$$H_1 \subset \mathring{H}_2 \subset H_2 \subset \mathring{H}_3 \subset H_3 \subset X.$$

Definition 2.1. A complex neighbourhood $\widetilde{U}_1$ of H_1 in $\widetilde{X}$ is called a *vertical neighbourhood* relative to H_2, H_3, if for any C^∞ function $\varphi\colon X \to \mathbb{R}$ such that $\operatorname{supp} \varphi \subset H_3$, $\varphi|_{H_2} \equiv 0$, and any $\varepsilon > 0$ and $p \in \mathbb{N}$, there exists an analytic function g on X such that:

(1) $\|g - \varphi\|^p_{H_3} < \varepsilon$, and

(2) g is the restriction of a holomorphic function $G\colon \widetilde{X} \to \mathbb{C}$ such that $|G(z)| < \varepsilon$ for $z \in \widetilde{U}_1$.

Remark 2.2. In the above situation, if $H_1 \subset \widetilde{U}_1' \subset \widetilde{U}_1$, $\widetilde{U}_1'$ is also a vertical neighbourhood.

Lemma 2.3. *Let* $(\widetilde{X}, \mathcal{O}_{\widetilde{X}})$, H_i, $i = 1, 2, 3$, $\widetilde{U}_1$ *as before. Let* $\widetilde{f}: (\widetilde{Y}, \mathcal{O}_{\widetilde{Y}}) \to (\widetilde{X}, \mathcal{O}_{\widetilde{X}})$ *be a complex analytic map. Assume that* $\widetilde{f}$ *and* $(\widetilde{Y}, \mathcal{O}_{\widetilde{Y}})$ *are defined over* $\mathbb{R}$ *and that* $(Y, \mathcal{O}_Y)$ *is the the real part of* $(\widetilde{Y}, \mathcal{O}_{\widetilde{Y}})$, *and consider* $f = \widetilde{f}|_Y$. *Assume there is an open subset* $Y' \subset Y$ *such that* f *defines an isomorphism between* Y' *and a closed analytic subset* $X' = f(Y')$ *of an open set* $W \supset H_3$ *of* X. *Define* $H_i' = f^{-1}(H_i \cap X')$ *for* $i = 1, 2, 3$. *Then* $\widetilde{f}^{-1}(\widetilde{U}_1)$ *is a vertical neighbourhood of* H_1' *relative to* H_2', H_3'.

Proof. Let $\varphi: Y \to \mathbb{R}$ be a C^∞ function such that $\operatorname{supp}\varphi \subset H_3'$, and $\varphi|_{H_2'} \equiv 0$. Clearly the function $\psi = \varphi \circ f^{-1}: X' \to \mathbb{R}$ can be extended to a C^∞ function on X (denoted also by ψ) such that $\operatorname{supp}\psi \subset H_3$ and $\psi|_{H_2} \equiv 0$. If $G: X \to \mathbb{R}$ is an analytic approximation of ψ and its holomorphic extension $\widetilde{G}$ is small on $\widetilde{U}_1$, then $\widetilde{G} \circ \widetilde{f}$ approximates φ and is "small" on $\widetilde{f}^{-1}(\widetilde{U}_1)$. □

Now we define vertical neighbourhoods for $\mathbb{R}^n$, considered as the real part of $\mathbb{C}^n$.

Lemma 2.4. *Let* Ω *be an open set in* $\mathbb{R}^n$. *Let* H_i, $i = 1, 2, 3$, *be three compact subsets of* Ω *such that* $H_i \subset \mathring{H}_{i+1}$ *for* $i = 1, 2$. *Define* $\delta = d(H_1, \Omega - H_2)$. *Then for any* $\alpha \in (0, 1)$ *the set*

$$\widetilde{U}_\alpha = \{z \in \mathbb{C}^n : \textit{for any } y \in \mathbb{R}^n - H_2, \quad |\Re(z - y)| > \alpha\delta\}$$

is a vertical neighbourhood of H_1 *relative to* H_2, H_3 *(where* $\Re(\ )$ *means the real part of* $(\ )$*).*

Proof. Let $\varphi: \Omega \to \mathbb{R}$ be a C^∞ function such that $\varphi|_{H_2} \equiv 0$ and $\operatorname{supp}\varphi \subset H_3$. For any $\lambda \in (0, +\infty)$, we define

$$I_\lambda(\varphi)(x) = c\lambda^{\frac{1}{2}n} \int_{\mathbb{R}^n} \varphi(y) \exp\{-\lambda \|x - y\|^2\}\, dy \tag{1}$$

where $c \cdot \int_{\mathbb{R}^n} \exp(-\|x^2\|)\, dx = 1$, that is $c = \pi^{-\frac{1}{2}n}$.

We have

$$I_\lambda(\varphi)(x) = c\lambda^{\frac{1}{2}n} \int_{\mathbb{R}^n} \varphi(x - y) \exp\{-\lambda \|y\|^2\}\, dy \tag{2}$$

and hence, for any $\alpha \in \mathbb{N}^n$

$$\begin{aligned} D^\alpha\left(I_\lambda(\varphi)\right)(x) &= c\lambda^{\frac{1}{2}n} \int_{\mathbb{R}^n} \left(D^\alpha\varphi\right)(x - y) \exp\{-\lambda \|y\|^2\}\, dy \\ &= c\lambda^{\frac{1}{2}n} \int_{\mathbb{R}^n} \left(D^\alpha\varphi\right)(y) \exp\{-\lambda \|x - y\|^2\}\, dy. \end{aligned} \tag{3}$$

From (2) and (3) we deduce:

$$D^{\alpha}\left(I_{\lambda}(\varphi)\right)(x) - D^{\alpha}(\varphi)(x) = c\lambda^{\frac{1}{2}n} \int_{\mathbb{R}^n} \left(\left(D^{\alpha}\varphi\right)(y) - \left(D^{\alpha}\varphi\right)(x)\right) \exp\{-\lambda \|x-y\|^2\}\, dy \quad (4)$$

and

$$\left|D^{\alpha}\left(I_{\lambda}(\varphi)\right)(x) - D^{\alpha}((\varphi))(x)\right| = \left|c\lambda^{\frac{1}{2}n} \int_{\|x-y\|<\delta} \left(\left(D^{\alpha}\varphi\right)(y) - \left(D^{\alpha}\varphi\right)(x)\right) \exp\{-\lambda \|x-y\|^2\}\, dy + c\lambda^{\frac{1}{2}n} \int_{\|x-y\|\geq\delta} \left(\left(D^{\alpha}\varphi\right)(y) - \left(D^{\alpha}\varphi\right)(x)\right) \exp\{-\lambda \|x-y\|^2\}\, dy\right|. \quad (5)$$

Relation (5) proves that for any $p \in \mathbb{N}$ we have

$$\lim_{\lambda\to\infty} \|I_{\lambda}(\varphi) - \varphi\|^{p}_{H_3} = 0. \quad (6)$$

In fact for any $\varepsilon > 0$ we may suppose δ small enough to ensure that the first integral in (5) is less then $\frac{\varepsilon}{2}$. (We use the fact that φ has compact support and hence $D^{\alpha}\varphi$ is uniformly continuous).

Given the positive number δ, we may find $\lambda \in \mathbb{R}$ big enough to ensure that the second integral in (5) has absolute value less than $\frac{\varepsilon}{2}$ (because of the nature of the "bump function" $\exp\{-\lambda \|x-y\|^2\}$).

So $I_{\lambda}(\varphi)$ approximates φ in the compact-open topology.

Coming back to the definition of $I_{\lambda}(\varphi)$, we remark that the variable x occurs only in $\exp\{-\lambda \|x-y\|^2\}$ which is holomorphic on $\mathbb{C}^n$. Moreover φ has compact support, hence we deduce that the function

$$I_{\lambda}(\varphi)(z) = c\lambda^{\frac{1}{2}n} \int_{\operatorname{supp}\varphi} \varphi(y) \exp\{-\lambda \|z-y\|^2\}\, dy \quad (7)$$

is holomorphic for any $z \in \mathbb{C}^n$, and in particular is analytic on Ω.

To complete the proof it is enough to verify the following: if $\varphi_{|H_2} \equiv 0$, then for any $\varepsilon > 0$, $\alpha \in (0, 1)$ there exists a λ_0 such that if $\lambda > \lambda_0$, we have

$$\left|(I_{\lambda}(\varphi))(z)\right| < \varepsilon \quad (8)$$

for any $z \in \widetilde{U}_{\alpha}$.

Fix $\alpha \in (0, 1)$. There exists $\sigma > 0$ such that for $z \in \widetilde{U}_{\alpha}$, if $d((\Re(z), H_1) < \sigma$ then $\varphi(\Re(z)) = 0$.

This implies that for $x \in \widetilde{U}_{\alpha} \cap \mathbb{R}^n$, we can evaluate $(I_{\lambda}(\varphi))(x)$ by the formula

$$(I_{\lambda}(\varphi))(x) = c\lambda^{\frac{1}{2}n} \int_{\operatorname{supp}\varphi\cap\{\|x-y\|>\sigma\}} \varphi(y) \exp\{-\lambda \|x-y\|^2\}\, dy, \quad (9)$$

and, as remarked before, for any ε, if λ is big enough, then $|(I_{\lambda}\varphi)(x)| < \varepsilon$.

Finally note that, since in (7) the variable z occours only in an exponential function, only its real part is significant for the norm of $I_\lambda(\varphi)(z)$. (We use here the fact that $|e^{a+ib}| = |e^a|$.)

The last remark ensures that the inequality $|I_\lambda(\varphi))(z)| < \varepsilon$ holds for any $z \in \widetilde{U}_\alpha$, and this completes the proof. □

Now we generalize Lemma 2.4 to real analytic subsets of $\mathbb{R}^n$.

Let $X \subset \Omega \subset \mathbb{R}^n$ be a real analytic set in the open set Ω of $\mathbb{R}^n$. We shall suppose X to be the real part of a complex space $\widetilde{X} \subset \widetilde{\Omega} \subset \mathbb{C}^n$ defined over $\mathbb{R}$.

Under these hypotheses we have:

Lemma 2.5. *Let H_i, $i = 1, 2, 3$ be three compact subsets of X such that $H_i \subset \mathring{H}_{i+1}$, $H_i = \overline{\mathring{H}_i}$, $\delta = d(H_1, X - H_2)$, $\alpha \in (0, 1)$ and*

$$\widetilde{U}_\alpha = \{x \in \widetilde{X} : |\Re(x - w)| > \alpha\delta \text{ for any } w \in X - H_2\}.$$

Then, for any $\alpha \in (0, 1)$, $\widetilde{U}_\alpha$ is a vertical neighbourhood of H_1, relative to H_2, H_3.

Proof. For $i = 1, 2, 3$, let us define

$$A_i = \{x \in \mathbb{R}^n - (X - \mathring{H}_i) : \text{ there exists } y \in H_i \text{ such that } d(x, y) < d(y, \partial H_i)\},$$

where d is the usual metric in $\mathbb{R}^n$. It is easy to verify that:

(1) *A_i is open and $A_i \cap X = \mathring{H}_i$.*
In fact A_i is union of the balls $B(y, \rho_y)$ with radius $\rho_y = d(y, \partial H_i)$ for $y \in \mathring{H}_i$. The condition $A_i \cap X = \mathring{H}_i$ follows from the definition of A_i.

(2) *$\overline{A}_i$ is compact and $\overline{A}_i \cap X = H_i$.*
This equality is an easy consequence of the hypothesis $H_i = \overline{\mathring{H}_i}$. The compactness follows from the fact that $\overline{A}_i$ is closed and bounded.

(3) $\overline{A}_i \subset A_{i+1}$, $i = 1, 2$.
This inclusion is a consequence of the hypothesis $H_i \subset \mathring{H}_{i+1}$; it implies that $d(y, \partial H_i) < d(y, \partial H_{i+1})$ if $y \in \mathring{H}_i$.

Let now $\varphi\colon X \longrightarrow \mathbb{R}$ be a C^∞ function such that $\operatorname{supp}\varphi \subset H_3$, $\varphi|_{H_2} \equiv 0$ and fix $\alpha \in (0, 1)$. We claim that there exists a C^∞ extension $\Phi\colon \mathbb{R}^n \to \mathbb{R}$ of φ such that $\operatorname{supp}\Phi \subset A_3$ and $\Phi|_{A_2} \equiv 0$.

The existence of such a Φ can be proved using a partion of unity or as a particular case of the Whitney extension theorem (see [W]).

Take $\alpha' \in (0, 1)$ and let δ' be the distance $d(\partial A_1, A_2)$; from Lemma 2.4 we get that the set

$$\widetilde{A}_1^{\alpha'} = \{x \in \mathbb{C}^n : |\Re(x - y)| > \alpha'\delta' \text{ for any } y \in \mathbb{R}^n - \overline{A}_2\}$$

is a vertical neighbourhood of $\overline{A}_1$ relative to $\overline{A}_2$, $\overline{A}_3$.

This implies that for any $\varepsilon > 0$, $p \in \mathbb{N}$ and $\alpha' \in (0, 1)$ there exists an analytic approximation $G: \mathbb{R}^n \to \mathbb{R}$ such that

a) $\|G - \Phi\|^{p}_{A_3} < \varepsilon$ and

b) G is the restriction of a holomorphic function $\widetilde{G}: \mathbb{C}^n \to \mathbb{C}$ such that $|\widetilde{G}(z)| < \varepsilon$ if $z \in \widetilde{A}_1^{\alpha'}$.

It is easy to verify that we can choose α' in such a way that $\widetilde{U}_\alpha \subset \widetilde{A}_1^{\alpha'}$ and hence $\widetilde{G}|_{\widetilde{X}}$ gives the approximation of φ. So $\widetilde{U}_\alpha$ is a vertical neighbourhood for H_1, relative to H_2, H_3. □

Definition 2.6. If $H_i, i = 1, 2, 3$, are compact sets satisfying the conditions of Lemma 2.5, the neighbourhoods $\widetilde{U}_\alpha$ of H_1 defined above shall be called *the canonical vertical neighbourhoods* of H_1.

Vertical neighbourhoods can be defined also for real analytic spaces which are not subsets of some $\mathbb{R}^n$.

Let $(X, \mathcal{O}_X)$ be a real analytic space and assume it is the real part of a Stein space $(\widetilde{X}, \mathcal{O}_{\widetilde{X}})$.

Let $\{K_n\}_{n\in\mathbb{N}}$ be a sequence of compact sets in X with the following properties:

(1) $K_n = \overline{\mathring{K}_n}$ and $\mathring{K}_{n+1} \supset K_n$ for any n.

(2) $\bigcup_n K_n = X$.

For any $p \in \mathbb{N}$ consider the compact sets K_p, K_{p+1}, K_{p+4}; they shall play the same role as H_1, H_2, H_3 before. We wish to prove the following

Lemma 2.7. *For any $p \in \mathbb{N}$ there exists a vertical neighbourhood $\widetilde{K}_p$ of K_p in $\widetilde{X}$, relative to K_{p+1}, K_{p+4} in such a way that $\widetilde{K}_{p+1} \supset \widetilde{K}_p$ for any $p \in \mathbb{N}$.*

Proof. From the general theory of analytic spaces we can easily deduce that for each $p \in \mathbb{N}$ there exists a holomorphic map $\widetilde{f}_p: \widetilde{X} \to \mathbb{C}^{n_p}$ such that

(1) $\widetilde{f}_p$ is defined over $\mathbb{R}$,

(2) $\widetilde{f}_p$ defines an isomorphism between an open neighbourhood $\widetilde{U}_p$ of K_{p+4} in $\widetilde{X}$ and a complex analytic subset $\widetilde{V}_p = \widetilde{f}_p(\widetilde{U}_p)$ of an open set Ω_p of $\mathbb{C}^{n_p}$,

(3) for any p, $\widetilde{U}_{p+1} \supset \widetilde{U}_p$, and

(4) $d(\widetilde{f}_p(\partial K_p), \widetilde{f}_p(\partial K_{p+1})) = \delta > 0$ for any $p \in \mathbb{N}$.

In fact, from [N] we know that for any compact set $H \subset \widetilde{X}$ there exists a holomorphic map $g: \widetilde{X} \to \mathbb{C}^n$ for some $n \in \mathbb{N}$, which is an isomorphism onto its image when restricted to a suitable open neighbourhood of H.

If $H \subset X$, it is easy to verify that g can be chosen so that it is defined over $\mathbb{R}$. Finally condition (4) is obtained by multiplying $\widetilde{f}_p$ by a suitable positive constant.

Define now $\widetilde{g}_q = (\widetilde{f}_1, \ldots, \widetilde{f}_q): \widetilde{X} \to \mathbb{C}^{n_1+\cdots+n_q}$. Denote by $\widetilde{U}^1_\alpha$ the canonical vertical neighbourhoods in $\widetilde{V}_1 = \widetilde{f}_1(\widetilde{U}_1)$ of $H_1 = \widetilde{f}_1(K_1)$ relative to $H_2 = \widetilde{f}_1(K_2)$ and $H_3 = \widetilde{f}_1(K_5)$ and define $\widetilde{K}^\alpha_1 = \widetilde{f}_1^{-1}(\widetilde{U}^1_\alpha)$. Using Lemma 2.3. it is easy to see that, for any $\alpha \in (0,1)$, $\widetilde{K}^\alpha_1$ is a vertical neighbourhood of K_1 in $\widetilde{X}$ relative to K_2, K_5.

Now we define the vertical neighbourhoods of K_2 relative to K_3, K_6.

Consider the map $\widetilde{g}_2 = (\widetilde{f}_1, \widetilde{f}_2): \widetilde{X} \to \mathbb{C}^{n_1+n_2}$; then $\widetilde{g}_2$ is an isomorphism on a neighbourhood of K_6 in $\widetilde{X}$, $\widetilde{U}_2 \supset \widetilde{U}_1$ and $\widetilde{V}_2 = \widetilde{g}_2(\widetilde{U}_2)$ is a complex analytic subset of an open set Ω_2 of $\mathbb{C}^{n_1+n_2}$. Take the compact sets $H^2_2 = g_2(K_2)$, $H^2_3 = g_2(K_3)$, $H^2_6 = g_2(K_6)$ and let $\widetilde{U}^2_\alpha$ be the canonical vertical neighbourhoods of H^2_2 relative to H^2_3, H^2_6 in $\widetilde{V}_2 = \widetilde{g}_2(\widetilde{U}_2)$.

Define $\widetilde{K}^\alpha_2 = \widetilde{g}_2^{-1}(\widetilde{U}^2_\alpha)$. As we remarked before, for any $\alpha \in (0,1)$, $\widetilde{K}^\alpha_2$ is a vertical neighbourhood of K_2 relative to K_3, K_6 in $\widetilde{X}$.

In a similar way we define the $\widetilde{K}^\alpha_3, \widetilde{K}^\alpha_4, \ldots$ as vertical neighbourhoods of $K_3, K_4, \ldots$. To complete the proof we have to verify that

$$\widetilde{K}^\alpha_{p+1} \supset \widetilde{K}^\alpha_p. \qquad (*)$$

It is enough to verify $(*)$ for $p = 1$, the general case follows from the same argument. Recall that:

$$\widetilde{U}^\alpha_1 = \left\{x \in \widetilde{f}_1(\widetilde{U}_1) : |\Re(x-w)| > \alpha\delta \text{ for any } w \in \widetilde{f}_1\left(\widetilde{U}_1 \cap (X - K_2)\right)\right\},$$
$$\widetilde{U}^\alpha_2 = \left\{x \in \widetilde{g}_2(\widetilde{U}_2) : |\Re(x-w)| > \alpha\delta \text{ for any } w \in \widetilde{g}_2\left(\widetilde{U}_2 \cap (X - K_3)\right)\right\}$$

and $\widetilde{K}^\alpha_1 = \widetilde{f}_1^{-1}(\widetilde{U}^\alpha_1)$, $\widetilde{K}^\alpha_2 = \widetilde{g}_2^{-1}(\widetilde{U}^\alpha_2)$.

We make the following easy observations.

a) If $x, y \in \widetilde{X}$ then $d(\widetilde{f}_1(x), \widetilde{f}_1(y)) \leq d(\widetilde{g}_2(x), \widetilde{g}_2(y))$ and $\left|\Re\left(\widetilde{f}_1(x) - \widetilde{f}_1(y)\right)\right| \leq \left|\Re\left(\widetilde{g}_2(x) - \widetilde{g}_2(y)\right)\right|$.

b) If $x \in K_2$ then $d\left(\widetilde{g}_2(x), \widetilde{g}_2(\partial K_2)\right) \leq d\left(\widetilde{g}_2(x), \widetilde{g}_2(\partial K_3)\right)$.

From the hypothesis $\delta = d(\widetilde{f}_1(\partial K_1), \widetilde{f}_2(\partial K_2)) = d(\widetilde{f}_2(\partial K_2), \widetilde{f}_2(\partial K_3))$ and these two remarks, $(*)$ follows. □

Note that Lemma 2.7 is trivial for $X = \mathbb{R}^n$: in this case we have the canonical vertical neighbourhoods.

Definition 2.8. A sequence $\left\{\widetilde{K}_p\right\}_{p \in \mathbb{N}}$ of neighbourhoods of compact sets as in Lemma 2.7 shall be called *a consistent sequence of vertical neighbourhoods*.

Theorem 2.9. *Let $(X, \mathcal{O}_X)$ be the real part of a complex space $(\widetilde{X}, \mathcal{O}_{\widetilde{X}})$. Denote by $\mathcal{E}_X$ the sheaf of germs of C^∞ functions on X. Then, for any open set $U \subset X$, $\Gamma(U, \mathcal{O}_X)$ is dense in $\Gamma(U, \mathcal{E}_X)$ for the Whitney topology, and hence for the weak topology.*

Proof. Consider a sequence $\{K_n\}_{n\in\mathbb{N}}$ of compact sets in X such that $\mathring{K}_{n+1} \supset K_n$, $\bigcup_n K_n = X$, $K_n = \overline{\mathring{K}_n}$.

Let $\{\widetilde{K}_n\}_{n\in\mathbb{N}}$ be a consistent sequence of vertical neigbourhoods as defined in Lemma 2.7; we recall that for each p, $\widetilde{K}_p$ is a vertical neighbourhood of K_p relative to K_{p+1}, K_{p+4}.

By a partition of unity we may construct a sequence of C^∞ functions $\varphi_p\colon X \to \mathbb{R}$, $p \geq 0$, such that

a) $\operatorname{supp}\varphi_p \subset K_{p+2}$

b) $\varphi_p(x) = 0$ in a neighbourhood of K_{p-1} and

c) $\varphi_p(x) = 1$ in a neighbourhood of $L_p = K_{p+1} - \mathring{K}_p$.

Take a C^∞ function $\gamma\colon X \to \mathbb{R}$, choose a sequence $\{\varepsilon_p\}$ of positive numbers and a sequence $\{m_p\}$ of natural numbers. We have to find an analytic function $g\colon X \to \mathbb{R}$ such that for any $p \geq 0$ one has

$$\|g - \gamma\|_{L_p}^{m_p} < \varepsilon_p. \tag{1}$$

We can assume $m_{p+1} \geq m_p$ for any $p \geq 0$. Define, for $p \geq 0$,

$$M_{p+1} = 1 + \|\varphi_p\|_{K_{p+2}}^{m_p}, \tag{2}$$

and choose positive numbers δ_p in such a way that

$$2\delta_{p+1} \leq \delta_p \quad \text{and} \quad \sum_{q\geq p} \delta_q M_{q+1} \leq \frac{1}{4}\,\varepsilon_p. \tag{3}$$

Now consider the compact sets

$$H_1 = H_2 = \emptyset,\ H_3 = K_2.$$

By definition of a vertical neighbourhood we can find a holomorphic function $g_0\colon \widetilde{X} \to \mathbb{C}$, defined over $\mathbb{R}$, such that

$$\|g_0 - \varphi_0\gamma\|_{K_1}^{m_0} = \|g_0 - \gamma\|_{K_1}^{m_0} \leq \|g_0 - \varphi_0\gamma\|_{K_2}^{m_0} < \delta_0.$$

Using the fact that $\widetilde{K}_p$ is a vertical neighbourhood of K_p relative to K_{p+1}, K_{p+4}, we may find inductively a sequence of holomorphic functions $\{g_p\}$ on $\widetilde{X}$, defined over $\mathbb{R}$, such that

$$\|g_p - \varphi_p(\gamma - g_0 - g_1 - \cdots - g_{p-1})\|_{K_{p+1}}^{m_p} < \delta_p \quad \text{and} \tag{4}$$

$$|g_p(z)| < \delta_p \text{ if } z \in \widetilde{K}_{p-2}. \tag{5}$$

(Condition (5) is empty for $p < 3$). From the conditions on φ_p we deduce

$$\|\gamma - \sum_{i=0}^{p} g_i\|_{L_p}^{m_p} < \delta_p \quad \text{and} \tag{6}$$

$$\|g_p\|_{K_{p-1}}^{m_p} < \delta_p. \tag{7}$$

If in (4) we replace p by $p+1$, we obtain

$$\begin{aligned} \|g_{p+1}\|_{L_p}^{m_p} &\leq \|\varphi_{p+1}(\gamma - \sum_{i=0}^{p} g_i)\|_{L_p}^{m_p} + \|g_{p+1} - \varphi_{p+1}(\gamma - \sum_{i=0}^{p} g_i)\|_{L_p}^{m_p} \\ &\leq \|\varphi_{p+1}\|_{K_{p+5}}^{m_p} \cdot \|\gamma - \sum_{i=1}^{p} g_i\|_{L_p}^{m_p} + \delta_{p+1} \leq M_{p+1}\delta_p + \delta_{p+1}. \end{aligned} \tag{8}$$

But (7) applied to g_{p+1} gives

$$\|g_{p+1}\|_{K_p}^{m_p} < \delta_{p+1} \tag{9}$$

since $m_p \leq m_{p+1}$. Finally, from (8) and (9), we deduce

$$\|g_{p+1}\|_{K_{p+1}}^{m_p} \leq M_{p+1}\delta_p + 2\delta_{p+1} \leq 2\delta_p M_{p+1},$$

and hence

$$\|\sum_{i\geq p} g_i\|_{K_{p+1}}^{m_p} \leq \|\sum_{i\geq p} g_i\|_{K_{p+1}}^{m_p} \leq \sum_{i\geq p} \|g_i\|_{K_{i+1}}^{m_i} \leq 2\sum_{i\geq p} \delta_i M_{i+1} < \frac{1}{2}\varepsilon_p. \tag{10}$$

Relations (6) and (10) prove that the series $\sum g_i$ converges on X to a C^∞ function g which approximates γ as wanted.

Moreover condition (5) proves that the series $\sum g_i$ in fact converges as a series of holomorphic functions on the union $\bigcup_p \widetilde{K}_p \subset \widetilde{X}$.

Since the space of holomorphic functions on a complex space is complete (see [GR]), the function $g = \sum g_i$ is the restriction of a holomorphic function, and hence it is analytic. The theorem is proved. □

Remark 2.10. Theorem 2.9 holds when $(X, \mathcal{O}_X)$ is the real part of a complex space. No hypothesis on the coherence or on the dimensions of the Zariski tangent spaces are necessary.

3. Approximation for sections of a sheaf

In the following $(X, \mathcal{O}_X)$ will be a real coherent reduced analytic space; no hypothesis on the dimensions of the Zariski tangent spaces are required.

Theorem 3.1. *Let $(X, \mathcal{O}_X)$ be a coherent real analytic space. Consider a coherent sheaf $\mathcal{F}$ of $\mathcal{O}_X$-modules and denote by $\mathcal{F}^\infty$ the sheaf $\mathcal{F} \otimes_{\mathcal{O}_X} \mathcal{E}_X$. For any open set $U \subset X$, $\Gamma(U, \mathcal{F})$ is dense in $\Gamma(U, \mathcal{F}^\infty)$ for the Whitney topology, hence also for the weak topology.*

Proof. Let $(\widetilde{X}, \mathcal{O}_{\widetilde{X}})$ be a complexification of $(X, \mathcal{O}_X)$. By a theorem of H. Cartan, there exists a neighbourhood (hence a Stein neighbourhood) of X in $\widetilde{X}$ and a coherent sheaf $\widetilde{\mathcal{F}}$ on it, such that $\widetilde{\mathcal{F}}|_X = \mathcal{F} \otimes_{\mathcal{O}_X} \mathcal{O}_{\widetilde{X}}$. (See [Ca] for the case $X \subset \mathbb{R}^n$; the proof is the same in the general case). So in the following we shall suppose that $(\widetilde{X}, \mathcal{O}_{\widetilde{X}})$ is a Stein space and that $\widetilde{\mathcal{F}}$ is defined over $\widetilde{X}$.

Take a sequence $\{K_n\}_{n\in\mathbb{N}}$ of compact sets such that $\overline{\mathring{K}_n} = K_n$, $K_n \subset \mathring{K}_{n+1}$, $\bigcup K_n = X$, and let $\{\widetilde{K}_n\}$ be a consistent sequence of vertical neighbourhoods in $\widetilde{X}$. More precisely, we assume that for any p, $\widetilde{K}_p$ is a vertical neighbourhood of K_p relative to K_{p+1}, K_{p+4}.

We know from Cartan's Theorem A, that for any $p \in \mathbb{N}$ there exists a finite set of global sections of $\widetilde{\mathcal{F}}$, say $\gamma_1^p, \dots, \gamma_{n_p}^p$, such that they generate the stalk $\widetilde{\mathcal{F}}_x$ for any x in an open neighbourhood $\widetilde{D}_p$ of K_p in $\widetilde{X}$.

We can assume $\gamma_i^p|_X \in \Gamma(X, \mathcal{F})$ for $i = 1, \dots, n_p$ and for any p.

Take now a C^∞ global section $\sigma \in \Gamma(X, \mathcal{F}^\infty)$. Using a suitable partition of unity, we can find C^∞ functions $\{\alpha_i^j\}_{i,j\in\mathbb{N}}$ on X such that

(1) for any j $\operatorname{supp} \alpha_i^j \subset K_{j+2} - K_{j-1}$ for $i = 1, \dots, n_j$, and

(2) $\sigma(x) = \sum_{j=1}^{\infty} \sum_{i=1}^{n_j} \alpha_i^j(x) \gamma_i^j(x)$.

(The family $\{\operatorname{supp} \alpha_i^j\}$ is locally finite, so (2) makes sense.) If we denote $\psi_j = \sum_{i=1}^{n_j} \alpha_i^j \gamma_i^j$ we can write $\sigma = \sum_{j=1}^{\infty} \psi_j$.

We have the following remarks.

(1) Take $\gamma_1, \dots, \gamma_q \in \Gamma(X, \mathcal{F})$ and a sequence of sections of $\mathcal{F}^\infty$ $\psi_n = \sum_{i=1}^{q} \alpha_i^n \gamma_i$.
Then, by using norms on $\Gamma(X, \mathcal{F})$ as in Example (4) of Section 1, it is not difficult to prove that $\lim_{n\to\infty} \|\alpha_i^n - a_i\|_K^m = 0$ implies
$$\lim_{n\to\infty} \|\psi_n - \sum_{i=1}^{q} a_i \gamma_i\|_K^m = 0.$$

(2) Remark (1) implies that it is possible to approximate on compact sets sections of $\mathcal{F}^\infty$ by sections of $\mathcal{F}$. Moreover if vertical neighbourhoods are defined and the original section has the required properties, the section of $\mathcal{F}$ extends to a section of $\widetilde{\mathcal{F}}$ which has small norm on the corresponding vertical neighbourhoods.

(3) The space $\Gamma(\widetilde{U}, \widetilde{\mathcal{F}})$ is a complete space (Example (3) of Section 1).

Now we can repeat the proof of Theorem 2.9 almost word for word. It is enough to replace $\varphi_p(\gamma - g_0 - \cdots - g_{p-1})$ by the section ψ_p, and to use seminorms for sections instead of seminorms for functions. This proves the theorem. □

Consider now a closed coherent subspace $Y \subset X$. The structure sheaf $\mathcal{O}_Y$ is defined by the exact sequence of coherent sheaves

$$0 \longrightarrow \mathcal{I}_Y \longrightarrow \mathcal{O}_X \longrightarrow \mathcal{O}_Y \longrightarrow 0.$$

If $\mathcal{F}$ is a coherent sheaf of $\mathcal{O}_X$-modules we can define the restriction of $\mathcal{F}$ to Y in the following way.

Definition 3.2. $\mathcal{F}|_Y = \mathcal{O}_Y \otimes_{\mathcal{O}_X} \mathcal{F}$.

From the above exact sequence we deduce the exactness of the sequence

$$\mathcal{I}_Y \otimes_{\mathcal{O}_X} \mathcal{F} \longrightarrow \mathcal{F} \xrightarrow{r} \mathcal{F}|_Y \longrightarrow 0,$$

where r is the restriction map; its kernel is the image in $\mathcal{F}$ of the sheaf $\mathcal{I}_Y \otimes_{\mathcal{O}_X} \mathcal{F}$ (the sheaf of germs of sections vanishing at Y.)

In this situation we can give a sort of relative version of Theorem 3.1.

Let g be an element in $\Gamma(Y, \mathcal{F}|_Y)$. Denote by $\Gamma(X, \mathcal{F})_g$ the set of sections which extend g to X. This set is not empty because of Cartan's Theorem B. Denote in the same way by $\Gamma(X, \mathcal{F}^\infty)_g$ the extensions of g to X in $\mathcal{F}^\infty$. Then we have:

Theorem 3.3. *Let Y be a closed coherent analytic subset of the coherent analytic space X and $g \in \Gamma(Y, \mathcal{F}|_Y)$; then $\Gamma(X, \mathcal{F})_g$ is dense in $\Gamma(X, \mathcal{F}^\infty)_g$ for the Whitney topology.*

Proof. Let φ be an element in $\Gamma(X, \mathcal{F}^\infty)_g$. Given a neighbourhood B_φ in the Whitney topology we have to find $h \in B_\varphi \cap \Gamma(X, \mathcal{F})_g$.

Let G be an element in $\Gamma(X, \mathcal{F})_g$. Then $(\varphi - G)|_Y = 0$ and so replacing φ by $\varphi - G$ we can suppose g to be the zero-section of $\Gamma(Y, \mathcal{F}|_Y)$.

By considering the exact sequence:

$$\mathcal{I}_Y \otimes_{\mathcal{O}_X} \mathcal{F}^\infty \xrightarrow{\beta} \mathcal{F}^\infty \xrightarrow{\alpha} \mathcal{O}_Y \otimes_{\mathcal{O}_X} \mathcal{F}^\infty \longrightarrow 0,$$

we see that φ is in the image of $\mathcal{I}_Y \otimes_{\mathcal{O}_X} \mathcal{F}^\infty = (\mathcal{I}_Y \otimes_{\mathcal{O}_X} \mathcal{F}) \otimes_{\mathcal{O}_X} \mathcal{E}_X$. Let $\psi \in \Gamma(X, \mathcal{I}_Y \otimes \mathcal{F}^\infty)$ be a preimage of φ. By Theorem 3.1 ψ can be approximated in the Whitney topology by a section f of $\Gamma(X, \mathcal{I}_Y \otimes_{\mathcal{O}_X} \mathcal{F})$. Then $\beta(f)$ approximates φ because β is continuous and $\beta(f)|_Y = 0$. □

Suppose now one has a sheaf homomorphism between two coherent analytic sheaves of $\mathcal{O}_X$-modules $\alpha\colon \mathcal{F} \to \mathcal{G}$. Then one has an exact sequence

$$0 \longrightarrow \ker \alpha \longrightarrow \mathcal{F} \xrightarrow{\alpha} \mathcal{G}.$$

Since $\mathcal{E}_X$ is flat over $\mathcal{O}_X$ we get an exact sequence

$$0 \longrightarrow (\ker\alpha)^\infty \longrightarrow \mathcal{F}^\infty \xrightarrow{\alpha^\infty} \mathcal{G}^\infty.$$

Remark 3.4. Any exact sequence of $\mathcal{O}_X$-modules is also locally an exact sequence of $\mathcal{O}_n$-modules, because X is locally isomorphic to an analytic subset of some $\mathbb{R}^n$. For $X \subset \mathbb{R}^n$, $\mathcal{E}_X$ is a faithfully flat $\mathcal{O}_n$-module (see [M], Cor.1.12, p. 88); so, by definition of $\mathcal{E}_X$, α is surjective if, and only if, α^∞ is surjective and $\alpha(\mathcal{F}) = \alpha^\infty(\mathcal{F}) \cap \mathcal{G}$.

Remark 3.5. If the sheaf $\mathcal{F}$ is $\mathcal{O}_X$, then the sheaf $\mathcal{I}_Y^\infty = \mathcal{I}_Y \otimes_{\mathcal{O}_X} \mathcal{E}_X \simeq \mathcal{I}_Y \cdot \mathcal{E}_X$ is the sheaf of germs of C^∞ functions vanishing at Y, because Y is coherent ([M], p. 95.)

Theorem 3.6. *Let X be a coherent analytic space, let $\mathcal{F}$, $\mathcal{G}$ be coherent sheaves of $\mathcal{O}_X$-modules and $\alpha: \mathcal{F} \to \mathcal{G}$ be a sheaf homomorphism; suppose $g \in \Gamma(X, \mathcal{G})$ is such that $g = \alpha^\infty(\eta)$ with $\eta \in \Gamma(X, \mathcal{F}^\infty)$. Then in each neighbourhood of η for the Whitney topology there exists $f \in \Gamma(X, \mathcal{F})$ such that $\alpha(f) = g$.*

Moreover if $Y \subset X$ is a closed coherent subspace and $\eta_{|Y}$ is in $\Gamma(Y, \mathcal{F}|_Y)$, we can find f such that $f|_Y = \eta_{|Y}$.

Proof. By Remark 3.4, $g = \alpha^\infty(\eta)$ is the image of some element $h \in \Gamma(X, \mathcal{F})$. Then $\eta - h \in \ker\alpha^\infty$ and, if $\eta_{|Y}$ is analytic, the same is true for $\eta - h|_Y$.

So we can apply Theorem 3.3 to the sheaf $\ker\alpha$ and find $h_1 \in \Gamma(X, \ker\alpha)$ very close to $\eta - h$ and such that $h_1|_Y = \eta - h|_Y$. Hence $f = h + h_1$ is very close to η, $\alpha(f) = g$ and, if $\eta_{|Y}$ is analytic, $f|_Y = \eta_{|Y}$. □

As an application of these results we obtain some approximation theorems for solutions of analytic linear systems.

Let U be an open set in $\mathbb{R}^n$ and $X \subset U$ be a coherent analytic set. Consider an analytic linear system on U:

$$\sum_{k=1}^{q} a_{hk}(x) y_k = g_h, \qquad h = 1, \dots, p, \tag{$*$}$$

where $a_{hk}, g_h \in C^\omega(U) = \Gamma(U, \mathcal{O})$. Then we have the following.

Theorem 3.7. *If $(*)$ has a C^∞ solution, i.e., if there exists $\varphi = (\varphi_1, \dots, \varphi_q) \in C^\infty(U)^q$ such that*

$$\sum_{k=1}^{q} a_{hk}(x)\varphi_k = g_h(x), \qquad h = 1, \dots, p,$$

then for each neighbourhood B_φ of φ in the product Whitney topology of $C^\infty(U)^q$, there exists a solution $f = (f_1, \dots, f_q) \in C^\omega(U)^q$ of $(*)$ that belongs to B_φ.

Moreover we have:

(1) *if $X \subset U$ is a coherent analytic set and $\varphi_{k|X} \in \Gamma(X, \mathcal{O}_X)$ for any k, then we can take*

$$f_k|_X = \varphi_{k|X} \quad \text{for } k = 1, \ldots, q;$$

(2) *if for some $l \leq q$, $\varphi_1, \ldots, \varphi_l \in C^\omega(U)$ then we can take $f_1 = \varphi_1, \ldots, f_l = \varphi_l$.*

Proof. Consider the sheaf homomorphism $\alpha\colon \mathcal{O}^q \longrightarrow \mathcal{O}^p$ defined by the matrix (a_{hk}).

If each g_h is the zero function the first statement is Theorem 3.1 applied to $\ker \alpha$.

We have $(g_1, \ldots, g_p) = \alpha^\infty(\varphi_1, \ldots, \varphi_q)$. By Remark 3.4, then $(g_1, \ldots, g_p) \in \Gamma(U, \operatorname{Im} \alpha)$. So the first statement and (1) are consequences of Theorem 3.6.

To prove (2) consider

$$g'_h = \sum_{k=1}^{l} a_{hk}\varphi_k.$$

This is an element in $\Gamma(U, \mathcal{O})$, so we have a system of the same type as $(*)$ given by

$$\sum_{k=l+1}^{q} a_{hk} y_k = g_h - g'_h, \qquad h = 1, \ldots, p, \tag{$**$}$$

and for any solution $(\psi_{l+1}, \ldots, \psi_q)$ of $(**)$, $(\varphi_1, \ldots, \varphi_l, \psi_{l+1}, \ldots, \psi_q)$ is a solution of $(*)$. So we reduce to the first statement. □

Remark 3.8. In Theorem 3.7 we can suppose (same proof) that U is any open set of a coherent real analytic space.

4. Approximation for sheaf homomorphisms

Let X be a coherent analytic space. If $\mathcal{F}$ and $\mathcal{G}$ are coherent sheaves of $\mathcal{O}_X$-modules, we know that the sheaf $\operatorname{Hom}(\mathcal{F}, \mathcal{G})$ is also a coherent sheaf (see [S]). The next proposition gives the relations between $\operatorname{Hom}(\mathcal{F}^\infty, \mathcal{G}^\infty)$ and $\operatorname{Hom}(\mathcal{F}, \mathcal{G}) \otimes_{\mathcal{O}_X} \mathcal{E}_X = \operatorname{Hom}(\mathcal{F}, \mathcal{G})^\infty$.

Proposition 4.1. $\operatorname{Hom}_{\mathcal{E}_X}(\mathcal{F}^\infty, \mathcal{G}^\infty) \simeq \operatorname{Hom}_{\mathcal{O}_X}(\mathcal{F}, \mathcal{G})^\infty$.

Proof. It is enough to prove the statement locally, since we have a natural map

$$\operatorname{Hom}_{\mathcal{O}_X}(\mathcal{F}, \mathcal{G})^\infty \to \operatorname{Hom}_{\mathcal{E}_X}(\mathcal{F}^\infty, \mathcal{G}^\infty).$$

Step 1. *The assertion is true for $\mathcal{F} = \mathcal{O}_X^p$ and $\mathcal{G} = \mathcal{O}_X^q$.*

In fact a sheaf homomorphism between $\mathcal{O}_X^p$ and $\mathcal{O}_X^q$ is given by a $p \times q$ matrix whose entries are analytic functions on X.

A similar result is true for $\text{Hom}_{\mathcal{E}_X}(\mathcal{E}_X^p, \mathcal{E}_X^q)$. In other words we have:

$$\text{Hom}_{\mathcal{O}_X}(\mathcal{O}_X^p, \mathcal{O}_X^q) \simeq \mathcal{O}_X^{p\times q}, \quad \text{Hom}_{\mathcal{E}_X}(\mathcal{E}_X^p, \mathcal{E}_X^q) \simeq \mathcal{E}_X^{p\times q}, \quad \mathcal{O}_X^{p\times q} \otimes_{\mathcal{O}_X} \mathcal{E}_X = \mathcal{E}_X^{p\times q}.$$

Step 2. *The assertion is true for* $\mathcal{F} = \mathcal{O}_X^p$ *and general* $\mathcal{G}$.

Take a local resolution of $\mathcal{G}$,

$$\mathcal{O}_X^n \longrightarrow \mathcal{O}_X^q \longrightarrow \mathcal{G} \longrightarrow 0,$$

and apply the functor Hom $(\mathcal{O}_X^p, -)$. Each homomorphism $\mathcal{O}_X^p \to \mathcal{G}$ can be lifted (see Proposition 6 of Chap. VIII in [G.R.]) to a morphism $\mathcal{O}_X^p \to \mathcal{O}_X^q$, hence we get an exact sequence

$$\text{Hom}(\mathcal{O}_X^p, \mathcal{O}_X^n) \longrightarrow \text{Hom}(\mathcal{O}_X^p, \mathcal{O}_X^q) \longrightarrow \text{Hom}(\mathcal{O}_X^p, \mathcal{G}) \longrightarrow 0.$$

Tensoring with $\mathcal{E}_X$, we get:

$$\big(\text{Hom}(\mathcal{O}_X^p, \mathcal{O}_X^n)\big)^\infty \longrightarrow \big(\text{Hom}(\mathcal{O}_X^p, \mathcal{O}_X^q)\big)^\infty \longrightarrow \big(\text{Hom}(\mathcal{O}_X^p, \mathcal{G})\big)^\infty \longrightarrow 0.$$

Doing the same but in opposite order (using again that homomorphisms can be lifted), from

$$\mathcal{E}_X^n \longrightarrow \mathcal{E}_X^q \longrightarrow \mathcal{G}^\infty \longrightarrow 0$$

we get

$$\text{Hom}(\mathcal{E}_X^p, \mathcal{E}_X^n) \longrightarrow \text{Hom}(\mathcal{E}_X^p, \mathcal{E}_X^q) \longrightarrow \text{Hom}(\mathcal{E}_X^p, \mathcal{G}^\infty) \longrightarrow 0.$$

The natural map $\text{Hom}(\mathcal{O}_X^p, \mathcal{O}_X^n) \otimes \mathcal{E}_X \longrightarrow \text{Hom}(\mathcal{E}_X^p, \mathcal{E}_X^n)$ gives the following commutative diagram

$$\begin{array}{ccccccc}
\text{Hom}(\mathcal{O}_X^p, \mathcal{O}_X^n) \otimes \mathcal{E}_X & \to & \text{Hom}(\mathcal{O}_X^p, \mathcal{O}_X^q) \otimes \mathcal{E}_X & \to & \text{Hom}(\mathcal{O}_X^p, \mathcal{G}) \otimes \mathcal{E}_X & \to & 0 \\
\downarrow & & \downarrow & & \downarrow & & \downarrow \\
\text{Hom}(\mathcal{E}_X^p, \mathcal{E}_X^n) & \to & \text{Hom}(\mathcal{E}_X^p, \mathcal{E}_X^q) & \to & \text{Hom}(\mathcal{E}_X^p, \mathcal{G}^\infty) & \to & 0
\end{array}$$

The first two vertical rows are isomorphisms by Step 1, so the third one is an isomorphism too.

Step 3. *The general case.*

Take a local resolution for $\mathcal{F}$

$$\mathcal{O}_X^m \longrightarrow \mathcal{O}_X^s \longrightarrow \mathcal{F} \longrightarrow 0,$$

Apply Hom $(-, \mathcal{G})$ to get

$$\text{Hom}(\mathcal{O}_X^m, \mathcal{G}) \longleftarrow \text{Hom}(\mathcal{O}_X^s, \mathcal{G}) \longleftarrow \text{Hom}(\mathcal{F}, \mathcal{G}) \longleftarrow 0.$$

Tensoring with $\mathcal{E}_X$ yields

$$\text{Hom}(\mathcal{O}_X^m, \mathcal{G})^\infty \longleftarrow \text{Hom}(\mathcal{O}_X^s, \mathcal{G})^\infty \longleftarrow \text{Hom}(\mathcal{F}, \mathcal{G})^\infty \longleftarrow 0.$$

Doing the same in opposite order yields

$$\operatorname{Hom}(\mathcal{E}_X^m, \mathcal{F}^\infty) \longleftarrow \operatorname{Hom}(\mathcal{E}_X^s, \mathcal{F}^\infty) \longleftarrow \operatorname{Hom}(\mathcal{F}^\infty, \mathcal{G}^\infty) \longleftarrow 0.$$

But the natural maps

$$\operatorname{Hom}(\mathcal{O}_X^m, \mathcal{F})^\infty \longrightarrow \operatorname{Hom}(\mathcal{E}_X^m, \mathcal{F}^\infty) \quad \text{and} \quad \operatorname{Hom}(\mathcal{O}_X^s, \mathcal{F})^\infty \longrightarrow \operatorname{Hom}(\mathcal{E}_X^s, \mathcal{F}^\infty)$$

are isomorphisms, hence $\operatorname{Hom}(\mathcal{F}, \mathcal{G})^\infty \longrightarrow \operatorname{Hom}(\mathcal{F}^\infty, \mathcal{G}^\infty)$ is an isomorphism too. □

Now we want to study the set of isomorphisms in $\operatorname{Hom}(\mathcal{F}, \mathcal{G})$. If α is such an isomorphism, then we have an isomorphism $\operatorname{Hom}(\mathcal{F}, \mathcal{G}) \to \operatorname{Hom}(\mathcal{F}, \mathcal{F})$, obtained by composition with α^{-1}, which is also an isomorphism. So we can assume $\mathcal{F} = \mathcal{G}$.

For each $x \in X$ we can consider a minimal resolution of $\mathcal{F}$ in a small neighbourhood U of x,

$$\mathcal{O}_U^n \xrightarrow{\tau} \mathcal{O}_U^p \xrightarrow{\sigma} \mathcal{F}|_U \longrightarrow 0.$$

This means:

(1) $\mathcal{F}_x$ is an $\mathcal{O}_{X,x}$-module generated by p sections $(a_1, \dots, a_p)$ and it cannot be generated by $p-1$ sections.

(2) The kernel of $\sigma: \mathcal{O}^p \to \mathcal{F}$ is a subsheaf of $\mathcal{O}^p$ generated by n p-tuples of analytic functions.

If $\alpha: \mathcal{F} \to \mathcal{F}$ is a given isomorphism, then $(\alpha(a_1) = b_1, \dots, \alpha(a_p) = b_p)$ is another system of generators of $\mathcal{F}_x$. Then we have:

Lemma 4.2. *α can be lifted to a morphism $\widetilde{\alpha}: \mathcal{O}^p \to \mathcal{O}^p$ (not in a unique way) and any lifting is an isomorphism on a neighbourhood of x.*

The proof is a consequence of the following more general fact.

Lemma 4.3. *Let $(A, \mathfrak{m})$ be a local ring and let M be an A-module of finite type. Let p be the minimal number of elements in M generating M over A. Then any A-homomorphism $f: M \to M$ can be lifted to a homomorphism $\widehat{f}: A^p \to A^p$. Moreover if f is an isomorphism, any lifting $\widehat{f}$ is an isomorphism.*

Proof. M is isomorphic to A^p / N, where N is a submodule and $N \subset \mathfrak{m}A^p$, since p is minimal. If $(v_1, \dots, v_p)$ is a system of generators for M over A, f can be expressed (not in a unique way) as a matrix F with entries in A; then $F: A^p \to A^p$ lifts f. If both F and G lift f then the columns of the difference $F - G$ are elements of N. In particular if $F = (f_{ij})$ and $G = (g_{ij})$, then $f_{ij} - g_{ij} \in \mathfrak{m}$ for any i, j. So, if f is invertible and $h = f^{-1}$, F lifts f and H lifts h, we have $f \circ h = \mathrm{id}$, hence $F \cdot H = (\delta_{ij} + m_{ij})$ with $m_{ij} \in \mathfrak{m}$. This implies that both $\det F$ and $\det H$ are units in A; in particular F is an isomorphism. □

Lemma 4.4. *Suppose $\alpha: \mathcal{F} \to \mathcal{F}$ is a sheaf isomorphism. For each point x in X there exist a compact neighbourhood H_x of x and a positive constant ε_x such that if $\beta: \mathcal{F} \to \mathcal{F}$ satisfies $\|\beta - \alpha\|^0_{H_x} < \varepsilon_x$, then β is an isomorphism on a neighbourhood of H_x.*

Proof. Let $\widetilde{\alpha}$ be a lifting of α on a neighbourhood U_x of x. By Lemma 4.2 we can suppose $\widetilde{\alpha}$ to be an isomorphism on U_x. Define

$$d_0 = \left|\det\left(a_{ij}(x)\right)\right|.$$

We can find a compact neighbourhood H_x of x such that for each $y \in H_x$ one has

$$\left|\det\left(a_{ij}(y)\right)\right| \geq \frac{d_0}{2}.$$

Then for each matrix of analytic functions $\left(b_{ij}(y)\right)$ sufficiently near to $\left(a_{ij}(y)\right)$ we have

$$\left|\det\left(b_{ij}(y)\right) - \det\left(a_{ij}(y)\right)\right| < \frac{d_0}{4}$$

for each $y \in H_x$. This means that there exists ε_x such that, if $\|\beta - \alpha\|^0_{H_x} < \varepsilon_x$, then β has a lifting $\left(b_{ij}(y)\right)$ such that $\left|\det\left(b_{ij}(y)\right)\right| > \dfrac{d_0}{4}$ for each $y \in H_x$; so $\left(b_{ij}(y)\right)$, and hence β, is an isomorphism over a neighbourhood of H_x. □

Corollary 4.5. *For any compact $K \subset X$ there exists a neighbourhood*

$$V(K, \varepsilon) = \{\beta \in \Gamma(X, \operatorname{Hom}(\mathcal{F}, \mathcal{F})) \mid \|\beta - \alpha\|^0_K < \varepsilon\}$$

such that any $\beta \in V(K, \varepsilon)$ is an isomorphism on a neighbourhood of K.

Proof. Cover K by a finite number of H_x and take $\varepsilon = \min\{\varepsilon_x\}$. □

Finally we have:

Theorem 4.6. *Let X be a coherent real analytic space and $\mathcal{F}$, $\mathcal{G}$ be two coherent sheaves of $\mathcal{O}_X$-modules. Then the set*

$$\operatorname{Iso}(\mathcal{F}, \mathcal{G}) = \{\beta \in \Gamma(X, \operatorname{Hom}(\mathcal{F}, \mathcal{G})) \mid \beta \textit{ is an isomorphism}\}$$

is an open set for the Whitney topology.

Proof. As before we can suppose $\mathcal{F} = \mathcal{G}$. Let α be an isomorphism. We have to show that $\operatorname{Iso}(\mathcal{F})$ contains a neighbourhood of α.

Let $\{K_i\}_{i \in \mathbb{N}}$ be an exhaustive sequence of compact sets. Define ε_i as follows:

— ε_0 is such that if $\|\beta - \alpha\|^0_{K_0} < \varepsilon_0$ then β is an isomorphism on a neighbourhood of K_0.

— ε_1 is such that if $\|\beta-\alpha\|^0_{K_1-\mathring{K}_0} < \varepsilon_1$ then β is an isomorphism on a neighbourhood of $K_1 - \mathring{K}_0$.
and so on.

Then $\{K_i, \varepsilon_i\}_{i\in\mathbb{N}}$ defines a neighbourhood of α in the Whitney topology, namely the set of β's such that for any i

$$\|\beta - \alpha\|^0_{K_i-\mathring{K}_{i-1}} < \varepsilon_i .$$

For any β in such a neighbourhood, β is an isomorphism on a neighbourhood of K_0 and on a neighbourhood of $K_i - \mathring{K}_{i-1}$ for each i. If all the ε_i are small enough, β is injective, hence $\beta \in \mathrm{Iso}(\mathcal{F})$. □

Theorem 4.7. *The set of isomorphisms in $\Gamma(X, \mathrm{Hom}\,(\mathcal{F}^\infty, \mathcal{G}^\infty))$ is an open set for the Whitney topology.*

Proof. The proofs of Lemmas 4.2 and 4.4, Corollary 4.5 and Theorem 4.6 can be repeated, almost without changes, with $\mathcal{F}^\infty$ instead of $\mathcal{F}$. □

Corollary 4.8. *Let $\varphi: \mathcal{F}^\infty \to \mathcal{G}^\infty$ be an isomorphism, then there exists an isomorphism $f: \mathcal{F} \to \mathcal{G}$ such that f^∞ is an isomorphism arbitrarily close to φ*

Proof. This is an application of Theorem 3.1 together with Proposition 4.1 and Theorems 4.6 and 4.7 below. □

II. The algebraic case

5. Preliminaries on real algebraic varieties

One of the main differences between the analytic case and the algebraic one is that Theorems A and B are not true for real algebraic coherent sheaves and hence some work is necessary to prove results similar to Theorem 3.1.

We start with some definitions and preliminaries that we shall use in the following.

Now we consider more specifically the case $\mathbb{K} = \mathbb{R}$.

Consider on $\mathbb{R}^n$ the sheaf $\mathcal{R}$ of germs of regular rational functions. We shall use in general the Zariski topology on $\mathbb{R}^n$, even though $\mathcal{R}$ can be defined also with the euclidean topology.

Let V be a real affine variety, $V \subset \mathbb{R}^n$, that is V is the zero set of finitely many polynomial functions. Denote by $\mathcal{R}_V$ the sheaf of germs of real functions that locally are restrictions to V of regular rational functions. $\mathcal{R}_V$ is called the *structure sheaf* of V, and its sections *regular rational functions*. So $(V, \mathcal{R}_V)$ is a ringed space.

If $V \subset \mathbb{R}^n$, by a *complexification* $\widetilde{V}$ of V we mean the smallest complex affine subvariety of $\mathbb{C}^n$ containing V.

If $(V, \mathcal{R}_V)$, $(W, \mathcal{R}_W)$ are real affine varieties, a continuous map $\phi: V \longrightarrow W$ is called an *algebraic (or regular) map* if locally ϕ is the restriction of a regular rational map between open sets of euclidean spaces. As usual ϕ shall be called an *algebraic isomorphism* if it is bijective and ϕ^{-1} is also algebraic.

Let $(V, \mathcal{R}_V)$ be a real affine variety; consider the ring

$$\Gamma_V = \Gamma(V, \mathcal{R}_V)$$

of regular rational functions on V.

Definition 5.1. The *affine scheme* associated to Γ_V is the ringed space $(\operatorname{Spec} \Gamma_V, \mathfrak{R}_V)$, where $\operatorname{Spec} \Gamma_V$ is the set of the prime ideals of Γ_V endowed with the spectral topology and $\mathfrak{R}_V$ is the *structure sheaf* on it (see [Ha]).

Now we give a list of known results; for details and proofs see [T11] or [B.C.R.].

(1) If $(V, \mathcal{R}_V)$ is a real affine subvariety of $\mathbb{R}^n$, then $\Gamma_V = (\mathcal{P}_V)_{N_V}$ is the localization of the polynomial ring $\mathcal{P}_V$ on V with respect to the multiplicative set N_V of polynomials without zeroes on V.

(2) The complexification $\widetilde{V}$ of an affine variety V depends on the embedding $V \hookrightarrow \mathbb{R}^n$. Nevertheless if $\widetilde{V}_1$, $\widetilde{V}_2$ are two complexification of V, then there exist complex neighbourhoods $\widetilde{U}_1, \widetilde{U}_2$ of V in $\widetilde{V}_1$, $\widetilde{V}_2$ which are isomorphic (as complex affine varieties defined over $\mathbb{R}$).

(3) Let $\phi: V \to W$ be an algebraic map between two real affine varieties. If $\widetilde{V}, \widetilde{W}$ are complexifications of V, W, then there exists a neighbourhood $\widetilde{U}$ of V in $\widetilde{V}$ such that ϕ extends to a complex algebraic map $\widetilde{\phi}: \widetilde{U} \to \widetilde{W}$. Moreover two such extensions coincide on a neighbourhood of V in $\widetilde{V}$.

(4) Let V be a real affine variety and $\widetilde{V}$ be a complexification of V. The points of $\operatorname{Spec} \Gamma_V$ are in one-to-one correspondence with those complex algebraic subvarieties $\widetilde{S}$ of $\widetilde{V}$ such that:
   a) $\widetilde{S}$ is defined over $\mathbb{R}$ and it is irreducible over $\mathbb{R}$.
   b) $\widetilde{S} \cap V \neq \emptyset$.

   In particular we may identify the point set V with the set of closed points of $\operatorname{Spec} \Gamma_V$.

(5) There is a canonical isomorphism between the sheaf $\mathcal{R}_V$ and the restriction to $V \hookrightarrow \operatorname{Spec} \Gamma_V$ of the sheaf $\mathfrak{R}_V$.

(6) The only open set of $\operatorname{Spec} \Gamma_V$ containing V is $\operatorname{Spec} \Gamma_V$.

(7) The real projective space with its usual structure as a real algebraic variety is isomorphic to an affine variety.

(8) Any open set of a real affine variety is isomorphic to an affine variety.

Remark 5.2. Results (1), (4), (5), (6) hold true for any field $\mathbb{K}$, if we replace $\mathbb{C}$ by the algebraic closure of $\mathbb{K}$.

Definition 5.3. A ringed space $(X, \mathcal{R}_X)$ is called an (*abstract*) *algebraic prevariety* over $\mathbb{K}$ if there exists a finite open covering $\mathfrak{A} = \{U_i\}_{i=1,\dots,q}$ of X such that any $(U_i, \mathcal{R}|_{U_i})$ is isomorphic to an affine variety defined over $\mathbb{K}$.

Definition 5.4. An algebraic prevariety $(X, \mathcal{R}_X)$ defined over $\mathbb{K}$ is called an *abstract variety* over $\mathbb{K}$ if the diagonal Δ_X is a closed subset of $X \times X$.

Remark 5.5. In the above definition $X \times X$ is endowed with the Zariski topology, which is different from the product topology, so the condition of 5.4 is not equivalent to the separation axiom.

Clearly any affine variety is an abstract variety.

Example 5.6. Consider $P = x^2(x-1)^2 + y^2$. It is easy to see that P is irreducible as an element of $\mathbb{R}[x, y]$ and its zero set is a pair of points, namely $X_1 = (0, 0)$ and $X_2 = (1, 0)$. Let $\tilde{U}_i = \mathbb{C}^2 - X_i$, $i = 1, 2$, and consider the following equivalence relation $\mathcal{R}$ on $\tilde{U}_1 \sqcup \tilde{U}_2$ (where $\sqcup$ means disjoint union). Let (x, y) be a point in $\tilde{U}_1$ and (x', y') be a point in $\tilde{U}_2$; then

$$(x, y)\mathcal{R}(x', y') \iff x = x',\ y = y' \text{ and } P(x, y) \neq 0.$$

It is easy to verify that $(\tilde{U}_1 \sqcup \tilde{U}_2)/\mathcal{R} = \mathbb{C}^2_P$ has a natural prevariety structure, but that $\mathbb{C}^2_P$ is not an abstract variety. Note also that $\mathbb{C}^2_P$ contains $\mathbb{R}^2$.

More generally, consider the following situation.

Let U_1, U_2 be two varieties defined over $\mathbb{K}$, $U_{12} \subset U_1$, $U_{21} \subset U_2$ two open sets and $\rho\colon U_{12} \longrightarrow U_{21}$ an isomorphism.

It is easy to see that the quotient space $X =: U_1 \sqcup U_2/_{\mathcal{R}_\rho}$, where $\mathcal{R}_\rho$ is the equivalence relation induced by ρ, has a natural prevariety structure over $\mathbb{K}$.

Finally let us denote:

$$\Gamma_\rho = \{(x, y) \in U_1 \times U_2 \mid y = \rho(x)\}.$$

Lemma 5.7. *In the above situation X is an abstract variety if, and only if, Γ_ρ is a closed subset of $U_1 \times U_2$.*

Proof. If $\pi\colon U_1 \sqcup U_2 \longrightarrow X$ is the natural projection and $U_i' = \pi(U_i)$, it is easy to see that $U_1' \times U_1'$, $U_1' \times U_2'$, $U_2' \times U_1'$, $U_2' \times U_2'$ is an open covering of $X \times X$. Moreover $\Delta_X = \cup \Delta'_{ik}$, where

$$\Delta'_{ik} = \Delta_X \cap U_i' \times U_k'.$$

Since Δ_X is closed in $X \times X$ if, and only if, Δ'_{ik} is closed in $U_i' \times U_k'$ for any i, k (because $\{U_i' \times U_k'\}$ is an open covering) and since U_1, U_2 are varieties, it follows that Δ'_{ii}, $i = 1, 2$, are closed and Δ'_{12} is precisely the graph of ρ. □

In the following sections we shall need also the notion of complexification for an abstract variety.

This type of result can also be generalized to any field $\mathbb{K}$. The topology is always the Zariski topology.

Definition 5.8. An abstract algebraic complex prevariety $(\widetilde{V}, \mathcal{R}_{\widetilde{V}})$ is called *a complexification of the real prevariety* $(V, \mathcal{R}_V)$ if $V \subset \widetilde{V}$ and there exists a finite open covering $\mathfrak{A} = \{\widetilde{U}_1, \dots, \widetilde{U}_q\}$ with the following properties:

1) there is an embedding $\rho_i : \widetilde{U}_i \longrightarrow \mathbb{C}^{n_i}$ such that $\widetilde{U}_i$ is affine,

2) $\rho_i(\widetilde{U}_i \cap V) = \rho_i(\widetilde{U}_i) \cap \mathbb{R}^{n_i}$, and

3) $\rho_i(\widetilde{U}_i)$ is a complexification of $\rho_i(\widetilde{U}_i) \cap \mathbb{R}^{n_i}$.

This means that $\{\widetilde{U}_i \cap V\}$ is an affine covering of V and that $\widetilde{U}_i$ is a complexification of $(\widetilde{U}_i \cap V)$ as an affine variety. In particular this definition implies that the maps

$$\rho_j \circ \rho_i^{-1} : \rho_i(\widetilde{U}_i \cap \widetilde{U}_j) \longrightarrow \rho_j(\widetilde{U}_i \cap \widetilde{U}_j)$$

are defined over $\mathbb{R}$.

Remark 5.9. If $(\widetilde{V}, \mathcal{R}_{\widetilde{V}})$ is a complexification of a variety $(V, \mathcal{R}_V)$, in general it is not a variety. For instance Example 5.6 yields a complexification $\mathbb{C}_p^2$ of $\mathbb{R}^2$ which is not a variety, since the diagonal in $\mathbb{C}_p^2 \times \mathbb{C}_p^2$ is not closed. This fact implies in particular that the complexification of $(V, \mathcal{R}_V)$ is not unique, up to isomorphism, even as a germ near V. Nevertheless uniqueness of such a germ does hold if both V and $\widetilde{V}$ are varieties.

Lemma 5.10. *Let $(V, \mathcal{R}_V)$ be an abstract real algebraic variety, $\{U_1, U_2\}$ an open covering of V and $\widetilde{U}_i$ a complexification of U_i which is a complex variety, $i = 1, 2$. Assume that the identity map* $\mathrm{id}: U_1 \cap U_2 \longrightarrow U_1 \cap U_2$ *extends to a complex algebraic isomorphism* $\widetilde{\mathrm{id}}$ *between complex neighbourhoods of $U_1 \cap U_2$ in $\widetilde{U}_1$ and $\widetilde{U}_2$.*

Consider the complex prevariety $\widetilde{V} = \widetilde{U}_1 \cup_{\widetilde{\mathrm{id}}} \widetilde{U}_2$ obtained by gluing together $\widetilde{U}_1$ and $\widetilde{U}_2$ by $\widetilde{\mathrm{id}}$, *and let $(\widehat{V}, \mathcal{R}_{\widehat{V}})$ be a complexification of $(V, \mathcal{R}_V)$. Assume the following:*

(i) *$(\widehat{V}, \mathcal{R}_{\widehat{V}})$ is a complex variety,*

(ii) *$\widetilde{U}_1$ is affine, and*

(iii) *the embedding $\rho_i : U_i \longrightarrow \widehat{V}$ extends to an embedding $\widetilde{\rho}_i : \widetilde{U}_i \longrightarrow \widehat{V}$, $i = 1, 2$.*

Then

1) *for any $x \in \widetilde{U}_1$, $\widetilde{\mathrm{id}}(x) = y$ implies $\widetilde{\rho}_1(x) = \widetilde{\rho}_2(y)$, and*

2) *the maps $\widetilde{\rho}_1$ and $\widetilde{\rho}_2$ glue together and define an algebraic map $\rho : \widetilde{V} \longrightarrow \widehat{V}$.*

Proof. It is enough to prove (1), since it implies (2). Without loss of generality, we can assume that all irreducible components V_i of V intersect the open set $U_1 \cap U_2$. So, $U_1 \cap U_2$ is Zariski dense in U_1, U_2, V. Since U_i is Zariski dense in $\widetilde{U}_i$ for any i, we have also that $U_1 \cap U_2$ is Zariski dense in $\widetilde{V}$; moreover, since $\widetilde{\rho}_i$ is an isomorphism, $\widetilde{\rho}_i(U_1 \cap U_2)$ is also dense in $\widetilde{\rho}_i(\widetilde{U}_i) = \widehat{U}_i$, $i = 1, 2$.

By (3) above, the map $\rho_2 \circ \rho_1^{-1}$, which is the identity map id when restricted to $\rho_1(U_1 \cap U_2)$, extends to an isomorphism $\widehat{\mathrm{id}}\colon \widehat{U}_{12} \longrightarrow \widehat{U}_{21}$, where $\widehat{U}_{12}$, $\widehat{U}_{21}$ are neighbourhoods of $U_1 \cap U_2$ resp. in $\widehat{U}_1$ and $\widehat{U}_2$; hence $U_1 \cap U_2$ is also Zariski dense in both $\widehat{U}_{12}$ and $\widehat{U}_{21}$, which are open and dense in $\widehat{U}_1$, resp. $\widehat{U}_2$.

Zariski open sets are also dense with respect to the usual euclidean topology of affine sets (see [S2]). So take points $x \in \widetilde{U}_1$, $y \in \widetilde{U}_2$ such that $\widetilde{\mathrm{id}}(x) = y$. We can find a sequence $\{x'_n\} \subset \widehat{U}_{12}$ which converges to $\widetilde{\rho}_1(x)$. Since $\widetilde{\rho}_1$ is an isomorphism, there is a sequence $\{x_n\} \subset \widetilde{U}_1$, $\widetilde{\rho}_1(x_n) = x'_n$, which converges to x. Consider the sequence $\{y'_n\}$, where $y'_n = \widehat{\mathrm{id}}(x'_n)$. There are two possibilities:

1) $\{y'_n\}$ converges to $\widetilde{\rho}_2(y)$, therefore $\widetilde{\rho}_1(x) = \widetilde{\rho}_2(y)$ since $\widehat{V}$ is a variety, and hence the graph of $\widehat{\mathrm{id}}$ is closed.

2) $\{y'_n\}$ does not converge to $\widetilde{\rho}_2(y)$, therefore the sequence $\{y_n\}$, with $\widetilde{\rho}_2(y_n) = y'_n$, does not converge to y.

But the second case is impossible because $\widetilde{\mathrm{id}}(x) = y$ and $\widetilde{\mathrm{id}}$ is continuous. □

Proposition 5.11. *Let $(V, \mathcal{R}_V)$ be a real abstract algebraic variety and U an open set. Let $(\widetilde{V}, \mathcal{R}_{\widetilde{V}})$, $(\widetilde{U}, \mathcal{R}_{\widetilde{U}})$ be complexifications of V, U such that $\widetilde{V}$, $\widetilde{U}$ are abstract complex algebraic varieties.*

Then the natural injection id: $U \hookrightarrow V$ *can be extended to a morphism* $\widetilde{\mathrm{id}}\colon \widetilde{U}' \hookrightarrow \widetilde{V}$ *where $\widetilde{U}'$ is an open neighbourhood of U in $\widetilde{U}$, $\widetilde{\mathrm{id}}(\widetilde{U}')$ is open and $\widetilde{\mathrm{id}}\colon \widetilde{U}' \longrightarrow \widetilde{\mathrm{id}}(\widetilde{U}')$ is an isomorphism.*

Moreover two such extensions coincide where both are defined.

Proof. Let $\{\widetilde{V}_1, \dots, \widetilde{V}_q\}$ be an open covering of $\widetilde{U}$ by charts, i.e., there exists an isomorphism ϕ_i, defined over $\mathbb{R}$, between $\widetilde{V}_i$ and an affine set in $\mathbb{C}^{n_i}$, $i = 1, \dots, q$.

The covering $\{U_i = \widetilde{V}_i \cap U\}$ can be extended to a covering $\{U_1, \dots, U_q, \dots, U_s\}$ of V in such a way that U_i is the real part of a complex chart $\widetilde{U}_i$, isomorphic to an affine set of $\mathbb{C}^{n_i}$ by an isomorphism ρ_i which is defined over $\mathbb{R}$, $i = 1, \dots, s$. We may assume that $\widetilde{U}_1 \cup \dots \cup \widetilde{U}_s = \widetilde{V}$.

Now for any $i = 1, \dots, s$, U_i is a real affine variety and $\widetilde{U}_i$ is a complexification of it; for $i \leq q$, $\widetilde{V}_i$ is another complexification of U_i. So the identity map of U_i can be extended to a complex isomorphism

$$g_i\colon \widetilde{V}_i^0 \longrightarrow \widetilde{U}_i^0$$

between neighbourhoods $\widetilde{V}_i^0$ of U_i in $\widetilde{V}_i$ and $\widetilde{U}_i^0$ of U_i in $\widetilde{U}_i$. $\widetilde{V}_1^0 \cup \dots \cup \widetilde{V}_q^0$ is a neighbourhood of U in $\widetilde{U}$.

We have to prove that $g_1, \dots, g_q$ glue together in a map g, possibly in a smaller neighbourhood. This can be done by induction on q, using Lemma 5.10. The fact that the gluing g is an isomorphism comes from the fact that, by the same arguments as before, to the inverses of the identity maps of U_i, $i = 1, \dots, q$, we may associate the inverses of the maps g_i, hence we find a gluing which is the inverse of g. □

Corollary 5.12. *Let $(V, \mathcal{R}_V)$ be a real algebraic variety and $\widetilde{V}'$, $\widetilde{V}''$ be two complexifications of V that are complex varieties. Then the identity map* id: $V \longrightarrow V$ *can be extended to an isomorphism* $\widetilde{\mathrm{id}}\colon \widetilde{U}' \longrightarrow \widetilde{U}''$ *between two open sets $\widetilde{U}'$, $\widetilde{U}''$ of $\widetilde{V}'$, $\widetilde{V}''$ containing V.* □

6. A- and B-coherent sheaves

Let V be an affine variety, $V \subset \mathbb{K}^n$, and let $\mathcal{F}$ be a coherent sheaf of $\mathcal{R}_V$-modules. If $\mathbb{K}$ is algebraically closed, $\mathcal{F}$ admits a global resolution (Theorem A):

$$\mathcal{R}_V^m \longrightarrow \mathcal{R}_V^n \longrightarrow \mathcal{F} \longrightarrow 0.$$

Moreover for any $p > 0$, $H^p(V, \mathcal{F}) = 0$ (Theorem B), see [Se1].

Theorems A and B do not hold if $\mathbb{K}$ is not algebraically closed. Consider the following example.

Example 6.1. Let $P = x^2(x-1)^2 + y^2$ be the polynomial of Example 5.6, it vanishes only at $X_1 = (0, 0)$ and $X_2 = (1, 0)$.

Consider the line bundle F defined on the covering $\mathfrak{U} = \{U_1, U_2\}$ of $\mathbb{R}^2$,

$$U_1 = \mathbb{R}^2 - X_1, \quad U_2 = \mathbb{R}^2 - X_2,$$

by the cocycle

$$\frac{1}{P}\colon U_1 \cap U_2 \longrightarrow \mathbb{R}^*.$$

We have:

(1) The cocycle $\dfrac{1}{P}$ is not a coboundary, hence: $H^1(\mathbb{R}^2, \mathcal{R}_{\mathbb{R}^2}) \neq 0$.

(2) Consider the sheaf $\mathcal{F}$ of germs of algebraic sections of F; it is locally free and it does not admit a global resolution.

In fact, if it were a coboundary, we could write $\dfrac{1}{P} = h_1 - h_2$; $h_i \in \Gamma(U_i, \mathcal{R}_{\mathbb{R}^2})$, $i = 1, 2$; hence, by (3) of Section 5, we could extend h_1, h_2 to complex regular functions $\widetilde{h}_1$, $\widetilde{h}_2$ on suitable complex neighbourhoods of $\widetilde{U}_1$ and $\widetilde{U}_2$. In particular $\widetilde{h}_i$ would be defined on a neighbourhood of $\widetilde{X}_i$ in $\mathbb{C}^2$. So $\widetilde{h}_1$, $\widetilde{h}_2$, $\widetilde{h}_1 - \widetilde{h}_2$ would be defined on some non-empty open set of $\widetilde{S}$, $\widetilde{S}$ being the $\mathbb{R}$-irreducible complex curve defined by P. This is a contradiction, since $\widetilde{h}_1 - \widetilde{h}_2$ extends $\dfrac{1}{P}$.

So $\frac{1}{P}$ is a nonvanishing element in $H^1(\mathfrak{A}, \mathcal{R}_{\mathbb{R}^2})$ and this implies, by standard arguments, that $H^1(\mathbb{R}^2, \mathcal{R}_{\mathbb{R}^2}) \neq 0$ (see [Go]).

If $\mathcal{F}$ admitted a global resolution, in particular its stalk $\mathcal{F}_x$, for any x, would be generated by global sections; but it is easy to prove that any $\gamma \in \Gamma(\mathbb{R}^2, \mathcal{F})$ vanishes at X_1, while $\mathcal{F}_{X_1}$ is not zero, hence it is not generated by global sections.

Remark 6.2. Example 6.1 is based on the existence of a prime ideal, the ideal (P), whose zero set is reducible.

There is also another fact: $U_1 \cap U_2 = \mathbb{R}^2 - \{X_1, X_2\}$ has a complexification $\widetilde{U}_{1,2} = \mathbb{C}^2 - \widetilde{S}$ and we cannot write $\widetilde{U}_{1,2}$ as intersection of complexifications of U_1 and U_2. In fact any complex open set $\widetilde{U}_i$ such that $\widetilde{U}_i \cap \mathbb{R}^2 = U_i$, $i = 1, 2$, contains an open subset of $\widetilde{S}$.

Any time we can reproduce such a situation, we claim that Theorems A and B do not hold. In particular it is easy to prove that $\dim H^1(\mathbb{R}^n, \mathcal{R}_{\mathbb{R}^n}) = \infty$ if $n > 1$ and $H^1(\mathbb{R}, \mathcal{R}_{\mathbb{R}}) = 0$. The last assertion follows from the identification $\mathbb{R} \cong \operatorname{Spec} \Gamma_{\mathbb{R}}$.

Definition 6.3. A coherent sheaf $\mathcal{F}$ of $\mathcal{R}_V$-modules is called *A-coherent* if there exists a coherent sheaf $\mathfrak{F}$ of $\mathfrak{R}$-modules such that $\mathfrak{F}|_V \cong \mathcal{F}$.

Lemma 6.4. *Let $\mathcal{F}$ be a sheaf of $\mathcal{R}_V$-modules, then the following conditions are equivalent.*

(i) *$\mathcal{F}$ is A-coherent.*

(ii) *There exists an exact sequence*

$$\mathcal{R}_V^m \longrightarrow \mathcal{R}_V^q \longrightarrow \mathcal{F} \longrightarrow 0.$$

Proof. (i) $\Rightarrow$ (ii). Let $\mathfrak{F}$ be an extension of $\mathcal{F}$ to $\operatorname{Spec} \Gamma_V$.

It is known (see [Ha]) that there exists an exact sequence

$$\mathfrak{R}_V^m \longrightarrow \mathfrak{R}_V^q \longrightarrow \mathfrak{F} \longrightarrow 0.$$

If we restrict this sequence to V we obtain the sequence in (ii).

(ii) $\Rightarrow$ (i). The morphism $\alpha\colon \mathcal{R}_V^m \longrightarrow \mathcal{R}_V^q$ of (ii) is given by a matrix (α_{ij}) with entries in Γ_V. It gives also a morphism

$$\mathfrak{a}\colon \mathfrak{R}_V^m \longrightarrow \mathfrak{R}_V^q$$

and its cokernel is a coherent sheaf $\mathfrak{F}$ that extends $\mathcal{F}$ to $\operatorname{Spec} \Gamma_V$. □

Example 6.1 proves that, in general, a coherent sheaf is not A-coherent.

Lemma 6.5. *Let $\mathcal{F}$ be an A-coherent sheaf of $\mathcal{R}_V$-modules and $\mathfrak{F}_i$, $i = 1, 2$, two coherent extensions of $\mathcal{F}$ to* $\operatorname{Spec} \Gamma_V$. *Then there exists an isomorphism $\mathfrak{f}\colon \mathfrak{F}_1 \longrightarrow \mathfrak{F}_2$ extending the identity $\mathcal{F} \longrightarrow \mathcal{F}$. Moreover, $\mathfrak{f}$ is uniquely determined.*

Proof. $\mathfrak{F}_{1x}$ and $\mathfrak{F}_{2x}$ are isomorphic, because they are isomorphic to $\mathcal{F}_x$. The coherence of $\mathfrak{F}_i$ implies that one can extend, in just one way, this isomorphism to an isomorphism $\mathfrak{F}_1|_{U_x} \longrightarrow \mathfrak{F}_2|_{U_x}$, where U_x is a open neighbourhood of x in Spec Γ_V. This implies that there exists a unique isomorphism that extends the identity $\mathcal{F} \longrightarrow \mathcal{F}$ to a neighbourhood U of V in Spec Γ_V.

By (6) of Section 5 we have $U = \operatorname{Spec} \Gamma_V$ and the lemma is proved. □

Let V be an affine subvariety of $\mathbb{R}^n$ and $\mathcal{F}$ an A-coherent sheaf on V.

Definition 6.6. The A-coherent sheaf $\mathcal{F}$ is called *B-coherent* if any $\gamma \in \Gamma(V, \mathcal{F})$ can be extended to a section $\tilde{\gamma}$ of $\Gamma(\operatorname{Spec} \Gamma_V, \mathfrak{F})$.

Definition 6.6 is well-posed by Lemma 6.5.

Lemma 6.7. *Let $\mathcal{F}$ be an A-coherent sheaf on the algebraic variety V and $\rho: \Gamma(\operatorname{Spec} \Gamma_V, \mathfrak{F}) \longrightarrow \Gamma(V, \mathcal{F})$ be the restriction map. $\mathcal{F}$ is B-coherent if and only if ρ is bijective.*

Proof. Let us suppose $\mathcal{F}$ is B-coherent, then by hypothesis ρ is surjective. Let $\tilde{\gamma}, \tilde{\gamma}'$ be two extensions of γ; then $\tilde{\gamma}, \tilde{\gamma}'$ coincide on a neighbourhood of V in Spec Γ_V. This implies, by (6) of Section 5, $\tilde{\gamma} = \tilde{\gamma}'$ and hence ρ is bijective.

If ρ is bijective in particular it is surjective and $\mathcal{F}$ is B-coherent. □

Corollary 6.8. *The category of B-coherent sheaves on $(V, \mathcal{R}_V)$ is isomorphic to the category of Γ_V-modules of finite type.*

Proof. To any B-coherent sheaf $\mathcal{F}$ we associate the unique extension $\mathfrak{F}$ over Spec Γ_V (see Lemma 6.5). The restriction map $\rho: \Gamma(\operatorname{Spec} \Gamma_V, \mathfrak{F}) \longrightarrow \Gamma(V, \mathcal{F})$ is bijective by Lemma 6.7.

So the corollary is a consequence of the fact that the category of coherent sheaves on Spec Γ_V is isomorphic to the category of Γ_V-modules of finite type ([Ha]). □

Lemma 6.9. *If $\mathcal{F}_i$, $i = 1, 2$, are B-coherent sheaves, then* Hom $(\mathcal{F}_1, \mathcal{F}_2)$ *is B-coherent.*

Proof. Let $\mathfrak{F}_i$ be the extension of $\mathcal{F}_i$ to Spec Γ_V, then Hom $(\mathfrak{F}_1, \mathfrak{F}_2)$ extends Hom $(\mathcal{F}_1, \mathcal{F}_2)$ and this proves that Hom $(\mathcal{F}_1, \mathcal{F}_2)$ is A-coherent.

$\mathcal{F}_1$ is A-coherent, hence there exist sections $\gamma_1, \ldots, \gamma_q \in \Gamma(V, \mathcal{F}_1))$ such that they generate $\mathcal{F}_{1x}$, for any $x \in V$. Let $\tilde{\gamma}_i$ be the unique extension of γ_i, then $\widetilde{\gamma_1}, \ldots, \widetilde{\gamma_q}$ generate $\mathfrak{F}_{1x}$ on a neighbourhood of V in Spec Γ_V. By (6) of Section 5, they generate $\mathfrak{F}_{1x}$ for any $x \in \operatorname{Spec} \Gamma_V$. If α belongs to $\Gamma(V, \operatorname{Hom}(\mathcal{F}_1, \mathcal{F}_2))$, then α is determined by the sections $\delta_i = \alpha(\gamma_i)$, $i = 1, \ldots, q$. $\mathcal{F}_2$ is B-coherent, so for each i there exists a unique extension $\tilde{\delta}_i \in \Gamma(\operatorname{Spec} \Gamma_V, \mathfrak{F}_2)$ of δ_i and the homomorphism $\tilde{\alpha}$ defined by $\tilde{\alpha}(\tilde{\gamma}_i) = \tilde{\delta}_i$, $i = 1, \ldots, q$, is an extension of α. This proves that Hom $(\mathcal{F}_1, \mathcal{F}_2)$ is B-coherent. □

Lemma 6.10. *Let*

$$0 \longrightarrow \mathcal{F}' \longrightarrow \mathcal{F} \longrightarrow \mathcal{F}'' \longrightarrow 0 \qquad (*)$$

be an exact sequence of A-coherent sheaves of $\mathcal{R}_V$-modules. Let us suppose $\mathcal{F}'$, $\mathcal{F}$ are B-coherent.

Then the sequence

$$0 \longrightarrow \Gamma(V, \mathcal{F}') \longrightarrow \Gamma(V, \mathcal{F}) \longrightarrow \Gamma(V, \mathcal{F}'') \longrightarrow 0 \qquad (**)$$

is exact if, and only if, $\mathcal{F}''$ is B-coherent.

Proof. From $(*)$ we obtain the sequence

$$0 \longrightarrow \mathfrak{F}' \longrightarrow \mathfrak{F} \longrightarrow \mathfrak{F}'' \longrightarrow 0, \qquad (***)$$

where $\mathfrak{F}'$, $\mathfrak{F}$, $\mathfrak{F}''$ are the extensions to Spec Γ_V of $\mathcal{F}'$, $\mathcal{F}$, $\mathcal{F}''$. In fact α has an extension $\tilde{\alpha}\colon \mathfrak{F}' \longrightarrow \mathfrak{F}$, see 6.9, and hence $\mathfrak{F}'' = \mathfrak{F}/\tilde{\alpha}(\mathfrak{F}')$ extends $\mathcal{F}'' \cong \mathcal{F}/\alpha(\mathcal{F}')$.

The sequence $(***)$ is exact on a neighbourhood of V in Spec Γ_V and hence, by the usual argument, on Spec Γ_V.

Thus we obtain the following commutative diagram

$$\begin{array}{ccccccccc} 0 & \longrightarrow & \Gamma(\operatorname{Spec}\Gamma_V, \mathfrak{F}') & \longrightarrow & \Gamma(\operatorname{Spec}\Gamma_V, \mathfrak{F}) & \longrightarrow & \Gamma(\operatorname{Spec}\Gamma_V, \mathfrak{F}'') & \longrightarrow & 0 \\ & & \downarrow{\scriptstyle\rho'} & & \downarrow{\scriptstyle\rho} & & \downarrow{\scriptstyle\rho''} & & \\ 0 & \longrightarrow & \Gamma(V, \mathcal{F}') & \longrightarrow & \Gamma(V, \mathcal{F}) & \longrightarrow & \Gamma(V, \mathcal{F}'') & \longrightarrow & 0 \end{array}$$

where ρ', ρ are bijective because $\mathcal{F}'$, $\mathcal{F}$ are B-coherent. Clearly the bottom row of the diagram is exact if, and only if, ρ'' is bijective, that is, precisely where $\mathcal{F}''$ is B-coherent. □

Now we shall state some conditions that are easy to verify and that imply B-coherence.

Definition 6.11. Let $\mathfrak{F}$ be a coherent sheaf over Spec Γ_V. We shall say that $\mathfrak{F}$ *satisfies condition (β)* if

> for any open set $U \subset \operatorname{Spec}\Gamma_V$ and for any $\gamma \in \Gamma(U, \mathfrak{F})$ we have $\gamma_{|U\cap V} = 0 \Rightarrow \gamma = 0$.

Lemma 6.12. *Let $\mathcal{F}$ be an A-coherent sheaf on V and $\mathfrak{F}$ its extension to* Spec Γ_V. *If $\mathfrak{F}$ satisfies condition (β), then $\mathcal{F}$ is B-coherent.*

Proof. Let $\gamma \in \Gamma(V, \mathcal{F})$. The condition $\mathfrak{F}|_V = \mathcal{F}$ implies that there exist a covering $\{\tilde{U}_i\}_{i=1,\dots,q}$ of V by open sets , $\tilde{U}_i$ of Spec Γ_V and $\tilde{\gamma}_i \in \Gamma(\tilde{U}_i, \mathfrak{F})$ such that $\tilde{\gamma}_i\big|_{\tilde{U}_i\cap V} = \gamma|_{\tilde{U}_i\cap V}$.

The sections $\tilde{\gamma}_{ik} = \tilde{\gamma}_i - \tilde{\gamma}_k$ are defined on $\tilde{U}_i \cap \tilde{U}_k$ and clearly $\tilde{\gamma}_{ik}\big|_{\tilde{U}_i\cap\tilde{U}_k\cap V} = 0$.

Condition (β) assures that the $\tilde{\gamma}_i$ glue together and define an extension $\tilde{\gamma}$ of γ defined on an open set $\tilde{U}$ containing V. By (6) of Section 5, $\tilde{U} = \operatorname{Spec} \Gamma_V$ and the lemma is proved. □

Let $(V, \mathcal{R}_V)$ be an affine variety and consider the closed set in Spec Γ_V defined by the equations $P_1, \ldots, P_q$. Put

$$T = \{\alpha \in \operatorname{Spec} \Gamma_V \mid \alpha \ni P_1, \ldots, \alpha \ni P_q,\ P_1, \ldots, P_q \in \Gamma_V\}.$$

If $\tilde{V}$ is a complexification of V and $\bar{P}_1, \ldots, \bar{P}_q$ are extensions of $P_1, \ldots, P_q$ to a neighbourhood $\tilde{V}'$ of V in $\tilde{V}$ (see (3) of Section 5), we associate to T the closed set $\check{T} = \{x \in \tilde{V}' \mid \bar{P}_1(x) = \cdots = \bar{P}_q(x) = 0\}$. It is easy to verify that T determines $\check{T}$ as a germ near V, and any closed set $\check{S}$ of $\tilde{V}$ defined over $\mathbb{R}$ defines a closed subset of Spec Γ_V.

In fact if $P_1, \ldots, P_q$ are the generators of the ideal associated to $\check{S}$, then we associate to $\check{S}$ the closed set

$$S = \{\alpha \in \operatorname{Spec} \Gamma_V \mid \alpha \ni P_1, \ldots, \alpha \ni P_q, \}.$$

Definition 6.13. If $\tilde{V}$ is a complexification of V and $\check{T}$ is a closed set of $\tilde{V}$ defined over $\mathbb{R}$, we shall say that $\check{T}$ is $\mathbb{R}$-*exceptional* if for some $x \in V$ the germ $\check{T}_x$ properly contains the complexification of the germ $(\check{T} \cap V)_x$.

This means that for any open set U of $\tilde{V}$, $U \supset V$, we have

$$\check{T} \cap U \supsetneq \widetilde{(\check{T} \cap V \cap U)}.$$

Definition 6.14. A closed set T of Spec Γ_V is called $\mathbb{R}$-*exceptional* if the corresponding set $\check{T}$ of a complexification (and hence of all complexifications) is $\mathbb{R}$-exceptional.

Lemma 6.15. *Let $\mathfrak{F}$ be a coherent sheaf on* Spec Γ_V, *then the following conditions are equivalent.*

(1) *$\mathfrak{F}$ does not satisfy condition* (β).

(2) *There exists $\gamma \in \Gamma(\operatorname{Spec} \Gamma_V, \mathfrak{F})$ such that* supp γ *is $\mathbb{R}$-exceptional.*

(Recall that $\operatorname{supp} \gamma = \{\alpha \in \operatorname{Spec} \Gamma_V \mid \gamma(\alpha) \neq 0\}$.*)*

Proof. (1) $\Rightarrow$ (2). Let U be an open set of Spec Γ_V and $\gamma \in \Gamma(U, \mathfrak{F})$ such that $\gamma \neq 0$ and $\gamma_{|U \cap V} = 0$. There exists $P \in \Gamma_V$ such that

(a) $\{P = 0\} = \{\alpha \in \operatorname{Spec} \Gamma_V \mid \alpha \ni P\} \supset \operatorname{Spec} \Gamma_V - U$, and

(b) $\gamma_{|\operatorname{Spec} \Gamma_V - \{P=0\}} \neq 0$.

This follows from the fact that the sets of the form

$$\operatorname{Spec} \Gamma_V - \{Q = 0\}$$

form a basis of open sets for Spec Γ_V when Q varies in Γ_V. So U is the union of the sets $\{\text{Spec}\,\Gamma_V - \{P_i = 0\}\}$ and for some i, $\gamma_{|\text{Spec}\,\Gamma_V - \{P_i=0\}} \neq 0$.

It is known that there exists $n_0 \in \mathbb{N}$, such that if $n > n_0$, then $P^n\gamma$ is the restriction of a global section $\gamma' \in \Gamma(\text{Spec}\,\Gamma_V, \mathfrak{F})$ (see [Ha]).

Let T be the support of γ'. T is a closed set of Spec Γ_V and relations (a) and (b) imply that T is not contained in $\{P = 0\}$; but $T \cap V \subset \{P = 0\} \cap V$, because $\gamma_{|U\cap V} = 0$, and $\{P = 0\} \supset \text{Spec}\,\Gamma_V \setminus U$. This proves that T is $\mathbb{R}$-exceptional as desired.

(2) $\Rightarrow$ (1). Let $\gamma \in \Gamma(\text{Spec}\,\Gamma_V, \mathfrak{F})$ be a section such that its support T is $\mathbb{R}$-exceptional.

If we take $U = \text{Spec}\,\Gamma_V - \{\alpha \in \text{Spec}\,\Gamma_V | \alpha \in \widetilde{T \cap V}\}$ and $\gamma' = \gamma_{|U}$, we have $\gamma' \neq 0$ but $\gamma'|_{U\cap V} = 0$. So we have proved that $\mathfrak{F}$ doesn't satisfy condition (β). □

Corollary 6.16. *Let $(V, \mathcal{R}_V)$ be an affine variety, S a closed subvariety and $\mathcal{R}_S = \mathcal{R}_V/\mathcal{I}_S$ the structure sheaf of S. Then $\mathcal{R}_V$ and $\mathcal{R}_S$ are B-coherent.*

Proof. It is clear that $\mathcal{R}_V$ and $\mathcal{R}_S$ are A-coherent. Let $\mathfrak{R}_V$, $\mathfrak{R}_S$ be the extensions; then $\mathfrak{R}_V$, $\mathfrak{R}_S$ do not satisfy condition (2) of 6.15 and hence, by 6.12, $\mathcal{R}_V$ and $\mathcal{R}_S$ are B-coherent. □

Corollary 6.17. *Let $\mathcal{F}$ be a B-coherent sheaf on V and $\mathcal{F}'$ an A-coherent subsheaf, then $\mathcal{F}'$ is B-coherent.*

Proof. Let $\gamma_1, \ldots, \gamma_s \in \Gamma(V, \mathcal{F}')$ be sections generating $\mathcal{F}'$ on V and $\tilde{\gamma}_1, \ldots, \tilde{\gamma}_s$ their extensions to Spec Γ_V. $\widetilde{\gamma}_1, \ldots, \widetilde{\gamma}_s$ generate the extension $\mathfrak{F}'$ of $\mathcal{F}'$, and $\mathfrak{F}'$ is a subsheaf of the extension $\mathfrak{F}$ of $\mathcal{F}$.

Take $\gamma \in \Gamma(V, \mathcal{F}')$ and let $\tilde{\gamma}$ be the extension of γ to $\mathfrak{F}$; $\tilde{\gamma}$ has values in $\mathcal{F}'$ over V and hence on an open set U containing V.

By (6) of Section 5, $U = \text{Spec}\,\Gamma_V$ and the corollary is proved. □

Corollary 6.18. *Let $\mathcal{F}$ be a locally free A-coherent sheaf on V, then $\mathcal{F}$ is B-coherent.*

Proof. Locally, $\mathcal{F}$ is isomorphic to $\mathcal{R}_V^q$, hence the extension $\mathfrak{F}$ to Spec Γ_V of $\mathcal{F}$ does not satisfy condition (2) of 6.15. Hence $\mathcal{F}$ is B-coherent by 6.12. □

Corollary 6.19. *Let V be an affine variety and S a closed subvariety, then the restriction map $\rho: \Gamma(V, \mathcal{R}_V) \longrightarrow \Gamma(S, \mathcal{R}_S)$ is surjective.*

Proof. We have an exact sequence of sheaves

$$0 \longrightarrow \mathcal{I}_S \longrightarrow \mathcal{R}_V \longrightarrow \frac{\mathcal{R}_V}{\mathcal{I}_S} = \mathcal{R}_S \longrightarrow 0.$$

By Corollaries 6.16 and 6.17, the sheaves $\mathcal{I}_S, \mathcal{R}_V, \mathcal{R}_S$ are B-coherent, hence, by 6.10, the sequence

$$0 \longrightarrow \Gamma(V, \mathcal{I}_S) \longrightarrow \Gamma(V, \mathcal{R}_V) \longrightarrow \Gamma(V, \mathcal{R}_S) \longrightarrow 0$$

is exact and the claim is proved. □

Remark 6.20. Let $P = x^2(x-1)^2 + y^2$ and $\mathcal{I}_P$ be the ideal subsheaf of $\mathcal{R}_{\mathbb{R}^2}$ generated by P. We have an exact sequence

$$0 \longrightarrow \mathcal{I}_P \longrightarrow \mathcal{R}_{\mathbb{R}^2} \longrightarrow \frac{\mathcal{R}_{\mathbb{R}^2}}{\mathcal{I}_P} \longrightarrow 0.$$

Clearly the sheaf $\mathcal{F} = \mathcal{R}_{\mathbb{R}^2}/\mathcal{I}_P$ is A-coherent (see 6.4.) We claim that $\mathcal{F}$ is not B-coherent.

The support of $\mathcal{F}$ is reduced to the set $S = (0,0) \cup (1,0)$; $\mathcal{I}_P$ and $\mathcal{R}_{\mathbb{R}^2}$ are B-coherent (see 6.16, 6.17) so, by 6.10, $\mathcal{F}$ is B-coherent if, and only if, the sequence:

$$0 \longrightarrow \Gamma(\mathbb{R}^2, \mathcal{I}_P) \longrightarrow \Gamma(\mathbb{R}^2, \mathcal{R}_{\mathbb{R}^2}) \longrightarrow \Gamma(\mathbb{R}^2, \mathcal{F}) \longrightarrow 0 \qquad (\diamond)$$

is exact. It is easy to verify that $(\diamond)$ is not exact For example $\Gamma(\mathbb{R}^2, \mathcal{F})$ has zero divisors and $\Gamma(\mathbb{R}^2, \mathcal{I}_P)$ is a prime ideal, so we must have

$$\Gamma(\mathbb{R}^2, \mathcal{F}) \neq \frac{\Gamma(\mathbb{R}^2, \mathcal{R}_{\mathbb{R}^2})}{\Gamma(\mathbb{R}^2, \mathcal{I}_P)}.$$

If we take $Q = x^2 + y^2$ and the usual sequence

$$0 \longrightarrow \mathcal{I}_Q \longrightarrow \mathcal{R}_{\mathbb{R}^2} \longrightarrow \frac{\mathcal{R}_{\mathbb{R}^2}}{\mathcal{I}_Q} = \mathcal{F} \longrightarrow 0,$$

we see that the support of $\mathcal{F}$ is $(0,0)$ and hence the sequence

$$0 \longrightarrow \Gamma(\mathbb{R}^2, \mathcal{I}_Q) \longrightarrow \Gamma(\mathbb{R}^2, \mathcal{R}_{\mathbb{R}^2}) \longrightarrow \Gamma(\mathbb{R}^2, \mathcal{F}) \longrightarrow 0$$

is exact. This implies (see 6.10) that $\mathcal{F}$ is B-coherent. If $\mathfrak{F}$ is the extension of $\mathcal{F}$ on Spec $\Gamma_{\mathbb{R}^2}$, then it is easy to verify that $\mathfrak{F}$ satisfies condition (2) of 6.15, hence it does not satisfy condition (β).

We remark that if we replace the notion of complexification with that of completion (see [To11]) the definitions and the results of this section can be proved (in the same way) for affine varieties defined over any field $\mathbb{K}$.

7. The approximation theorems in the algebraic case

In this paragraph $\mathbb{R}^n$ shall be considered endowed with the euclidean topology.

Let $V \subset \mathbb{R}^n$ be a real affine variety. Denote by $\mathcal{O}_V$ the structure sheaf of V as a real analytic set.

Definition 7.1. Let $\widetilde{V}$ be a complexification of V as an affine variety. We say that V is *almost regular* if for each $x \in V$ the germ $\widetilde{V}_x$ coincides with the analytic complexification of the germ V_x.

For an A-coherent sheaf of $\mathcal{R}_V$-modules $\mathcal{F}$ denote by $\mathcal{F}^\circ$ the sheaf $\mathcal{F} \otimes_{\mathcal{R}_V} \mathcal{O}_V$.

Remark 7.2. If V is almost regular in particular it is coherent as an analytic space ([T7]); in this case $\mathcal{O}_V$ is a faithfully flat $\mathcal{R}_V$-module (see [Se2]).

If $U \subset V$ is an open set, we can endow $\Gamma(U, \mathcal{F})$ with a local system of seminorms by considering the usual weak topology on $\mathcal{R}(U)$ as in Examples (2) and (3) in § 1. Then we have:

Theorem 7.3. *Let V be a real affine variety and $Y \subset V$ be an almost regular algebraic subset. Let $\mathcal{F}$ be an A-coherent sheaf, then $\Gamma(U, \mathcal{F})$ is dense in $\Gamma(U, \mathcal{F}^\circ)$ in the weak topology.*

Moreover if $f \in \Gamma(V, \mathcal{F}^\circ)$ and $f|_Y = g \in \Gamma(Y, \mathcal{F}|_Y)$, then the set $\Gamma(V, \mathcal{F})_g$ of regular sections extending g is dense in the corresponding set $\Gamma(V, \mathcal{F}^\circ)_g$.

Proof. This is the same as the proofs of Theorems 3.1 and 3.3, using Stone–Weierstrass instead of the Whitney approximation theorem. Remark 7.2 and the fact that $\mathcal{F}$ is A-coherent give us the necessary ingredients to repeat the proofs. □

Now consider a sheaf homomorphism $\alpha: \mathcal{F} \to \mathcal{G}$ between two A-coherent sheaves of $\mathcal{R}_V$-modules. We can define $\alpha^\circ: \mathcal{F}^\circ \to \mathcal{G}^\circ$ by tensoring with the sheaf $\mathcal{O}_V$. By Remark 7.2 we have

$$\alpha^\circ(\mathcal{F}^\circ) \cap \mathcal{G} = \alpha(\mathcal{F}),$$

and hence the following result:

Theorem 7.4. *Let U be a Zariski open set in $\mathbb{R}^n$. Consider the linear system*

$$\sum_{k=1}^{q} a_{hk}(x) y_k = g_h, \quad h = 1, \ldots, p, \tag{$*$}$$

where $a_{hk}(x)$ and $g_h(x)$ are regular functions on U for $h = 1, \ldots, p$ and $k = 1, \ldots, q$. Then any differentiable solution $(f_1, \ldots, f_q)$ of $(*)$ can be approximated in the weak topology by a regular solution $(g_1, \ldots, g_q)$.

Moreover:

(1) *If for an almost regular affine variety $V \subset U$ we have $f_k|_V \in \Gamma(V, \mathcal{R}_V)$ for $k = 1, \ldots, q$, we can take $g_1, \ldots, g_q$ in such a way that $g_k|_V = f_k|_V$ for $k = 1, \ldots, q$.*

(2) *If the first $l < q$ components $f_1, \ldots, f_l$ are regular, then we can take $g_1 = f_1, \ldots, g_l = f_l$.*

Proof. By Theorem 3.7 we may suppose $(f_1, \ldots, f_q)$ is an analytic solution of $(*)$. As we remarked before, the matrix (a_{hk}) defines an exact sequence of coherent sheaves

$$\begin{array}{ccccccc} 0 & \longrightarrow & \ker\alpha & \longrightarrow & \mathcal{R}_U^q & \xrightarrow{\alpha} & \mathcal{R}_U^p \\ & & & & \downarrow & & \\ & & & & \operatorname{Im}\alpha & & \\ & & & & \downarrow & & \\ & & & & 0 & & \end{array}$$

where $\ker\alpha$ and $\operatorname{Im}\alpha$ are A-coherent because they admit a complexification. Consider the corresponding exact sequence obtained by applying $\otimes_{\mathcal{R}_U}\mathcal{O}$. We can apply Theorem 7.3 to $\ker\alpha$ and $\ker\alpha^\circ$. So any analytic solution of the homogeneous system has a regular approximation.

For the general case we can use the fact that $\alpha^\circ(\mathcal{F}^\circ) \cap \mathcal{G} = \alpha(\mathcal{F})$, and we have surjectivity for sections because $\mathcal{R}_U^p$ is B-coherent and so also $\operatorname{Im}\alpha$ is B-coherent, being a subsheaf of $\mathcal{R}_U^p$.

So if $(f_1, \ldots, f_q)$ is an analytic solution of $(*)$ we have $(g_1, \ldots, g_p) = \alpha^\circ(f_1, \ldots, f_q)$ and hence $(g_1, \ldots, g_p) = \alpha(h_1, \ldots, h_q)$ with $h_1, \ldots, h_q \in \Gamma(U, \mathcal{R}_U^q)$, because $\mathcal{O}_V$ is faithfully flat on $\mathcal{R}_V$ (see Remark 7.2).

So $(f_1 - h_1, \ldots, f_q - h_q) \in \ker\alpha^\circ$ and we conclude as before.

If $f_1, \ldots, f_l$ are regular, then $(f_{l+1}, \ldots, f_q)$ is a solution of the system

$$\sum_{k=l+1}^{q} a_{hk} y_k = g_h - \sum_{k=1}^{l} a_{hk} f_k, \quad h = 1, \ldots, p, \tag{$**$}$$

and this has a regular approximation. This proves that (2) may be satisfied.

For assertion (1) we can use a more direct argument instead of repeating the proof of Theorem 3.7.

Suppose $f_k|_V$ to be regular for each k and let F_k be a regular function on U which extends $f_k|_V$. Then $f_k - F_k$ is an analytic function vanishing on V. Since V is almost regular, if $p_1, \ldots, p_\nu$ are generators for the ideal $I(V) \subset \mathcal{R}_U$, we can write

$$f_k - F_k = \sum_{j=1}^{\nu} \beta_{jk} p_j, \quad k = 1, \ldots, q.$$

Then by applying α° to the vector $(f_1 - F_1, \ldots, f_q - F_q))$ we find that the set $\{\beta_{jk}\}_{\substack{j=1,\ldots,\nu \\ k=1,\ldots,q}}$ is an analytic solution of the system

$$\sum_{k=1}^{q}\sum_{j=1}^{\nu} a_{hk} p_j \beta_{jk} = g_h - \sum_{k=1}^{q} a_{hk} F_k \qquad h = 1, \ldots, p. \tag{$***$}$$

Hence the β_{jk} can be approximated in the weak topology by regular functions b_{jk} such that the set $\{b_{jk}\}$ is a regular solution of $(***)$. Consider for $k = 1, \ldots, q$ the regular function $G_k = \sum_{j=1}^{\nu} b_{jk} p_j$. It vanishes on V , approximates $f_k - F_k$ and by construction

$$\sum_{k=1}^{q} a_{hk} G_k = g_h - \sum_{k=1}^{q} a_{hk} F_k, \quad h = 1, \ldots, p$$

so $(G_1 + F_1, \ldots, G_q + F_q)$ is the required approximation of $\left(f_1, \ldots, f_q\right)$. □

III. Algebraic and analytic bundles

8. Duality theory

The purpose of this section is to state the duality between the category of B-coherent sheaves and a subcategory of the category of (not necessarily locally trivial) vector bundles over a real affine variey V. The whole category of bundles corresponds to the category of A-coherent sheaves, but in this case the correspondence depends on the resolution of the sheaf.

In this section the base field $\mathbb{K}$ will be always $\mathbb{R}$ or $\mathbb{C}$, even though almost all the results can be stated and proved, essentially with the same proofs, for any field.

Let $(V, \mathcal{R}_V)$ be an affine variety defined over $\mathbb{K}$ and $\Gamma_V = \Gamma(V, \mathcal{R}_V)$ be the ring of regular functions of V. Let

$$\alpha = (\alpha_{ik})_{\substack{i=1,\ldots,n \\ k=1,\ldots,q}}$$

be a matrix of elements of Γ_V and $\alpha \colon V \times \mathbb{K}^q \longrightarrow V \times \mathbb{K}^n$ be the map defined by

$$\alpha(x, t_1, \ldots, t_q) = (x, \sum_j \alpha_{1j}(x) t_j, \ldots, \sum_j \alpha_{nj}(x) t_j).$$

Definition 8.1. The set

$$F = \{(x, t) \in V \times \mathbb{K}^q \mid \alpha(x, t) = (x, 0)\} = \ker \alpha$$

is called *a linear algebraic subbundle* of $V \times \mathbb{K}^q$ or *a linear bundle.*

If $\pi_F \colon F \longrightarrow V$ is the restriction of the first projection then, for any $x \in V, \pi_F^{-1}(x)$ is a linear subspace of $\mathbb{K}^q$ and

$$\dim_{\mathbb{K}} \pi_F^{-1}(x) = q - \operatorname{rank} \alpha(x).$$

The function $q - \operatorname{rank} \alpha(x)$ is upper semicontinuous but, in general, not locally constant. It follows that, in general, F is not a (locally trivial) vector bundle.

Definition 8.2. Let F, G be two linear bundles on V. An algebraic map $\phi: F \longrightarrow G$ is called a *morphism (of linear bundles)* if

(1) the following diagram is commutative

$$\begin{array}{ccc} F & \xrightarrow{\phi} & G \\ \downarrow{\scriptstyle \pi_F} & & \downarrow{\scriptstyle \pi_G} \\ V & \xrightarrow{\text{id}} & V \end{array}$$

(2) ϕ is linear on each fiber.

ϕ is called an *isomorphism* if it is bijective and ϕ^{-1} is a morphism.

We shall denote by $\mathcal{L}(V)$ the category of linear bundles F on V, such that $F = \ker \alpha$ for some $\alpha: V \times \mathbb{K}^q \longrightarrow V \times \mathbb{K}^n$.

Definition 8.3. Let $\pi: F \to V$ be a map of abstract varieties defined over $\mathbb{K}$. We shall say that F is *an abstract linear bundle* if there exist an open covering $\{U_i\}_{i=1,\dots,p}$ of V, a family of linear bundles $F_i \in \mathcal{L}(U_i)$ and algebraic isomorphisms

$$\phi_i: \pi^{-1}(U_i) \longrightarrow F_i$$

such that, if $U_i \cap U_j \neq \emptyset$, $\phi_i \circ \phi_j^{-1}$ is an isomorphisms of linear bundles.

We remark that the fibers of F have a natural structure of vector space.

Definition 8.4. Let F, G be two abstract linear bundles over the abstract variety V. An algebraic map $\phi: F \longrightarrow G$ is called a *morphism (of abstract linear bundles)* if $\pi_G \circ \phi = \pi_F$ and it is linear on the fibers.

Again ϕ is an *isomorphism* if it is bijective and ϕ^{-1} is a morphism

We shall denote by $\mathcal{L}_a(V)$ the category of abstract linear bundles over V.

The next proposition shows that the notion of linear bundle over an affine variety reduces to the usual one when the dimension of its fibers is constant.

Proposition 8.5. *Let F be a linear subbundle of $V \times \mathbb{K}^q$ defined by the equations*

$$\sum_j \alpha_{ij}(x) t_j = 0,\ i = 1, \dots, n, \tag{$*$}$$

where $t_1, \dots, t_q$ are coordinates of $\mathbb{K}^q$.

Let us suppose

$$\operatorname{rank}(\alpha_{ij}(x)) = r \quad \textit{for all } x \in V.$$

Then for any $x^0 \in V$ there exists a neighbourhood U_{x^0} such that the pair $\big(U_{x^0} \times \mathbb{K}^q, (U_{x^0} \times \mathbb{K}^q) \cap F\big)$ is isomorphic to the pair $\big(U_{x^0} \times \mathbb{K}^q, U_{x^0} \times \mathbb{K}^{q-r}\big)$.

Proof. We can assume that the minor α' formed by the first r rows and columns of the matrix α has rank r at the point x^0; hence α' has rank r at any point in a neighbourhood U_{x^0} of x^0. This means that on U_{x^0} we can remove from $(*)$ the last $n-r$ equations. Consider the matrix α'^{-1}. Its entries are regular functions on U_{x^0}. So we can consider the automomorphism of $U_{x^0} \times \mathbb{K}^q$ defined by $\phi(x,t) = (x, \phi_x(t))$, where ϕ_x is the invertible matrix

$$\begin{pmatrix} \alpha'^{-1} & 0 \\ 0 & I_{q-r} \end{pmatrix}.$$

It is a straightforward verification that the system $(*)$ in the new coordinates becomes

$$t_i + \sum_{j>r} \beta_{ij}(x) t_j = 0, \qquad i = 1, \dots, r; \tag{$**$}$$

hence its kernel is isomorphic to $U_{x^0} \times \mathbb{K}^{q-r}$. So the pair $\left(U_{x^0} \times \mathbb{K}^q, F|_{U_{x^0}}\right)$ is isomorphic to the pair $\left(U_{x^0} \times \mathbb{K}^q, U_{x^0} \times \mathbb{K}^{q-r}\right)$. □

A linear bundle $F = \ker \alpha \subset V \times \mathbb{K}^q$ is clearly an affine subvariety of $V \times \mathbb{K}^q$. The next theorem proves the converse under the hypothesis that $\pi^{-1}(x) \cap F$ is a linear subspace of $\mathbb{K}^q$.

Theorem 8.6. *Let F be an affine subvariety of $V \times \mathbb{K}^q$. Assume that, for each $x \in V$, $\pi^{-1}(x) \cap F$ is a vector subspace of $\{x\} \times \mathbb{K}^q$, where $\pi: V \times \mathbb{K}^q \longrightarrow V$ is the natural projection. Then F is a linear bundle on V.*

Proof. We have to find regular functions $\alpha_1, \dots, \alpha_n \in \Gamma_{V \times \mathbb{K}^q}$ such that

(1) for any i, α_i is linear in the coordinates $t_1, \dots, t_q$ of $\mathbb{K}^q$, and

(2) $F = \{x \in V \times \mathbb{K}^q \mid \alpha_1(x) = \cdots = \alpha_n(x) = 0\}$.

Fix an embedding $V \hookrightarrow \mathbb{K}^l$. Let $u = (u_1, \dots, u_l), t = (t_1, \dots, t_q)$ be coordinates in $\mathbb{K}^l$, $\mathbb{K}^q$ and consider the set $\mathcal{I}_F^{\mathcal{L}}$ of polynomials $P \in \mathbb{K}[u,t]$ of the form

$$P(u,t) = P_1(u)t_1 + \cdots + P_q(u)t_q$$

and belonging to the ideal $\mathcal{I}_F$ of polynomials vanishing on F

The set $\mathcal{I}_F^{\mathcal{L}}$ generates an ideal in $\mathbb{K}[u,t]$ whose zero set we denote by $F_{\mathcal{L}}$.

We can write

$$F_{\mathcal{L}} = \{(u,t) \in \mathbb{K}^l \times \mathbb{K}^q \mid \beta_1(u,t) = \cdots = \beta_p(u,t) = 0\}$$

with $\beta_1, \dots, \beta_p \in \mathcal{I}_F^{\mathcal{L}}$.

So $F_{\mathcal{L}}$ is a linear bundle, it is an affine variety and clearly contains F. We wish to prove $F_{\mathcal{L}} = F$.

If $F_{\mathcal{L}}$ properly contains F, then at least one irreducible component $F_{\mathcal{L}}^i$ of $F_{\mathcal{L}}$ does not lie in F, and so it satisfies

$$\dim F_{\mathcal{L}}^i > \dim(F_{\mathcal{L}}^i \cap F).$$

Let $F'_{\mathcal{L}}$ be the union of the irreducible components $F^h_{\mathcal{L}}$, $h \neq i$, of $F_{\mathcal{L}}$. $F'_{\mathcal{L}}$ is closed, hence $F^i_{\mathcal{L}} - F'_{\mathcal{L}}$ is open non empty.

The coefficients of $\beta_1(u,t), \dots, \beta_p(u,t)$, as linear functions on $\mathbb{K}^q$, give a matrix $\beta(u)$. We consider the sequence of closed varieties

$$V_i = \{x \in V \mid \operatorname{rank} \beta(x) \leq i\}.$$

Clearly, for $i \geq 0$, $V_i \subseteq V_{i+1}$ and β has constant rank on $V_{i+1} - V_i$.

By Proposition 8.5, $F_{\mathcal{L}}|_{V_{i+1}-V_i}$ is a locally trivial vector bundle.

Let now j be the last index such that

$$W = (F_{\mathcal{L}}|_{V_j - V_{j-1}} \cap F^i_{\mathcal{L}}) - F$$

is not empty and contains an open set of $F^i_{\mathcal{L}}$. In particular we can find a regular point of $F^i_{\mathcal{L}}$ belonging to W. Call it x_0 and consider the fibers F_{x_0}, $F_{\mathcal{L}x_0}$. They are not equal, hence there is a line in $F_{\mathcal{L}x_0}$ which is not contained in F_{x_0}; we can assume it is the t_1-axis and that x_0 is the point $(0,0) \in \mathbb{K}^l \times \mathbb{K}^q$. We can construct a polynomial R with the following properties:

— R vanishes on F, but does not vanish on the t_1-axis,

— $\dfrac{\partial R}{\partial t_1}(0,0) \neq 0$.

We can decompose R into a sum of homogeneous polynomials R_i of degree i in the variables $t_1, \dots, t_q$. The assumptions on R imply in particular that R_1 is not the zero polynomial. By a standard argument, from $R|_F = 0$ we deduce $R_i|_F = 0$, since if $(u_0, t_0) \in F$ then $(u_0, \lambda t_0) \in F$ for any $\lambda \in \mathbb{K}$.

In particular R_1 vanishes on F and this contradicts the hypothesis: the t_1-axis is not contained in F and $\dfrac{\partial R}{\partial t_1}(0,0) \neq 0$. So we have proved $F = F_{\mathcal{L}}$. □

Corollary 8.7. *A linear bundle $F \subset V \times \mathbb{K}^n$ is a locally trivial algebraic vector bundle on V if it is an affine subvariety of $V \times \mathbb{K}^n$ and the dimension of the fibers is locally constant.*

Let $F \subset V \times \mathbb{K}^n$, $G \subset V \times \mathbb{K}^q$ be two elements of $\mathcal{L}(V)$ and $\phi \colon F \longrightarrow G$ a morphism.

In next theorem we prove that ϕ extends to a neighbourhood of F in $V \times \mathbb{K}^n$. This extension property is the main technical result to state the duality theory.

Theorem 8.8. *For any $x \in V$ there exists an open set $U_x \ni x$ and a morphism $U_x \times \mathbb{K}^n \longrightarrow U_x \times \mathbb{K}^q$ that extends $\phi \colon \pi_F^{-1}(U_x) \longrightarrow \pi_G^{-1}(U_x)$.*

To prove the above theorem we need several lemmas and definitions.

Definition 8.9. Let $F \subset V \times \mathbb{K}^q$ be an element of $\mathcal{L}(V)$, $x \in V$ and $n = \dim \pi_F^{-1}(x)$. A *tangent space to F at x* is a trivial bundle $U_x \times \mathbb{K}^n$ on a neighbourhood of x toghether with an injective bundle morphism $\pi_F^{-1}(U_x) \hookrightarrow U_x \times \mathbb{K}^n$.

The above definition makes sense because $\dim \pi_F^{-1}(y) \leq \dim \pi_F^{-1}(x)$ in a neighbourhood of x.

Lemma 8.10. *Let $F \subset V \times \mathbb{K}^q$ be an element of $\mathcal{L}(V)$. Then for any $x_0 \in V$ there exists a tangent space to F at x_0.*

Proof. Let $\alpha(x) = \{\alpha_{ij}(x)\}$ be the matrix of regular functions which defines F as a subbundle of $V \times \mathbb{K}^q$. Assume $\dim \pi_F^{-1}(x_0) = p$. Then

$$\operatorname{rank}(\alpha(x_0)) = q - p.$$

Take $q - p$ rows of α which are linearly independent at x_0. They form a matrix α' of rank $q - p$ on a neighbourhood of x_0; so α' defines a locally trivial bundle F' which is tangent to F at x_0 □

Lemma 8.11. *Let F, G be two elements of $\mathcal{L}(V)$, $F \subset V \times \mathbb{K}^n$, $G \subset V \times \mathbb{K}^q$ and $\phi: F \longrightarrow G$ a morphism.*

Let us suppose

(1) *$V \times \mathbb{K}^n$ is tangent to F at $x \in V$,*

(2) *$\phi\big|_{\pi_F^{-1}(x)}$ is injective, and*

(3) *there exists a morphism $\hat{\phi}: V \times \mathbb{K}^n \longrightarrow V \times \mathbb{K}^q$ that extends ϕ.*

Then there exists a neighbourhood U_x of x such that $\hat{\phi}\big|_{U_x \times \mathbb{K}^n}$ is an isomorphism onto the image.

Proof. $\hat{\phi}$ is a morphism of usual vector bundles and it is injective on the fiber over x, hence it is injective on a neighbourhood $U_x \times \mathbb{K}^n$. An injective bundle morphism is an isomorphism onto the image. □

Lemma 8.12. *Let F and G be two elements of $\mathcal{L}(V)$, $F \subset V \times \mathbb{K}^n$, $G \subset V \times \mathbb{K}^q$ and $\phi: F \longrightarrow G$ a morphism. Assume $V \times \mathbb{K}^n$ is tangent to F at x.*

Then there exists a neighbourhood U_x of x such that ϕ extends to a morphism $\hat{\phi}: U_x \times \mathbb{K}^n \longrightarrow U_x \times \mathbb{K}^q$

Proof. Consider the set

$$\Upsilon_\phi = \{(x, t, \tau) \in V \times \mathbb{K}^n \times \mathbb{K}^q \mid (x, t) \in F,\ \phi(x, t) = (x, \tau)\}.$$

It is a linear subbundle of $V \times \mathbb{K}^n \times \mathbb{K}^q$ by Theorem 8.6.

Denote by π_n and π_q the projections of $V \times \mathbb{K}^n \times \mathbb{K}^q$ on $V \times \mathbb{K}^n$, resp. on $V \times \mathbb{K}^q$. Then $\pi_n|_{\Upsilon_\phi}$ is an isomorphism between Υ_ϕ and F; let $\omega: F \longrightarrow \Upsilon_\phi$ be its inverse. Then we have:

— π_q extends $\phi \circ \omega^{-1} = \phi \circ \pi_n|_{\Upsilon_\phi}: \omega(F) \longrightarrow G$,

— π_n extends $\pi_n|_{\Upsilon_\phi}: \Upsilon_\phi \longrightarrow F$.

Now ω is injective on the fiber over x, hence by Lemma 8.11 it extends to an injective morphism $\widehat{\omega}: U_x \times \mathbb{K}^n \longrightarrow U_x \times \mathbb{K}^n \times \mathbb{K}^q$. Consider the composition

$$\pi_q \circ \widehat{\omega}: U_x \times \mathbb{K}^n \longrightarrow U_x \times \mathbb{K}^q.$$

It is the required extension of ϕ; in fact $\widehat{\omega}$ is the inverse of π_n when restricted to a tangent space to Υ_ϕ. □

Proof of Theorem 8.8. Lemma 8.12 shows that ϕ extends on a neighbourhood of x to a suitable tangent space to F at x. Any tangent space to F is locally a direct factor of $V \times \mathbb{K}^n$, so locally, ϕ extends. □

We are ready to define the duality. Let $(V, \mathcal{R}_V)$ be an affine variety over $\mathbb{K}$ and Γ_V be the ring of regular functions.

Consider a $(q \times n)$-matrix $\alpha = (\alpha_{ij})$ and a $(n \times q)$-matrix $\beta = (\beta_{ij})$ both with entries in Γ_V; they induce, in a natural way, two maps which we shall denote respectively by α and β:

$$\alpha: V \times \mathbb{K}^n \longrightarrow V \times \mathbb{K}^q \quad \text{given by } \alpha(x, t_1, \dots, t_n) = (x, \sum_{j=1}^{n} \alpha_{1j} t_j, \dots, \sum_{j=1}^{n} \alpha_{qj} t_j)$$

$$\beta: \mathcal{R}_V^q \longrightarrow \mathcal{R}_V^n \quad \text{given by } \beta(\sigma_1, \dots, \sigma_q) = (\sum_{j=1}^{q} \beta_{1j}\sigma_j, \dots, \sum_{j=1}^{q} \beta_{nj}\sigma_j).$$

So α determines a linear bundle $F = \ker \alpha$ and β determines an A-coherent sheaf $\mathcal{F} = \operatorname{coker} \beta$.

Definition 8.13. The pairs (F, α), $(\mathcal{F}, \beta)$ are called *dual* if

(1) $F = \ker \alpha, \quad \mathcal{F} = \operatorname{coker} \beta$

(2) $\beta = {}^t\alpha$.

Let F be a linear bundle on V; we denote by $\mathcal{D}(F)$ the sheaf associated to the presheaf $U \longrightarrow \operatorname{Hom}(F|_U, U \times K)$.

Proposition 8.14. *Let F be in $\mathcal{L}(V)$ and let $\alpha: V \times \mathbb{K}^n \longrightarrow V \times \mathbb{K}^q$ be such that $F = \ker \alpha$. Then the dual sheaf $(\mathcal{F}, {}^t\alpha)$ of (F, α) is canonically isomorphic to $\mathcal{D}(F)$.*

Proof. By hypothesis there is an exact sequence

$$0 \longrightarrow F \longrightarrow V \times \mathbb{K}^n \longrightarrow V \times \mathbb{K}^q.$$

For any open set $U \subset V$, it yields

$$0 \longrightarrow F|_U \longrightarrow U \times \mathbb{K}^n \longrightarrow U \times \mathbb{K}^q.$$

If we apply the functor Hom $(-, U \times \mathbb{K})$ we obtain

$$\mathrm{Hom}\,(U \times \mathbb{K}^q, U \times \mathbb{K}) \xrightarrow{{}^t\alpha} \mathrm{Hom}\,(U \times \mathbb{K}^n, U \times \mathbb{K}) \xrightarrow{\beta} \mathrm{Hom}\,(F|_U, U \times \mathbb{K}).$$

The standard properties of the functor Hom $(-, -)$ and Theorem 8.8 prove that the last sequence defines the following exact sequence of sheaves:

$$\mathcal{R}_V^{q*} \xrightarrow{{}^t\alpha} \mathcal{R}_V^{n*} \longrightarrow \mathcal{D}(F) \longrightarrow 0,$$

where $\mathcal{R}_V^{j*} = \mathrm{Hom}\,(\mathcal{R}_V^j, \mathcal{R}_V)$.

$\mathcal{F}$ is defined by the exact sequence

$$\mathcal{R}_V^q \xrightarrow{{}^t\alpha} \mathcal{R}_V^n \longrightarrow \mathcal{F} \longrightarrow 0$$

and $\mathcal{R}_V^j \cong \mathcal{R}_V^{j*}$; so $\mathcal{F} \cong \mathcal{D}(F)$. □

Corollary 8.15. *Let* $F = \ker\alpha$ *be in* $\mathcal{L}(V)$*; then the dual sheaf* $(\mathcal{F}, {}^t\alpha)$ *does not depend on the resolution* α *of* F.

Proof. By Proposition 8.14 $(\mathcal{F}, {}^t\alpha)$ is isomorphic to $\mathcal{D}(F)$. □

Lemma 8.16. *Let*

$$\mathcal{R}_V^n \xrightarrow{\alpha} \mathcal{R}_V^q \longrightarrow \mathcal{F} \longrightarrow 0$$

$$\mathcal{R}_V^{n'} \xrightarrow{\alpha'} \mathcal{R}_V^{q'} \longrightarrow \mathcal{F}' \longrightarrow 0$$

be two resolutions of the A-coherent sheaves $\mathcal{F}$, $\mathcal{F}'$ *and* $\phi\colon \mathcal{F} \longrightarrow \mathcal{F}'$ *a morphism.*

If $\mathcal{F}'$ *is B-coherent there exist morphisms* β_1, β_2 *such that the following diagram is commutative:*

$$\begin{array}{ccccccc}
\mathcal{R}_V^n & \xrightarrow{\alpha} & \mathcal{R}_V^q & \xrightarrow{\pi} & \mathcal{F} & \longrightarrow & 0 \\
\downarrow{\scriptstyle \beta_2} & & \downarrow{\scriptstyle \beta_1} & & \downarrow{\scriptstyle \phi} & & \\
\mathcal{R}_V^{n'} & \xrightarrow{\alpha'} & \mathcal{R}_V^{q'} & \xrightarrow{\pi'} & \mathcal{F}' & \longrightarrow & 0.
\end{array}$$

Proof. Let $\{\sigma_i\}, i = 1, \dots, q$, be the canonical basis of $\Gamma(V, \mathcal{R}_V^q) = (\Gamma_V)^q$ and $\delta_i = (\phi \circ \pi)(\sigma_i) \in \Gamma(V, \mathcal{F}')$; $\mathcal{F}'$ is B-coherent, hence there exists $\eta_i \in \Gamma(V, \mathcal{R}_V^{q'})$

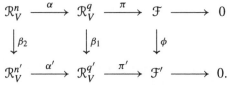

such that $\pi'(\eta_i) = \delta_i, i = 1, \dots, q$. We may define for each i

$$\beta_1(\sigma_i) = \eta_i.$$

$\beta_1(\mathcal{R}^q_V)$ is an A-coherent subsheaf of $\mathcal{R}^{q'}_V$, hence it is B-coherent.

So we may repeat the above argument and define β_2. □

Corollary 8.17. *Let $\phi: \mathcal{F} \longrightarrow \mathcal{F}'$ be a morphism of A-coherent sheaves. If $\mathcal{F}'$ is B-coherent then for any resolution α of $\mathcal{F}$, resp. α' of $\mathcal{F}'$, ϕ induces a dual morphism $\phi^*: F' \longrightarrow F$, where F, F' are the dual linear bundles.*

Proof. If we dualize the diagram in Lemma 8.16 we obtain:

$$\begin{array}{ccccccc} 0 & \longrightarrow & \ker^t \alpha & \longrightarrow & V \times \mathbb{K}^q & \longrightarrow & V \times \mathbb{K}^n \\ & & \uparrow{\scriptstyle \phi''} & & \uparrow{\scriptstyle {}^t\beta_1} & & \uparrow{\scriptstyle {}^t\beta_2} \\ 0 & \longrightarrow & \ker^t \alpha' & \longrightarrow & V \times \mathbb{K}^{q'} & \longrightarrow & V \times \mathbb{K}^{n'} \end{array}$$

and the sought-after morphism ϕ^* is the restriction of ${}^t\beta_1$. □

Corollary 8.18. *Let $\mathcal{F}$ be a B-coherent sheaf and F, F' be two dual linear bundles, associated to different resolutions of $\mathcal{F}$, then F is isomorphic to F'.*

Proof. The identity map id: $\mathcal{F} \longrightarrow \mathcal{F}$ defines, by Corollary 8.17, an invertible morphism id*: $F \longrightarrow F'$.

Hence F is isomorphic to F'. □

We can summarize the previous results in the following:

Theorem 8.19. *The duality between B-coherent sheaves and corresponding linear bundles is independent of the resolution.*

If $\mathcal{F}$, $\mathcal{G}$ are B-coherent sheaves, to any morphism $\phi: \mathcal{F} \longrightarrow \mathcal{G}$, a dual morphism $\phi^*: G \longrightarrow F$ is associated, and conversely to any morphism $\psi: G \longrightarrow F$, a dual morphism $\psi^*: \mathcal{F} \longrightarrow \mathcal{G}$ is associated.

Moreover $\phi^{**} = \phi$ and $\psi^{**} = \psi$.

Proof. The fact that the duality is independent of the resolution is proved in 8.15 and 8.18. Any $\phi: \mathcal{F} \longrightarrow \mathcal{G}$ induces $\phi^*: G \longrightarrow F$ by 8.17. Let $\psi: G \longrightarrow F$ be a morphism, then we can define

$$\psi^*_U: \operatorname{Hom}(F|_U, U \times \mathbb{K}) \longrightarrow \operatorname{Hom}(G|_U, U \times \mathbb{K})$$

by $\psi^*_U(\gamma)(\eta) = \gamma(\psi(\eta))$, $\gamma \in \operatorname{Hom}(F|_U, U \times \mathbb{K})$.

ψ^*_U defines $\psi^*: \mathcal{D}(F) \longrightarrow \mathcal{D}(G)$ and hence, by 8.14, a morphism $\psi^*: \mathcal{F} \longrightarrow \mathcal{G}$.

The relations $\phi^{**} = \phi$ and $\psi^{**} = \psi$ are a direct consequence of the identity ${}^t({}^t\alpha) = \alpha$. □

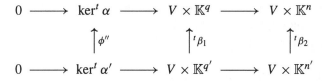

The above theorem shows that the duality between B-coherent sheaves $\mathcal{F}$ on V and the corresponding elements in $\mathcal{L}(V)$ is well-defined.

It should be possible to define a duality between the category of coherent sheaves that are locally B-coherent (defined over any abstract variety V) and abstract linear bundles.

9. Strongly algebraic vector bundles

In this section we shall study some properties of real algebraic vector bundles. After the first definitions we shall recall some basic constructions and results from the classical theory of fiber bundles, for which we refer to [St].

Let V be an affine variety over $\mathbb{K}$, $\mathbb{K} = \mathbb{R}$ or $\mathbb{K} = \mathbb{C}$.

Definition 9.1. An abstract linear bundle (see 8.3) F on V, is called *an algebraic vector bundle* if the dimension of its fibers is locally constant on V.

An algebraic vector bundle F on V is called *strongly algebraic* if it is a linear bundle, i.e., a subbundle of $V \times \mathbb{K}^n$ for some n. In particular, by 8.5, it is locally trivial.

Here *locally* means *locally with respect to the Zariski topology*. Nevertheless, if V is compact in the euclidean topology, then the notion of local triviality is the same in both topologies.

Remark 9.2. Let F be an algebraic vector bundle on the affine variety V. We call *the dual sheaf of* F the sheaf $\mathcal{D}(F)$ associated to the presheaf

$$\mathcal{D}(F)(U) = \mathrm{Hom}(F|_U, U \times \mathbb{K}).$$

Note that if F is also a linear bundle the dual sheaf $\mathcal{F}$ as defined in Section 8 is canonically isomorphic to $\mathcal{D}(F)$.

Denote by $\mathcal{S}(F)$ the sheaf of germs of algebraic sections of F. The following proposition gives some relations among these sheaves.

Proposition 9.3. *Let F be an algebraic vector bundle on V, then the sheaf $\mathcal{S}(F)$ is locally isomorphic to $\mathcal{D}(F)$.*

Proof. F is locally trivial, i.e., for each $x \in V$ there exists an open neighbourhood U_x of x such that $F|_{U_x}$ is isomorphic to $U_x \times \mathbb{K}^n$, with $n = \dim \pi_F^{-1}(x)$. So $\mathcal{D}(F)$ is locally isomorphic to

$$\mathrm{Hom}\,(U_x \times \mathbb{K}^n, U_x \times \mathbb{K}) = \mathrm{Hom}\,(\mathcal{R}_V^n, \mathcal{R})\big|_{U_x}.$$

$\mathrm{Hom}\,(\mathcal{R}^n_V, \mathcal{R}) = (\mathcal{R}^n_V)^*$ is isomorphic to $\mathcal{R}^n_V$, via the canonical basis $\{\delta_i\}_{i=1,\dots,n}$ of $\mathcal{R}^n_V$, where δ_i is the n-tuple of regular functions $\{(0,\dots.\underset{\text{place } i}{0,1,0},\dots,0)\}$. But also $\mathcal{S}(F)(U) = \Gamma(U, F)$ is isomorphic to $\mathcal{R}^n_V(U)$ if $F|_U \cong U \times \mathbb{K}^n$. □

Remark 9.4. We cannot find in general a global isomorphism between $\mathcal{S}(F)$ and $\mathcal{D}(F)$.

Let us now recall some facts from the theory of *topological vector bundles*. For details we refer to [St].

Definition 9.5. A *topological vector bundle* is a triple (E, π, X) where E, X are topological spaces, $\pi: E \longrightarrow X$ is a surjective continuous map and

(1) for any $x \in X$, $\pi^{-1}(x)$ has the structure of a finite dimensional $\mathbb{K}$-vector space,

(2) π is locally equivalent to the projection $U_x \times \mathbb{K}^n \longrightarrow U_x$.

A *morphism* of vector bundles $\phi: E_1 \longrightarrow E_2$ is a continuous map such that $\pi_{E_2} \circ \phi = \pi_{E_1}$ and $\phi\big|_{\pi^{-1}_{E_1}(x)}: \pi^{-1}_{E_1}(x) \longrightarrow \pi^{-1}_{E_2}(x)$ is linear for each $x \in X$.

If $\phi: X \longrightarrow Y$ is a continous map and (E, π, Y) is a vector bundle, the *pull-back* $\phi^*(E)$ on X is defined as the vector bundle

$$\phi^*(E) = \{(x, y) \in X \times E \mid \phi(x) = \pi_E(y)\}$$

with the natural projection on X.

Moreover we have a continous map $\phi^*: \phi^*(E) \longrightarrow E$, induced by the projection $X \times E \longrightarrow E$, such that ϕ^* is an isomorphism on each fiber and the following diagram commutes

$$\begin{array}{ccc} \phi^*(E) & \xrightarrow{\phi^*} & E \\ \downarrow & & \downarrow{\scriptstyle \pi} \\ X & \xrightarrow{\phi} & Y. \end{array}$$

Of course if ϕ is an algebraic map between affine algebraic varieties and E is algebraic then also $\phi^*(E)$ is an algebraic vector bundle.

Consider now the Grassmann manifold $G_{n,q}$ of q-linear subspaces of $\mathbb{R}^n$. $G_{n,q}$ carries the structure of a projective (hence affine) real variety, which can briefly be defined as follows. Linear subspaces L of the euclidean space $\mathbb{R}^n$ correspond one-to-one to linear orthogonal projections $\mathbb{R}^n \longrightarrow L$, i.e., to the associated matrices A_L. It is easy to see that $n \times n$ matrices associated to such projections are characterized by the following conditions:

$$A^2 = A \qquad \mathrm{tr}\, A = q.$$

These equations define $G_{n,q}$ as an affine subvariety of $\mathbb{R}^{n^2}$; this variety is homogeneous, hence nonsingular.

Definition 9.6. The vector bundle

$$\gamma_{n,q} = \{(x, y) \in G_{n,q} \times \mathbb{R}^n \mid x \ni y\}$$

is called the *tautological vector bundle.*

We shall use the following results.

Theorem 9.7. *$(\gamma_{n,q}, \pi, G_{n,q})$ is a strongly algebraic vector bundle.*

Proof. $\gamma_{n,q}$ has fiber of constant dimension q, and it is an affine variety by definition. □

Theorem 9.8. *Let X be a metric space of dimension n.*

There exists $N \in \mathbb{N}$ such that for any topological vector bundle (F, π, X) there exists a continuous map $\varphi_F\colon X \longrightarrow G_{N,q}$ such that we have:

(1) *F is isomorphic to $\varphi_F^*(\gamma_{N,q})$.*

(2) *If $\varphi^*(\gamma_{N,q})$ is isomorphic to $\psi^*(\gamma_{N,q})$ then φ is homotopic to ψ.*

Proof. See [St]. □

In other words Theorem 9.8 says that vector bundles on X are classified by homotopy classes of maps $X \longrightarrow G_{N,q}$.

Consider now a real affine variety V and an algebraic vector bundle F on V. The following theorem, which is the main result of this section, characterizes the strongly algebraic vector bundles.

Theorem 9.9. *Let V be a real affine variety and F be an algebraic vector bundle on V. The following conditions are equivalent:*

(1) *There exists a regular rational map $\varphi\colon V \longrightarrow G_{N,q}$ such that F is algebraically isomorphic to $\varphi^*(\gamma_{N,q})$.*

(2) *F is strongly algebraic.*

(3) *There exists a strongly algebraic vector bundle F' on V that $F \oplus F'$ is algebraically isomorphic to $V \times \mathbb{R}^n$.*

(4) *The dual sheaf $\mathcal{F}$ of F is A-coherent.*

(5) *The dual sheaf $\mathcal{F}$ of F is B-coherent.*

The proof of the theorem is a consequence of the following lemmas.

Lemma 9.10. *Let V, W be two real affine varieties, F a strongly algebraic vector bundle on W and $\varphi: V \longrightarrow W$ an algebraic map.*

Then $\varphi^*(F)$ is a strongly algebraic vector bundle on V.

Proof. By hypothesis F is a linear subbundle of some $W \times \mathbb{R}^n$.

It is easy to verify that $\varphi^*(W \times \mathbb{R}^n) = V \times \mathbb{R}^n$ and hence $\varphi^*(F)$ is a linear subbundle of $V \times \mathbb{R}^n$. □

Let us now consider on $\mathbb{R}^n$ the usual scalar product. For any vector subbundle F of $V \times \mathbb{R}^n$ we can define

$$F^\perp = \{(x, y) \in V \times \mathbb{R}^n \mid y \perp \pi^{-1}(x)\}.$$

Lemma 9.11. *Let (F, π_F, V) be an algebraic subbundle of $V \times \mathbb{R}^n$, then $(F^\perp, \pi', V)$ is a strongly algebraic vector subbundle of $V \times \mathbb{R}^n$.*

Proof. We have to prove that $F^\perp$ is an algebraic vector subbundle of $V \times \mathbb{R}^n$. It is clear that $\dim \pi'^{-1}(x)$ is locally constant, hence the claim follows from Proposition 8.5 and Theorem 8.6 if we verify that $F^\perp$ is a linear algebraic subbundle of $V \times \mathbb{R}^n$.

By Proposition 8.5, locally, the pair $(V \times \mathbb{R}^n,\ F)$ is algebraically isomorphic to $(U \times \mathbb{R}^n,\ U \times \mathbb{R}^q)$. Hence for any $x_0 \in V$ there exists a neighbourhood U_{x_0} and algebraic sections $\sigma_1, \dots, \sigma_q$ such that $\sigma_1(x), \dots, \sigma_q(x)$ is a basis of $\pi_F^{-1}(x)$ if $x \in U_{x_0}$. Hence there exist $\sigma_{q+1}, \dots, \sigma_n \in \Gamma(U_{x_0}, \mathcal{R}_V^n)$ such that $\sigma_1(x), \dots, \sigma_n(x)$ is a basis of $\mathbb{R}^n$, $x \in U_{x_0}$.

Let $\sigma_1', \dots, \sigma_n'$ be an orthogonal basis constructed from $\sigma_1, \dots, \sigma_n$ by the Gram–Schmidt process. Clearly $\sigma_1', \dots, \sigma_n'$ are algebraic and $\sigma_{q+1}', \dots, \sigma_n'$ generate $F^\perp|_{U_{x_0}}$.

This proves that $F^\perp$ is algebraic. □

Proof of Theorem 9.9.

(1)⇒(2) Since $\gamma_{N,q}$ is strongly algebraic; this follows from Lemma 9.10.

(2)⇒(3) This is a consequence of Lemma 9.11.

(3)⇒(1) In fact if $F \subset V \times \mathbb{K}^N$, the map $\rho: V \longrightarrow G_{N,q}$ defined by $\rho(x) = \pi_F^{-1}(x)$ is algebraic and F is isomorphic to $\rho^*(\gamma_{N,q})$.

(2)⇒(4) This is a consequence of the duality.

(4)⇒(2) If $\mathcal{D}(F)$ is A-coherent, it has a resolution

$$\mathcal{R}_V^n \longrightarrow \mathcal{R}_V^q \longrightarrow \mathcal{D}(F) \longrightarrow 0.$$

From Lemma 6.5, we deduce that $\mathcal{D}(F)$ is also B-coherent, because it is locally free. By duality we obtain the sequence

$$0 \longrightarrow F \longrightarrow V \times \mathbb{K}^q \longrightarrow V \times \mathbb{K}^n.$$

This proves that F is strongly algebraic.

(4)⇔(5) This follows from Corollary 6.18, since $\mathcal{F}$ is locally free.

The theorem is proved. □

Corollary 9.12. *Let F be an algebraic vector bundle on a real affine variety. F is strongly algebraic if and only if its sheaf of germs of algebraic sections $\mathcal{S}(F)$ is B-coherent.*

Proof. If F is strongly algebraic, then it has a resolution

$$0 \longrightarrow F \longrightarrow V \times \mathbb{K}^q \xrightarrow{\alpha} V \times \mathbb{K}^n,$$

and $F = \ker \alpha$. The dual sheaf in this case is $\mathcal{F} = \operatorname{coker}{}^t\alpha$ given by

$$\mathcal{R}_V^n \xrightarrow{{}^t\alpha} \mathcal{R}_V^q \longrightarrow \mathcal{F} \longrightarrow 0,$$

and it is B-coherent by Theorem 9.9. The sheaf $\mathcal{F}^* = \operatorname{Hom}(\mathcal{F}, \mathcal{R}_V)$ is B-coherent by Lemma 6.9. By construction $\mathcal{F}^*$ is isomorphic to $\mathcal{S}(F)$.

Conversely if $\mathcal{S}(F)$ is B-coherent, by the same argument as before, we get an isomorphism between $\operatorname{Hom}(\mathcal{S}(F), \mathcal{R})$ and $\mathcal{D}(F)$. Hence $\mathcal{D}(F)$ is B-coherent and F is strongly algebraic. □

Remark 9.13. The notion of strongly algebraic vector bundle can be given for any field $\mathbb{K}$.

Theorem 9.9 can be partially restated in the general situation and (2) ⇔ (4) ⇔ (5) are true, with the same proof.

Finally we want to study the complexifications of real algebraic vector bundles and their properties.

Theorem 9.14. *Let (F, π, V) be a real algebraic vector bundle, where V and F are affine varieties.*

Then there exist an affine complexification $\widetilde{V}$ of V and a complex algebraic vector bundle $(\widetilde{F}, \widetilde{\pi}, \widetilde{V})$ such that:

(1) *$\widetilde{F}$ is a complexification of F,*

(2) *$\widetilde{F}$ is an affine variety, and*

(3) *$\widetilde{F}|_V \cong F \otimes_{\mathbb{R}} \mathbb{C}$.*

Proof. Let $\widetilde{V}'$, $\widetilde{F}'$ be affine complexifications of the real affine varieties V, F and $\widetilde{\pi}'\colon \widetilde{F}' \longrightarrow \widetilde{V}'$ an extension of π.

F is locally trivial, so we can find an open covering $\{U_1, \ldots, U_q\}$ of V and for each i an isomorphism

$$\rho_i\colon F|_{U_i} \longrightarrow U_i \times \mathbb{R}^n.$$

Clearly we may suppose there exist open sets $\widetilde{U}_i$ of $\widetilde{V}'$ such that $\widetilde{U}_i \cap V = U_i$. Consider now the maps

$$\rho'_{ik} = \rho_k \circ \rho_i^{-1} : (U_i \cap U_k) \times \mathbb{R}^n \longrightarrow (U_i \cap U_k) \times \mathbb{R}^n$$
$$\rho'_{ik}(x, t) = (x, g_{ik}(x) \cdot t),$$

where $g_{ik}(x) \in GL(n, \mathbb{R})$ is the cocycle of the vector bundle F.

From (3) of §5 there exist algebraic extensions $\widetilde{g}_{ik}: \widetilde{U}_{ik} \longrightarrow GL(n, \mathbb{C})$ of g_{ik}, where $\widetilde{U}_{ik}$ is a neighbourhood of $U_{ik} = U_i \cap U_k$ in $\widetilde{V}'$. Moreover we may suppose that, where they are defined, these maps satisfy the cocycle condition

$$\widetilde{g}_{ik} \cdot \widetilde{g}_{kj} = \widetilde{g}_{ij}.$$

If we can prove that there exist open neighbourhoods $\widetilde{U}'_i$ of U_i in $\widetilde{V}'$ such that $\widetilde{U}'_i \cap \widetilde{U}'_k \subset \widetilde{U}_{ik}$ then the theorem is proved because the $\widetilde{g}_{ik}$ define the desired complexification on the covering $\{\widetilde{U}'_i\}$.

Now we wish to prove the existence of such $\widetilde{U}'_i$; the covering is finite, so we may consider only two open sets U_i and U_k.

Consider the set

$$T = \{x \in \widetilde{U}_i \cup \widetilde{U}_k \mid \text{ the algebraic map } g_{ik}: U_i \cap U_k \longrightarrow GL(n, \mathbb{C}) \text{ cannot be extended at } x\}$$

We have two possibilities:

(1) In a neighbourhood $\widetilde{U}$ of $U_i \cup U_k$ in $\widetilde{V}'$, T is contained in the union of two closed sets T_i, T_k such that $U_i \cap T_k = U_k \cap T_i = \emptyset$. In this case the open set $\widetilde{U}'_i = \widetilde{U} - T_k$, $\widetilde{U}'_k = \widetilde{U} - T_i$ solves our problem.

(2) Condition (1) is not satisfied.

Let now recall the following facts:

(a) $\widetilde{F}'$ is a complexification of F and it is an affine variety. From (3) of §5 it follows that, perhaps taking a smaller neighbourhood $\widetilde{U}''_i$ of U_i in $\widetilde{V}'$, we have an embedding $\widetilde{h}_i: (\widetilde{U}''_i \times \mathbb{C}^n) - Z_i \hookrightarrow \widetilde{F}'$, where Z_i is a closed set such that $Z_i \cap (\widetilde{U}''_i \times \mathbb{R}^n) = \emptyset$.

(b) Clearly the complex algebraic maps $\widetilde{h}_k^{-1} \circ \widetilde{h}_i$ are the gluing functions that define $\widetilde{F}'$ and hence they satisfy the cocycle relations. Moreover they are linear on the fibers, because this is true on the real part and $\widetilde{F}'$ is a complexification of F. More precisely the maps $\widetilde{h}_k^{-1} \circ \widetilde{h}_i$ are determined by $\widetilde{h}_k^{-1} \circ \widetilde{h}_i\big|_{U_i \times \mathbb{R}^n}$ and $\widetilde{h}_k^{-1} \circ \widetilde{h}_i\big|_{U_i \times \mathbb{R}^n}$, and $\rho_k \circ \rho_i^{-1}$ coincide where both are defined.

(c) The validity of property (1) does not depend on the chosen trivialization of F.

(d) Any open set $\widetilde{U}''_i \times \mathbb{C}^n - Z_i$ has the property that for any $y \in \widetilde{U}''_i$ the set $\widetilde{\pi}_i^{-1}(y) \cap ((\widetilde{U}''_i \times \mathbb{C}^n) - Z_i)$ contains a basis of the fiber $\widetilde{\pi}_i^{-1}(y)$, where $\widetilde{\pi}_i: \widetilde{U}''_i \times \mathbb{C}^n \longrightarrow \widetilde{U}''_i$ is the projection.

Finally we note that from (b) and (d), it follows that $\widetilde{h}_k \circ \widetilde{h}_i^{-1}$ are isomorphisms on the fibers and this proves that (2) cannot happen.

The theorem is now proved. □

As a corollary we have an improvement of Corollary 8.7.

Corollary 9.15. *Let (F, π, V) be a real algebraic vector bundle such that V, F are affine varieties, then F is strongly algebraic.*

Proof. Let $(\widetilde{F}, \widetilde{\pi}, \widetilde{V})$ be a complexification of F. If γ is a global section of $\widetilde{F}$, we can decompose locally it as $\gamma = \gamma' + i\gamma''$. But $\widetilde{F}$ is given by a cocycle $\{\widetilde{g}_{ij}\}$ which is defined over $\mathbb{R}$, hence globally $\gamma = \gamma' + i\gamma''$, with $\gamma'|_V, \gamma''|_V$ global sections of F.

This proves that any fiber of F is generated by global sections, since the same is true for $\widetilde{F}$. This means that the dual sheaf $\mathcal{D}(F)$ is A-coherent and so, by Theorem 9.9, F is strongly algebraic. □

This corollary is essentially proved in [MR].

Remark 9.16. The above results and the existence of vector bundles that are not strongly algebraic prove that there exist real algebraic varieties that are not affine. Moreover let F be an algebraic vector bundle over an affine variety V. If F is not strongly algebraic, then any complexification $\widetilde{F}$ of F is not a complex algebraic variety.

10. Approximation for sections of vector bundles

Before giving some consequences of the theorems in Sections 3 and 4, we give the definitions of generalized analytic vector bundle, as we did for the algebraic case in Section 8. We refer to [F1], [F2] and [P] for the complex case and to [Ct] for the real analytic one.

Let $\mathbb{K}$ be the field $\mathbb{C}$ or $\mathbb{R}$, and $(X, \mathcal{O}_X)$ be an analytic set in $\mathbb{K}^m$. If $\mathbb{K} = \mathbb{R}$, assume X to be coherent.

Given a matrix $\alpha(x) = \big(a_{ij}(x)\big)_{\substack{i=1,\dots,q\\ j=1,\dots,p}}$ with entries in $\Gamma(X, \mathcal{O}_X)$, we can think of it as a map

$$\begin{aligned} X \times \mathbb{K}^p &\longrightarrow X \times \mathbb{K}^q \\ (x\ ,\ t\ ) &\longrightarrow (x\ ,\ \alpha(x)t). \end{aligned}$$

Definition 10.1. A *linear analytic bundle* is the set

$$F = \ker \alpha = \big\{(x, t) \in X \times \mathbb{K}^p \,|\, t \in \ker \alpha(x)\big\}.$$

Let π_F be the projection $F \to X$. A *morphism* of linear analytic bundles $F \subset X \times \mathbb{K}^p$, $G \subset X \times \mathbb{K}^r$ is an analytic map $\varphi: F \to G$ such that:

(1) The diagram

$$\begin{array}{ccc} F & \xrightarrow{\varphi} & G \\ {\scriptstyle \pi_F}\downarrow & & \downarrow{\scriptstyle \pi_G} \\ X & \xrightarrow{\text{id}} & X \end{array}$$

commutes.

(2) For any $x \in X$, $\varphi_{|\pi_F^{-1}(x)}: \pi_F^{-1}(x) \to \pi_G^{-1}(x)$ is linear.

Denote by $\mathcal{L}(X)$ the category of linear analytic bundles over X.

More generally, we could define an abstract notion of analytic bundle as a triple (F, π, X) locally isomorphic to a bundle as in Definition 10.1.

Remark 10.2. If $\alpha(x)$ has constant rank we have the usual notion of locally trivial vector bundle.

A matrix $\big(b_{ij}(x)\big)$ with entries in $\Gamma(X, \mathcal{O}_X)$ defines also a morphism $\beta: \mathcal{O}_X^p \to \mathcal{O}_X^q$. Its cokernel $\mathcal{F}$ is a coherent sheaf over X.

Proposition 10.3. *There is a duality associating to each linear analytic bundle* $F = \ker \alpha$ *the coherent sheaf* $\mathcal{F} = \operatorname{coker}^t \alpha$. *If* $\mathcal{D}(F)$ *is the sheaf associated to the presheaf*

$$X \supset U \longrightarrow \operatorname{Hom}(F|_U, U \times \mathbb{K})$$

then $\mathcal{D}(F)$ *is canonically isomorphic to* $\operatorname{coker}^t \alpha$

Now we can apply the results of Sections 3 and 4.

In the following by analytic (C^∞) sections of F we shall mean sections of $\mathcal{D}(F)$ ($\mathcal{D}(F)^\infty$).

Let Y be a coherent analytic subset of X. We can consider smooth sections of $\mathcal{D}(F)$ vanishing on Y; again by Malgrange's theorem these are precisely the sections in the image of $\mathcal{I}^\infty \otimes \mathcal{D}(F)$. Thus we have:

Theorem 10.4. *Let X be a coherent real analytic space, Y a coherent subspace and F a linear analytic bundle; let σ be a smooth section of F which is analytic on Y.*

Then in each neighbourhood of σ for the Whitney topology, one can find an analytic section s such that $s_{|Y} = \sigma_{|Y}$.

Theorem 10.5. *Let X be a compact real affine variety, Y an almost regular subvariety and F a linear algebraic bundle; let σ be a smooth section of F which is regular on Y.*

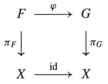

Then in each neighbourhood of σ in the weak topology, one can find a regular section s such that $s_{|Y} = \sigma_{|Y}$.

Remark 10.6. Theorems 10.4 and 10.5 are generalizations of the results about approximations of sections in a locally trivial analytic or algebraic vector bundle (see [BT1], [BT2], [BT3]).

Finally, by duality, Corollary 4.8 gives a classification theorem for linear analytic bundles, which extends the classical results for the complex case (see [G1], [G2], [G3]) and for the real case (see [T1], [T2], [T3]).

In the case of strongly algebraic vector bundles, all the sheaves involved are B-coherent, in particular Hom $(\mathcal{F}, \mathcal{G})$ is B-coherent for $\mathcal{F}$, $\mathcal{G}$ B-coherent.

So Theorems 4.6, 4.7 and Corollary 4.8 hold true when X is a real affine compact variety and $\mathcal{F}$, $\mathcal{G}$ are B-coherent sheaves of $\mathcal{R}_X$-modules.

Hence we have:

Theorem 10.7. *Let F, G be linear analytic (algebraic) bundles over the coherent set (compact affine variety) $X \subset \mathbb{R}^n$. Let $\varphi\colon F \to G$ be a smooth isomorphism, i.e., there is a commutative diagram*

$$\begin{array}{ccc} F & \xrightarrow{\varphi} & G \\ {\scriptstyle \pi_F}\big\downarrow & & \big\downarrow{\scriptstyle \pi_G} \\ X & \xrightarrow{\text{id}} & X \end{array}$$

where φ is bijective, C^∞, invertible and linear on the fibers. Then there exists $f\colon F \to G$ which is an analytic (algebraic) bundle isomorphism and is arbitrarily close to φ.

References

[ABT] F. Acquistapace, F. Broglia, A. Tognoli, Smooth and analytic solutions for analytic linear systems, to appear in Rev. Mat. Univ. Complut. Madrid.

[BT1] R. Benedetti, A. Tognoli, Teoremi di approssimazione in topologia differenziale I, Boll. U.M.I. 14–B (1977), 866–887.

[BT2] R. Benedetti, A. Tognoli, Approximation Theorems in Real Algebraic Geometry, Supplemento Boll. U.M.I Algebra e Geometria 2 (1980), 209–228.

[BT3] R. Benedetti, A. Tognoli, On real algebraic vector bundles, Bull. Sci. Math. 104 (1980), 89–112.

[BKS] J. Bochnak, W. Kucharz, M. Shiota, On equivalence of ideals of real global analytic functions and the 17th Hilbert problem, Invent. Math. 63 (1981), 403–421.

[Ca] H. Cartan, Varietés analytiques réelles et varietés analytiques complexes, Bull. Soc. Math. France 85 (1957), 77–99.

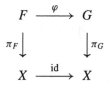

[Ct] F. Catanese, Un teorema di dualità per i fasci coerenti su uno spazio analitico reale, Ann. Sc. Norm. Sup. Classe di Scienze XXVII fasc IV (1973), 845–871.

[F1] G. Fischer, Eine Charakterisierung von holomorphen Vektorraumbündeln, Bayer. Akad. d. Wiss. Math.-Natur. Kl. S. B. 1966, 101–107.

[F2] G. Fischer, Lineare Faserräume und kohärente Modulgarben über komplexen Räumen, Arch. Math. 18 (1967), 609–617.

[Go] R. Godement, Théorie des faisceaux, Hermann, Paris 1958.

[Gr1] H. Grauert, Approximationssätze für holomorphe Funktionen mit Werten in komplexen Räumen, Math. Ann. 133 (1957), 139–159.

[Gr2] H. Grauert, Holomorphe Funktionen mit Werten in komplexen Lieschen Gruppen, Math. Ann. 133 (1957), 450–472.

[Gr3] H. Grauert, Analytische Faserungen über holomorph-vollständingen Räumen, Math. Ann. 135 (1958), 263–273.

[GaRo] R. C. Gunning, H. Rossi, Analytic functions of several complex variables, Prentice-Hall, London 1965.

[Ha] R. Hartshorne, Algebraic Geometry, Springer-Verlag, New York 1977.

[M] B. Malgrange, Ideals of Differentiable Functions, Tata Institute of Fundamental Research, Bombay, Oxford University Press, Oxford 1966.

[MR] M. G. Marinari, M. Raimondo, Fibrati vettoriali su varietà algebriche definite su corpi non algebricamente chiusi, Bull. U.M.I. (5) 16-A (1979), 128–136.

[N] R. Narasimhan, Analysis on Real and Complex Manifolds, Masson & Cie, Paris; North-Holland, Amsterdam 1968.

[P] D. Prill, Über lineare Faserräume und schwach negative holomorphe Geradenbündel Math. Z. 105 (1968), 313–326.

[Se1] J. P. Serre, Faisceaux Algébriques Coherents, Ann. of Math. (2) 61 (1955), 197–278, 1–42

[Se2] J. P. Serre, Géométrie Algébrique et Géométrie Analytique, Ann. Ist. Fourier IV (1955–56), 1–42.

[St] N. Steenrod, The topology of fibre bundles, Princeton University Press 1951.

[T1] A. Tognoli, Sulla classificazione dei fibrati analitici reali, Ann. Scuola Norm. Sup. XXI (1967), 709–744.

[T2] A. Tognoli, L'analogo del teorema delle matrici olomorfe invertibili nel caso analitico reale, Ann. Scuola Norm. Sup. XXII (1968), 537–558.

[T3] A. Tognoli, Sulla classificazione dei fibrati analitici E-principali, Ann. Scuola Norm. Sup. XXIII (1969), 75–86.

[T4] A. Tognoli, Teoremi di approssimazione per gli spazi analitici reali coerenti non immergibili in $\mathbb{R}^n$, Ann. Scuola Norm. Sup. XXV (1971), 507–516.

[T5] A. Tognoli, Su una congettura di Nash, Ann. Scuola Sup. di Pisa XXVII (1973), 167–185.

[T6] A. Tognoli, Un teorema di approssimazione relativo, Atti Acc. Naz. Lincei Ser. 8 LIV (1973), 496–502.

[T7] A. Tognoli, Algebraic approximation of manifolds and spaces, in: Sém. Bourbaki 1979/80 Exposé 548, 73–94, Lectures Notes in Math. 842 (1981).

[T8] A. Tognoli, Problèmes d'approximation pour espaces analytiques réels, Ann. Univ. Ferrara sez. VII Sci. Mat. XXVIII (1982), 55–66.

[T9] A. Tognoli, Coherent Sheaves and Algebraic Geometry, Ann. Univ. Ferrara sez. VII Sci. Mat. XXXI (1985), 99–123.

[T10] A. Tognoli, Proprietà globali degli spazi analitici reali, Ann. Mat. Pura e Appl. (IV) LXXV (1967), 143–218.

[T11] A. Tognoli, Introduzione alla teoria degli spazi analitici reali, Accademia Nazionale dei Lincei Roma 1976.

[To] J. C. Tougeron, Ideaux de fonctions differentiables, Springer-Verlag, Berlin–Heidelberg–New York 1972.

[W] H. Whitney, Analytic extensions of differentiable functions defined in closed sets, Trans. Amer. Math. Soc. 36 (1934), 63–89.

Dipartimento di Matematica, Università di Trento, v. Sommarive 14, 38050 Povo, Trento, Italy

Real abelian varieties and real algebraic curves

Ciro Ciliberto and Claudio Pedrini

Introduction

The main object of this paper is the theory of real abelian varieties with applications to curves. Most of the theory presented here is contained in two papers by Comessatti (see [C2]) which appeared in 1924–1926. In the modern literature they have been rediscovered or reworked by various authors, among which we mention here first of all M. Raimondo, who first pointed out in his preprint [Ra] Comessatti's important contribution, then Silhol and Gross–Harris (see the list of references).

Whenever possible we have followed Comessatti's ideas as close as possible, in view of their strength and simplicity. Indeed it seems to us that Comessatti's viewpoint, consisting in applying to real algebraic geometry ideas and techniques coming from the complex projective case, has been extremely fruitful and can still lead to further applications. Furthermore it is a stimulating source of new ideas and open questions.

In particular we prove Theorem 3.1.4 which describes the structure of the group of fixed points on a complex torus with a real structure, using a fundamental result (see Theorem 2.5.3) by Comessatti, on the existence of a normal basis for the action of an involutory automorphism on a free $\mathbb{Z}$-module. This result, which leads to the definition of the so-called *Comessatti character* of a real structure, is fundamental also in the classification of real rational surfaces, as given by Comessatti himself (for modern references, see [S1], [CP]).

The theory of real abelian varieties is related with important areas of research, which however we do not treat in this paper, such as the so-called real Hodge problem, and the problem of determining the number of connected components of real abelian varieties with complex multiplication (see [H] also for a list of references). A still open question is to analyse the image of the Torelli mapping (see Theorem 4.4.2) for the moduli space of real curves of genus g, in the case $g \geq 3$. Here we describe, following Comessatti, the case $g = 2$ (see §4).

Because of the expository character of this paper, we tried to make it self-contained. Therefore we put one preliminary chapter (Chapter 1) where, with no aim to completeness, and sometimes without proofs, we list general facts about complex tori.

In Chapter 2 we prove several results on real structures on complex varieties, including Harnack's theorem, which will be used later. Moreover we introduce the

Comessatti character and relate it to group cohomology. These results will be applied in Chapters 3 and 4.

Chapter 4 is devoted to the applications to real curves. After recalling some classical results on complex curves and their Jacobians, we relate the algebraic invariants of a real structure on a compact Riemann surface, with topological invariants of the real locus. Most of the results given here go back to Comessatti, if not to older authors as Klein (see [K]) and Weichold (see [Wh]). A modern approach to the same results is contained in a paper by Gross and Harris (see [GrHa]). We have also included in Chapter 4 some recent results (see [PW]) on singular real curves, which extend the classical ones for smooth projective curves. By doing so we show how modern tools such as étale cohomology can be usefully applied to real algebraic geometry.

We assume the reader is well acquainted with some basic notions in algebraic geometry (see [GH], [Hr]), in the theory of real and complex manifolds (see [GH]), homology and cohomology of complexes (see [HS], [HW]), sheaves and their cohomology (see [Hr]). We will often use, without any further reference, the so called Serre's GAGA principle (see [Se2]), switching from the analytic category to the algebraic one; we assume that this principle is well known to the reader.

The authors would like to thank the organizers of the Winter School on Real Algebraic Geometry for the invitation to lecture on this beautiful subject, and for the great job they did during the days in which the School was held. Special thanks go to M. G. Marinari for many useful comments and criticisms to the first draft of these notes. Finally the authors want to thank the referee whose report has greatly helped them to improve this paper. In particular he suggested a shorter proof of Comessatti's Theorem (see Theorem 2.5.3), which has been included here.

1. Generalities on complex tori

1.1. Complex tori

Let V be an $\mathbb{R}$-vector space of dimension q and let Λ be a *discrete subgroup* of the additive structure of V, i.e., a subgroup of $V(+)$ which is discrete for the euclidean topology. It is a well known fact, which is easy to prove, that a subgroup Λ of $V(+)$ is discrete if and only if it has no limit point in V. Such a subgroup Λ of $V(+)$ is called a *lattice* in V.

The typical example of a lattice is the following: let $S = \{x_1, \ldots, x_r\}$ be linearly independent elements of V and let Λ be the set of all vectors which are linearly dependent on S with coefficients in $\mathbb{Z}$. Then Λ is a finitely generated free subgroup of $V(+)$ of rank r. It is not difficult to prove that any lattice is of this sort. Lattices of maximal rank q are called *maximal* lattices of V.

Given a lattice Λ in V, one can consider the abelian group $X := V/\Lambda$, and the canonical projection map $p: V \to X$. The group X is naturally endowed with a structure of differentiable manifold which is compatible with the group structure, i.e., it is a Lie group. Furthermore the map p is a homomorphism of Lie groups and it is a local diffeomorphism. If Λ has rank r one has the following obvious chain of isomorphisms of real Lie groups:

$$X = V/\Lambda \simeq \mathbb{R}^q/\mathbb{Z}^r \simeq (\mathbb{R}/\mathbb{Z})^r \times \mathbb{R}^{q-r} \simeq (S_1)^r \times \mathbb{R}^{q-r},$$

where $S_1 \simeq \mathbb{R}/\mathbb{Z}$ is the circle. Hence X is compact if and only if $r = q$, i.e., if and only if Λ is a maximal lattice, in which case $X \simeq S_1^q$ is a *q-(real) torus*.

Next we consider the case in which V is a $\mathbb{C}$-vector space of dimension q. If Λ is a maximal lattice in V, the real manifold $X = V/\Lambda$ inherits, via the local diffeomorphism $p: V \to X$, the complex structure of V. Hence X is a complex, compact, connected Lie group, which is called a q-dimensional *complex torus*. It is a basic result, which we are not going to prove here, that, conversely, any complex, compact, connected Lie group of dimension q is a q-dimensional complex torus.

Let $X = V/\Lambda$ be a complex torus of dimension q. If we fix a basis $\{e_1, \dots, e_q\}$ of V, we have complex coordinates in V, i.e., we may identify V with $\mathbb{C}^q$, whose elements we will regard as column vectors of order q. Let $\{\gamma_1, \dots, \gamma_{2q}\}$ be a basis of Λ as a free abelian group of rank $2q$. We denote by $\omega^1, \dots, \omega^{2q}$ the column vectors corresponding to $\gamma_1, \dots, \gamma_{2q}$ respectively, and we consider the $q \times 2q$ matrix Ω whose columns are $\omega^1, \dots, \omega^{2q}$. Then $\Omega = (\omega_{ij})_{i=1,\dots,q;\, j=1,\dots,2q}$ where $\gamma_i = \sum_{j=1}^{q} \omega_{ij} e_i$ $(1 \le j \le 2q)$. The matrix Ω is called a *period matrix* of the complex torus X. Notice that such a period matrix is not uniquely determined. It changes by changing coordinates in V and by changing the basis of Λ. Accordingly Ω changes by multiplication to the left by a matrix in $\mathrm{GL}(q, \mathbb{C})$, and by multiplication to the right by a matrix in $\mathrm{GL}(2q, \mathbb{Z})$.

Given a period matrix Ω for the complex torus X of dimension q, one can form the complex $2q \times 2q$ matrix

$$\begin{pmatrix} \Omega \\ \overline{\Omega} \end{pmatrix}$$

and the real $2q \times 2q$ matrix

$$\begin{pmatrix} \mathrm{Re}(\Omega) \\ \mathrm{Im}(\Omega) \end{pmatrix}.$$

It is quite easy to see that

$$\det \begin{pmatrix} \Omega \\ \overline{\Omega} \end{pmatrix} = (-2i)^q \det \begin{pmatrix} \mathrm{Re}(\Omega) \\ \mathrm{Im}(\Omega) \end{pmatrix}. \tag{1.1.1}$$

Since $\{\gamma_1, \dots, \gamma_{2q}\}$ spans V as a real vector space, we have that

$$\det \begin{pmatrix} \mathrm{Re}(\Omega) \\ \mathrm{Im}(\Omega) \end{pmatrix} \neq 0.$$

Hence we have

$$\mathrm{rk} \begin{pmatrix} \Omega \\ \overline{\Omega} \end{pmatrix} = 2q. \tag{1.1.2}$$

It is then clear that (1.1.2) is a necessary and sufficient condition in order that Ω is a period matrix of a q-dimensional complex torus.

1.2. Homology and cohomology of tori

Let $X := V/\Lambda$ be a real torus of dimension q. Since $V \simeq \mathbb{R}^q$ is simply connected, it is clear that $p\colon V \to X$ is the universal cover of X, hence $\pi_1(X) \simeq \Lambda \simeq \mathbb{Z}^q$. Therefore also $H_1(X, \mathbb{Z}) \simeq \Lambda \simeq \mathbb{Z}^q$. On the other hand, we know that $X \simeq S_1^q$. The Künneth formula and an easy induction argument show that

$$H_n(X, \mathbb{Z}) \simeq \wedge^n H_1(X, \mathbb{Z})$$

for all positive integers n.

The theorem of universal coefficients yields that

$$H^n(X, \mathbb{Z}) \simeq \mathrm{Hom}(H_n(X, \mathbb{Z}), \mathbb{Z})$$

for all integers n. Therefore one has

$$H^n(X, \mathbb{Z}) \simeq \wedge^n H^1(X, \mathbb{Z}) \simeq \wedge^n H_1(X, \mathbb{Z}) \simeq H_n(X, \mathbb{Z})$$

for all positive integers n. Furthermore one has

$$H_n(X, \mathbb{C}) \simeq H_n(X, \mathbb{Z}) \otimes_{\mathbb{Z}} \mathbb{C},$$

$$H^n(X, \mathbb{C}) \simeq H^n(X, \mathbb{Z}) \otimes_{\mathbb{Z}} \mathbb{C}.$$

So far only the topology of X came into the picture. Now let us look at X as a differentiable manifold. Then we can consider the *De Rham cohomology* of X:

$$H^n_{\mathrm{DR}}(X) := \frac{\{d - \text{closed } n - \text{forms on } X\}}{d\{(n-1) - \text{forms on } X\}},$$

where d is the differential operator acting on forms on X. De Rham's theorem says that

$$H^n_{\mathrm{DR}}(X) \simeq H^n(X, \mathbb{C}).$$

This isomorphism can be made very explicit in the present case. Indeed, let $x_1, \ldots, x_{2q}$ be coordinates on V. The 1-forms $dx_1, \ldots, dx_{2q}$ are *translation invariants*, and therefore descend to 1-forms $\xi_1, \ldots, \xi_{2q}$ on X, which are clearly closed. Their classes in $H^1_{\mathrm{DR}}(X)$ are of course independent, hence they span $H^1_{\mathrm{DR}}(X)$. Then $H^n_{\mathrm{DR}}(X)$ is generated by forms of type $\xi_{i_1} \wedge \cdots \wedge \xi_{i_n}$ with $1 \leq i_1 < \cdots < i_n \leq q$. The *De Rham pairing*

$$\mu_n\colon (\omega, \gamma) \in H^n_{\mathrm{DR}}(X) \times H_n(X, \mathbb{C}) \to \int_\gamma \omega \in \mathbb{C}$$

which is well defined because of Stokes' theorem, is an isomorphism, inasmuch as

$$\mu_n(\xi_{i_1} \wedge \cdots \wedge \xi_{i_n}, \phi_{j_1,\ldots,j_n}) = \delta^{i_1,\ldots,i_n}_{j_1,\ldots,j_n}$$

where $1 \le i_1 < \cdots < i_n \le q$, $1 \le j_1 < \cdots < j_n \le q$, and δ is, as usual, the *Kronecker symbol*.

Finally, let us put the complex structure of X into the picture. Let $z_1, \dots, z_q$ be complex coordinates on V, so that $z_1, \dots, z_q, \overline{z}_1, \dots, \overline{z}_q$ give real coordinates. The 1-forms $dz_1, \dots, dz_q$ are translation invariants and therefore give us holomorphic 1-forms $\zeta_1, \dots, \zeta_q$ on X. Hence they span a q-dimensional subspace Z of $H^0(X, \Omega^1_X)$, where, as usual, Ω^p_X is the locally free sheaf of holomorphic p-forms on the variety X. One has

$$H^1_{\mathrm{DR}} = Z \oplus \overline{Z}$$

and accordingly

$$H^n_{\mathrm{DR}}(X) \simeq \oplus_{a+b=n} Z^{a,b}, \quad Z^{a,b} := \wedge^a Z \otimes \wedge^b \overline{Z}$$

for all positive integers n. In this setting, recall that *Dolbeault's theorem* says that

$$H^b(X, \Omega^a_X) \simeq H^{a,b}(X) := \frac{\{\overline{\partial} - \text{closed } (a,b) - \text{forms on } X\}}{\overline{\partial}\{(a, b-1) - \text{forms on } X\}}$$

and the *theorem of Hodge* (see [LB], p. 17) says that

$$H^{a,b}(X) \simeq Z^{a,b}.$$

Hence we have the Hodge decomposition

$$H^n_{\mathrm{DR}}(X) \simeq \oplus_{b+a=n} H^b(X, \Omega^a_X).$$

In particular one has $H^0(X, \Omega^1_X) \simeq Z$, i.e., $\dim_{\mathbb{C}} H^0(X, \Omega^1_X) = q$, and

$$H^1(X, \mathcal{O}_X) \simeq \overline{H^0(X, \Omega^1_X)}.$$

Finally let us point out a different interpretation of a period matrix of a complex torus. We keep the above notation. Let γ be an element of Λ with coordinate vector $\alpha = (\alpha_1, \dots \alpha_q)^t \in \mathbb{C}^q$. One has

$$\alpha_i = \int_\gamma \zeta_i.$$

This remark clearly yields that any period matrix Ω of X is obtained by fixing bases $\{\omega_1, \dots, \omega_q\}$ of $H^0(X, \Omega^1_X)$ and $\{\gamma_1, \dots, \gamma_{2q}\}$ of $\Lambda \simeq H_1(X, \mathbb{C})$, and by taking

$$\Omega := (\int_{\gamma_j} \omega_i)_{i=1,\dots,q;\, j=1,\dots,2q}.$$

1.3. Morphisms of complex tori

Let $X := V/\Lambda$ and $X' := V'/\Lambda'$ be complex tori, with the natural maps $p\colon V \to X$, $p'\colon V' \to X'$. One has the following basic result.

Proposition 1.3.1. *Let $f: X \to X'$ be a holomorphic map. Then there is an affine map $\phi: V \to V'$ such that $p' \circ \phi = f \circ p$. The map ϕ is uniquely defined up to translations by elements of Λ'.*

The idea of the proof of this proposition, which we only sketch here (see [LB], p. 9), is that $f: X \to X'$ lifts to a holomorphic map $\phi: V \to V'$ of the universal covers of X and X', such that $\phi(x + \Lambda) - \phi(x) \in \Lambda'$ for all $x \in V$ and $\lambda \in \Lambda$. Using this information is not difficult to see that ϕ must be affine, i.e., a linear map followed by a translation.

Let us denote now by $\mathrm{Hom}(X, X')$ the abelian group of all *homomorphisms* of X to X', i.e., of all holomorphic maps $X \to X'$ which are also group homomorphisms. In particular we have the group $\mathrm{End}(X)$ of all the *endomorphisms* of X, which is a ring with the multiplication given by composition of endomorphisms. We will let $\mathrm{End}(X)^*$ be the group of invertible endomorphisms of X, i.e., the elements of $\mathrm{End}(X)^*$ are the complex Lie group automorphisms of X.

As an immediate consequence of Proposition 1.3.1, one finds the following:

Corollary 1.3.2.

(i) *Let $f: X \to X'$ be a homomorphism. Then there is a unique linear map $\phi: V \to V'$ such that $p' \circ \phi = f \circ p$. In particular ϕ maps Λ to Λ'.*

(ii) *Let $f: X \to X'$ be a holomorphic map and set $f(0) := -y$. Then $f' := \tau_y \circ f: X \to X'$ is a homomorphism, where τ_y denotes the translation by y.*

(iii) *The map $(y, f) \in X \times \mathrm{End}(X)^* \to \tau_y \circ f \in \mathrm{Aut}(X)$ is a group isomorphism, where the group structure on $X \times \mathrm{End}(X)^*$ is given by*

$$(y, f) \cdot (y', f') = (y + f(y'), f \circ f').$$

The above corollary tells us that there are two injective group homomorphism

$$\rho_a: \mathrm{Hom}(X, X') \to \mathrm{Hom}_{\mathbb{C}}(V, V')$$

$$\rho_r: \mathrm{Hom}(X, X') \to \mathrm{Hom}_{\mathbb{Z}}(\Lambda, \Lambda')$$

which are called the *analytic* and the *rational* representation of $\mathrm{Hom}(X, X')$. Since $\mathrm{Hom}_{\mathbb{Z}}(\Lambda, \Lambda')$ is a free abelian group of rank equal to $4 \cdot \dim X \cdot \dim X'$, we have the following theorem.

Theorem 1.3.3. *The group $\mathrm{Hom}(X, X')$ is a free abelian group of rank $m \leq 4 \cdot \dim X \cdot \dim X'$.*

In the case $X = X'$ the analytic and rational representations

$$\rho_a: \mathrm{End}(X) \to \mathrm{End}_{\mathbb{C}}(V)$$

$$\rho_r: \mathrm{End}(X) \to \mathrm{End}_{\mathbb{Z}}(\Lambda)$$

are easily seen to be also ring-homomorphisms. One sets

$$\mathrm{Hom}_{\mathbb{Q}}(X, X') := \mathrm{Hom}(X, X') \otimes_{\mathbb{Z}} \mathbb{Q}$$

$$\mathrm{End}_{\mathbb{Q}}(X) := \mathrm{End}(X) \otimes_{\mathbb{Z}} \mathbb{Q}$$

the latter being a $\mathbb{Q}$-algebra, which is a sub-algebra of $\mathrm{End}_{\mathbb{Q}}(\Lambda) := \mathrm{End}_{\mathbb{Z}}(\Lambda) \otimes_{\mathbb{Z}} \mathbb{Q}$.

Let us now fix bases in V and V'. Let $q := \dim X$, $q' := \dim X'$, and let Ω, Ω' be period matrices of X and X' respectively. If $f: X \to X'$ is a homomorphism, we have that $\rho_a(f) := A$ can be interpreted as a $q' \times q$ matrix over $\mathbb{C}$, whereas $\rho_r(f) := B$ is a $2q' \times 2q$ integral matrix. The content of Corollary 1.3.2, (i), can be rephrased by saying that

$$A \cdot \Omega = \Omega' \cdot B. \tag{1.3.4}$$

This is called a *Hurwitz relation* for the homomorphism f.

The relation between the analytic and the rational representations is indicated in the following result.

Theorem 1.3.5. *Let us set*

$$\rho_r \otimes 1: f \otimes z \in \mathrm{End}_{\mathbb{Q}}(X) \otimes \mathbb{C} \to \rho_r(f) \cdot z \in \mathrm{End}_{\mathbb{C}}(\Lambda \otimes \mathbb{C}) \simeq \mathrm{End}_{\mathbb{C}}(V \times V).$$

One has

$$\rho_r \otimes 1 = \rho_a \oplus \overline{\rho}_a$$

where $\overline{\rho}_a$ is the complex conjugate representation of ρ_a.

Proof. Let (1.3.4) be a Hurwitz relation for f. Then one also has

$$\overline{A} \cdot \overline{\Omega} = \overline{\Omega}' \cdot B,$$

whence

$$\begin{pmatrix} A & 0_q \\ 0_q & \overline{A} \end{pmatrix} \cdot \begin{pmatrix} \Omega \\ \overline{\Omega} \end{pmatrix} = \begin{pmatrix} \Omega \\ \overline{\Omega} \end{pmatrix} \cdot B$$

where 0_q is the zero square matrix of order q. From this relation the assertion easily follows. □

A homomorphism $f: X \to X'$ which is surjective and with finite kernel is said to be an *isogeny* between X and X'. Note that f is an isogeny if and only if $\rho_a(f)$ is a isomorphism (whence $q = q'$) and $\rho_r(f)$ is injective with finite cokernel. The order of the kernel of an isogeny f is called the *degree* of the isogeny, and denoted by $\deg(f)$. If (1.3.4) is a Hurwitz relation for the isogeny f, it is not difficult to see that $\deg(f) = |\det B|$ (see [LB], p. 12).

A relevant feature of isogenies is that they are $\mathbb{Q}$-*invertible*. Indeed, let $f: X \to X'$ be an isogeny. We denote by $e(f)$ the minimum positive integer, called the *exponent* of f, such that for all elements $x \in \ker(f)$ one has $e(f) \cdot x = 0$. It is then easy to see (see [LB], p. 12) that the following theorem holds.

Theorem 1.3.6. *Let $f: X \to X'$ be an isogeny. There is an isogeny $g: X' \to X$ such that $g \circ f$ [resp. $f \circ g$] is the multiplication by $e(f)$ in X [resp. in X'].*

In particular the isogenies of a complex torus X to itself are invertible in $\mathrm{End}_{\mathbb{Q}}(X)$. Since the image of a homomorphism is clearly a complex torus, we have the following result.

Proposition 1.3.7. *Let X be a complex torus which has no non trivial complex subtori (in which case X is said to be simple). Then every non zero element in $\mathrm{End}_{\mathbb{Q}}(X)$ is invertible.*

1.4. The Albanese and the Picard variety

Let X be any compact Kähler variety. We denote by q its *irregularity*, i.e., $q := h^0(X, \Omega_X^1)$. We let $H_1(X, \mathbb{Z})_0$ be $H_1(X, \mathbb{Z})$ modulo torsion and we consider the map

$$f: \sigma \in H_1(X, \mathbb{Z})_0 \to \{\omega \to \int_\sigma \omega\} \in H^0(X, \Omega_X^1)^*.$$

Using the Hodge decomposition

$$H^1_{\mathrm{DR}}(X) \simeq H^0(X, \Omega_X^1) \oplus \overline{H^0(X, \Omega_X^1)}$$

(see [GH], p. 116), one sees that f is injective. Furthermore De Rham's theorem yields that $f(H_1(X, \mathbb{Z})_0)$ is a maximal lattice in $H^0(X, \Omega_X^1)^*$. Therefore one can consider the complex torus of dimension q

$$\mathrm{Alb}(X) := \frac{H^0(X, \Omega_X^1)^*}{f(H_1(X, \mathbb{Z})_0)}$$

which is called the *Albanese variety* of X.

If p is a point in X, one can define the *Albanese map*

$$\alpha_p: x \in X \to \{\omega \in H^0(X, \Omega_X^1) \to \int_p^x \omega \in \mathbb{C}\} \in \frac{H^0(X, \Omega_X^1)^*}{f(H_1(X, \mathbb{Z})_0)} = \mathrm{Alb}(X)$$

which depends on p, and is defined up to translations as p varies in X. We notice the following important theorem, whose proof we omit (see [LB], chapter 11).

Theorem 1.4.1 (Universal property of the Albanese variety). *Let X be a compact Kähler manifold, and let $\phi: X \to T$ be a holomorphic map of X to a complex torus T. Then there is a unique homomorphism ϕ': $\mathrm{Alb}(X) \to T$ such that $\tau_{-\phi(p)} \circ \phi = \phi' \circ \alpha_p$.*

Next we consider period matrices for $\mathrm{Alb}(X)$, which are also called *period matrices* for X. In order to do so, we introduce coordinates in $\mathrm{Alb}(X)$ by fixing a basis $\{\omega_i\}_{i=1,\dots,q}$ of $H^0(X, \Omega_X^1)$. If $\{\gamma_1, \dots, \gamma_{2q}\}$ is a basis of $H_1(X, \mathbb{Z})_0$, the coordinate vector of $f(\gamma_j)$ with respect to the dual basis to $\{\omega_i\}_{i=1,\dots,q}$ is clearly

$$\omega^j = \begin{pmatrix} \int_{\gamma_j} \omega_1 \\ \cdots \\ \cdots \\ \int_{\gamma_j} \omega_q \end{pmatrix}.$$

Therefore the related period matrix of $\mathrm{Alb}(X)$ is

$$\Omega = (\int_{\gamma_j} \omega_i)_{i=1,\dots,q;\, j=1,\dots,2q}.$$

Furthermore the Albanese map can be written as

$$\alpha_p \colon x \in X \to (\int_p^x \omega_1, \dots, \int_p^x \omega_q) + \Lambda \in \mathrm{Alb}(X)$$

where $\Lambda = f(H_1(X, \mathbb{Z})_0)$.

Note that from the above discussion and from §3 it follows that if X is a complex torus, then $\mathrm{Alb}(X) \simeq X$.

There is another natural complex torus related to a complex compact Kähler manifold X, namely its *Picard variety*. We are now going to recall its definition. Consider the *exponential* exact sequence of sheaves

$$0 \to \mathbb{Z} \xrightarrow{\times i} \mathcal{O}_X \to \mathcal{O}_X^* \to 0 \tag{1.4.2}$$

where, as usual, $\mathcal{O}_X$ [resp. $\mathcal{O}_X^*$] is the sheaf of holomorphic functions [resp. of never vanishing holomorphic functions] on X. Accordingly one has the long exact cohomology sequence

$$0 \to H^1(X, \mathbb{Z}) \to H^1(X, \mathcal{O}_X) \to H^1(X, \mathcal{O}_X^*) \xrightarrow{c_1} H^2(X, \mathbb{Z}). \tag{1.4.3}$$

It is well know that $\mathrm{Pic}(X) := H^1(X, \mathcal{O}_X^*)$ is the *Picard group* of X, whose elements are the isomorphism classes of all holomorphic line bundles on X (see [GH], p. 132). The map c_1 is the *first Chern map* associating to the class of a line bundle its *first Chern class*. One defines the *Picard variety* of X to be

$$\mathrm{Pic}^0(X) := \ker c_1$$

i.e., $\mathrm{Pic}^0(X)$ is the group of *topologically trivial* line bundles on X. One also defines the *Néron–Severi group* of X as

$$NS(X) := \mathrm{Im}\, c_1.$$

One has

$$\mathrm{Pic}^0(X) \simeq \frac{H^1(X, \mathcal{O}_X)}{H^1(X, \mathbb{Z})}$$

and this shows that $\mathrm{Pic}^0(X)$ is a complex torus of dimension q. Indeed we have

$$H^1(X,\mathbb{Z}) \subset H^1(X,\mathbb{C}) \simeq H^1_{\mathrm{DR}}(X) \simeq H^0(X,\Omega^1_X) \oplus \overline{H^0(X,\Omega^1_X)}. \tag{1.4.4}$$

On the other hand by Dolbeault's theorem one has

$$H^1(X,\mathcal{O}_X) \simeq \overline{H^0(X,\Omega^1_X)}$$

and the inclusion $H^1(X,\mathbb{Z}) \to H^1(X,\mathcal{O}_X)$ given by the sequence (1.4.3) is easily seen to be deduced from (1.4.4) by projecting $H^1(X,\mathbb{Z})$ to the summand $\overline{H^0(X,\Omega^1_X)}$. In conclusion it is quite easy to see that $H^1(X,\mathbb{Z})$ is a maximal lattice in $\overline{H^0(X,\Omega^1_X)}$, which proves that $\mathrm{Pic}^0(X)$ is a complex torus of dimension q.

Given a complex torus X, one defines its *dual torus* to be $\hat{X} := \mathrm{Pic}^0(X)$, which has the same dimension q as X. It is not difficult to prove that $\hat{\hat{X}} = X$, and that the Picard and Albanese varieties of a given variety are dual to each other (see [LB], p. 362).

1.5. Line bundles on complex tori

Let $X = V/\Lambda$ be a complex torus of dimension q. We want to indicate how to construct line bundles on X.

Let $H\colon V \times V \to \mathbb{C}$ be a hermitian form on V. We will denote by E its imaginary part. Note that E is antisymmetric and such that

$$E(ix,iy) = E(x,y), \quad \text{for all } x,y \in V. \tag{1.5.1}$$

Conversely, it is easily seen that, given an antisymmetric real form E on V such that (1.5.1) holds, there is a unique hermitian form H on V such that $E = \mathrm{Im}(H)$. Precisely one has

$$H(x,y) = -E(x,iy) + iE(x,y).$$

From now on we will assume that H is such that E takes integral values on Λ. Given the form H, or equivalently the form E, we define a map $\alpha\colon \Lambda \to S_1 = \{z \in \mathbb{C} : |z| = 1\}$ to be a *quasi-character* related to H (or to E), if

$$\alpha(\lambda+\mu) = (-1)^{E(\lambda,\mu)}\alpha(\lambda)\cdot\alpha(\mu) \tag{1.5.2}$$

for all λ, $\mu \in \Lambda$. Notice that, once H is given, certainly there are quasi-characters α related to H. Indeed one can arbitrarily define α on a basis of Λ and then extend it by using the formula (1.5.2).

A pair (H,α) where H is a hermitian form on V whose imaginary part is integral on Λ, and α is a quasi-character for H, is called an *Appèll–Humbert datum* for X. We denote by $\mathcal{AH}(X)$ the set of all Appèll–Humbert data for X. Note that $\mathcal{AH}(X)$ has a natural group structure defined by

$$(H,\alpha)\cdot(H',\alpha') := (H+H',\alpha\cdot\alpha').$$

Note the exact sequence

$$0 \to \mathrm{Hom}(\Lambda, S_1) \simeq S_1^q \to \mathcal{AH}(X) \to \mathcal{H}(V, \Lambda) \to 0$$

where $\mathcal{H}(V, \Lambda)$ is the group of all hermitian forms on V whose imaginary part is integral on Λ, and the maps in the sequence are the obvious ones.

Now we relate Appèll–Humbert data to line bundles. Precisely one defines a map

$$L\colon (H, \alpha) \in \mathcal{AH}(X) \to L(H, \alpha) \in \mathrm{Pic}(X)$$

as follows. First, given $(H, \alpha) \in \mathcal{AH}(X)$ one defines a function

$$e\colon (\lambda, x) \in \Lambda \times V \to e_\lambda(x) \in \mathbb{C}^*$$

called the *automorphy factor* of the Appèll–Humbert datum (H, α), via the formula

$$e_\lambda(x) := \alpha(\lambda) \cdot \exp(\pi H(x, \lambda) + \frac{\pi}{2} H(\lambda, \lambda)).$$

Then one defines a Λ-action on $V \times \mathbb{C}$ via the formula

$$\lambda.(x, z) := (x + \lambda, e_\lambda(x) \cdot z).$$

One sets $L(H, \alpha) := V \times \mathbb{C}/\Lambda$, and it is quite easy to see that $L(H, \alpha)$ is a line bundle on X. The main result in this setting, which we are not going to prove (see [LB], p. 29–32), is the following:

Theorem 1.5.3 (Appèll–Humbert). *The map L is a group isomorphism.*

As a consequence one finds the following exact commutative diagram

$$\begin{array}{ccccccc} 0 & \to & \mathrm{Hom}(\Lambda, S_1) \simeq S_1^q & \to & \mathcal{AH}(X) & \to & \mathcal{H}(V, \Lambda) & \to 0 \\ & & L \updownarrow & & L \updownarrow & & \updownarrow & \\ 0 & \to & \mathrm{Pic}^0(X) & \to & \mathrm{Pic}(X) & \to & NS(X) & \to 0 \end{array}$$

which yields an isomorphism $\mathcal{H}(V, \Lambda) \to NS(X) \subset H^2(X, \mathbb{Z})$. To be explicit $c_1(L(H, \alpha))$ can be identified, as an element of $NS(X) \subset H^2(X, \mathbb{Z}) = \wedge^2 H^1(X, \mathbb{Z})$, with $E = \mathrm{Im}(H)$ which is an antisymmetric form on $\Lambda \simeq H^1(X, \mathbb{Z})$ with integral values. Hence the isomorphism $\mathcal{H}(V, \Lambda) \to NS(X)$ maps $H \in \mathcal{H}(V, \Lambda)$ to its imaginary part which sits in $NS(X) \subset H^2(X, \mathbb{Z})$.

1.6. Polarizations

Let $X = V/\Lambda$ be a complex torus. One has the following result (see [LB], chapter 3).

Theorem 1.6.1. *The line bundle $L = L(H, \alpha)$ is ample on $X = V/\Lambda$ if and only if H is positive definite.*

A *polarization* on X is defined to be an element $c_1(L) \in NS(X)$ such that L is ample. In other words, a polarization is a hermitian form in $\mathcal{H}(V, \Lambda)$ which is positive definite. A complex torus X on which there is a polarization is called an *abelian variety*. By Theorem 1.6.1, a complex torus is an abelian variety if and only if it is isomorphic to a complex projective variety. A pair $(X, c_1(L))$, where X is a complex torus and $c_1(L)$ is a polarization on X is said to be a *polarized abelian variety*. If there is no danger of confusion, one refers to X itself as the polarized abelian variety.

Given two polarized abelian varieties $(X, c_1(L))$ and $(X', c_1(L'))$, one says that they are *isomorphic* if there is an isomorphism $f: X \to X'$ of complex tori such that the associated map in cohomology $f^*: H^2(X', \mathbb{Z}) \to H^2(X, \mathbb{Z})$ carries $c_1(L')$ to $c_1(L)$.

Examples 1.6.2. There are natural polarizations on the Picard variety of a smooth projective variety X of dimension n. Indeed, let $X \subset \mathbb{P}^m$ and let $\omega \in H^{1,1}(X) \cap H^2(X, \mathbb{Z})$ be the Chern class of the line bundle $\mathcal{O}_X(1)$ on X (see [GH], p. 139). Furthermore, recall the Dolbeault isomorphism $H^1(X, \mathcal{O}_X) \simeq \overline{H^0(X, \Omega^1_X)}$. Then one has the following hermitian form:

$$H: (\phi_1, \phi_2) \in H^1(X, \mathcal{O}_X) \times H^1(X, \mathcal{O}_X) \to \int_X \wedge^{n-1}\omega \wedge \phi_1 \wedge \overline{\phi}_2 \in \mathbb{C}$$

which is clearly positive definite and sits in $\mathcal{H}(H^1(X, \mathcal{O}_X), H^1(X, \mathbb{Z}))$. Hence it gives a polarization on $\mathrm{Pic}^0(X)$. In particular:

(a) if X is an abelian variety, then also $\hat{X} = \mathrm{Pic}^0(X)$ is an abelian variety;

(b) if X is a smooth projective variety, then $\mathrm{Alb}(X)$, which is dual to $\mathrm{Pic}^0(X)$, is an abelian variety.

In order to analyse polarizations on a complex torus, it is useful to recall the following.

Theorem 1.6.3 (Frobenius). *Let M be a finitely generated free $\mathbb{Z}$-module and let E be an antisymmetric, non-degenerate integral valued form on M. Then M has even rank $2n$ and there exists a basis $\{\lambda_1, \ldots, \lambda_n, \mu_1, \ldots, \mu_n\}$ of M in which E has matrix of the form*

$$A = \begin{pmatrix} 0_n & \Delta \\ -\Delta & 0_n \end{pmatrix}$$

where Δ is a diagonal matrix whose entries are positive integers $(d_1, \ldots, d_n)$ such that d_i divides d_{i+1}, for all $i = 1, \ldots, n-1$. These integers are uniquely determined.

Any basis of M with the above properties is called a *symplectic basis* of M with respect to the form E and the integers $(d_1, \ldots, d_n)$ are called the *elementary divisors* of E.

Now, given a polarization H on $X = V/\Lambda$, it determines an antisymmetric form $E = \mathrm{Im}(H)$ on Λ, which is non-degenerate since H is positive definite. Hence one can apply Frobenius' theorem to determine the elementary divisors $(d_1, \dots, d_n)$ of E. They are invariants of the polarization, which is therefore said to be of *type*, or *level*, $\boldsymbol{d} = (d_1, \dots, d_n)$. A polarization of level $(1, \dots, 1)$ is said to be a *principal* polarization.

Now we can recall the basic theorem (see [LB], chapter 3).

Theorem 1.6.4 (Riemann–Roch). *Let $L = L(H, \alpha)$ be an ample line bundle on X determining a polarization of type $(d_1, \dots, d_n)$. Then one has $h^0(X, L) = d_1 \cdot d_2 \dots d_n$, where $h^0(X, L)$ is the dimension of $H^0(X, L)$.*

Note that, according to the descriptions of line bundles given in §5, an element $\theta \in H^0(X, L(H, \alpha))$ can be regarded as a holomorphic map $\theta \colon V \to \mathbb{C}$ such that, for all $x \in V$ and all $\lambda \in \Lambda$ one has

$$\theta(x + \lambda) = e_\lambda(x) \cdot \theta(x) \tag{1.6.5}$$

where e is the automorphy factor of (H, α). Such a function is called a *theta function* related to the Appèll–Humbert datum (H, α). Just to give an idea of the proof of Theorem 1.6.4, let us observe that formula (1.6.5) leads to a Fourier expansion for θ. The coefficients of this expansion can be recursively uniquely determined, once the first $d := d_1 \cdot d_2 \cdot \dots \cdot d_n$ are arbitrarily given. The positivity assumption on H insures convergence for every choice of these d coefficients. It follows that the dimension of $H^0(X, L(H, \alpha))$ is exactly d.

1.7. Riemann's bilinear relations and moduli spaces

The existence of a polarization on a complex torus $X = V/\Lambda$ of dimension q is a restriction on X, or equivalently on its period matrices, as soon as $q \geq 2$. Indeed one has the following theorem (see [LB], p. 75).

Theorem 1.7.1 (Riemann's bilinear relations). *Let X be a complex torus of dimension q and let Ω be a period matrix of X. There is a polarization on X if and only if there is a non degenerate, antisymmetric $2q \times 2q$, integral matrix A such that*

(i) $\Omega \cdot A^{-1} \cdot \Omega^t = 0_q$,

(ii) $i\Omega \cdot A^{-1} \cdot \overline{\Omega}^t$ *is positive definite.*

Without entering into details, we remark that condition (i) of the theorem says that there is a non degenerate, antisymmetric, integral form E on Λ which can be interpreted as the imaginary part of a hermitian form on V, and (ii) yields that this hermitian form is positive definite.

In particular, one can take $\{\lambda_1, \dots, \lambda_q, \mu_1, \dots, \mu_q\}$ to be a symplectic basis of Λ with respect to E. Then the matrix A of E in this basis is of the form indicated in

Frobenius' theorem. Furthermore it is not difficult to see that $\{\lambda_1, \ldots, \lambda_q\}$ are linearly independent in V. Therefore, by performing a suitable change of coordinates in V, one may find a period matrix of X of the form

$$\Omega = (\Delta \quad Z).$$

Then Riemann's bilinear relations become

$$Z = Z^t, \quad \mathrm{Im}(Z) \text{ is positive definite.} \tag{1.7.2}$$

We will say that Ω is a period matrix of X *corresponding* to the given polarization.

Next we want to indicate how to give a structure of variety to the set $\mathcal{A}_{(q,\boldsymbol{d})}$ of all isomorphism classes of abelian varieties of dimension q with a polarization of type $\boldsymbol{d} = (d_1, \ldots, d_q)$. This construction is quite natural if one takes into account Theorem 1.7.1 and particularly Riemann's bilinear relations in the form (1.7.2).

Let us take $\mathbb{H}_q$ to be the open subset of $\mathbb{C}^{\frac{q(q+1)}{2}}$ formed by all $q \times q$ symmetric complex matrices Z such that $\mathrm{Im}(Z)$ is positive definite. The complex variety $\mathbb{H}_q$ is called the q-th *Siegel upper-half-space*. Let Z be a point in $\mathbb{H}_q$. We can associate in a natural way to Z a polarized abelian variety in $\mathcal{A}_{(q,\boldsymbol{d})}$ in the following way. The lattice Λ_Z generated by the columns of $\Omega_Z = (\Delta \quad Z)$ in $\mathbb{C}^q$ is maximal, and, according to Theorem 1.7.1 and to (1.7.2), the complex torus $X_Z := \mathbb{C}^q/\Lambda_Z$ possesses a polarization L_Z of level $\boldsymbol{d}$. Thus we have a map $\pi : \mathbb{H}_q \to \mathcal{A}_{(q,\boldsymbol{d})}$, which sends $Z \in \mathbb{H}_q$ to the isomorphism class of the polarized abelian variety (X_Z, L_Z). According to Theorem 1.7.1 and to the comments we made after it, the map π is surjective.

What we want to do now is to realize $\mathcal{A}_{(q,\boldsymbol{d})}$ as the *quotient* of $\mathbb{H}_q$ by the action of a suitable group. Precisely we consider the so called *real symplectic group of level* $\boldsymbol{d} = (d_1, \ldots, d_q)$, which is the subgroup $\mathrm{Sp}(\boldsymbol{d}, \mathbb{R})$ of $\mathrm{GL}(2q, \mathbb{R})$ formed by all matrices $N \in \mathrm{GL}(2q, \mathbb{R})$ such that

$$N^t \cdot \begin{pmatrix} 0_q & \Delta \\ -\Delta & 0_q \end{pmatrix} \cdot N = \begin{pmatrix} 0_q & \Delta \\ -\Delta & 0_q \end{pmatrix}$$

where Δ is as in the statement of Frobenius' theorem. We denote by $\mathrm{Sp}(\boldsymbol{d}, \mathbb{Q})$ [resp. $\mathrm{Sp}(\boldsymbol{d}, \mathbb{Z})$] the subgroup of $\mathrm{Sp}(\boldsymbol{d}, \mathbb{R})$ formed by rational [resp. integral] matrices and by $\mathrm{Sp}(2q, \mathbb{R})$, $\mathrm{Sp}(2q, \mathbb{Q})$, $\mathrm{Sp}(2q, \mathbb{Z})$ the symplectic groups of level $(1, \ldots, 1)$.

The real symplectic group acts on $\mathbb{H}_q$ in the following way. If

$$N = \begin{pmatrix} a & b \\ c & d \end{pmatrix} \in \mathrm{Sp}(2q, \mathbb{R})$$

where a, b, c, d are $q \times q$ matrices, and if Z is in $\mathbb{H}_q$, then $Z \cdot b + \Delta \cdot d$ is invertible and we define the action of $\mathrm{Sp}(d, \mathbb{R})$ by

$$N.Z := \Delta \cdot (\Delta \cdot a + Z \cdot c)^{-1} \cdot (\Delta \cdot b + Z \cdot d).$$

By restricting this action to the *integral symplectic group of level* $\boldsymbol{d} = (d_1, \ldots, d_q)$, i.e., the group $\mathrm{Sp}(\boldsymbol{d}, \mathbb{Z}) := \mathrm{Sp}(\boldsymbol{d}, \mathbb{R}) \cap \mathrm{GL}(2q, \mathbb{Z})$, one has a $\mathrm{Sp}(\boldsymbol{d}, \mathbb{Z})$ action on $\mathbb{H}_q$. By using Theorem 1.7.1 it is not difficult to prove the next result.

Theorem 1.7.3. *Let Z, Z' be points in $\mathbb{H}_q$. There is an isomorphism $f\colon X_{Z'} \to X_Z$ of the corresponding polarized abelian varieties if and only if there is a matrix*

$$N = \begin{pmatrix} a & b \\ c & d \end{pmatrix}$$

in $\mathrm{Sp}(\boldsymbol{d}, \mathbb{Z})$ such that $N.Z = Z'$. In such a case the analytic representation of f is given by

$$\rho_a(f)\colon x \in \mathbb{C}^q \to (\Delta \cdot a + Z \cdot c) \cdot \Delta^{-1} \cdot x \in \mathbb{C}^q.$$

As a consequence we have, as asserted before, that $\mathcal{A}_{(q,\boldsymbol{d})}$ appears as the quotient $\mathbb{H}_q / \mathrm{Sp}(\boldsymbol{d}, \mathbb{Z})$. More precisely, the following holds.

Theorem 1.7.4. *There is a unique structure of complex analytic space of dimension $\frac{q(q+1)}{2}$ on $\mathcal{A}_{(q,\boldsymbol{d})} = \mathbb{H}_q / \mathrm{Sp}(\boldsymbol{d}, \mathbb{Z})$ such that the natural projection map $\pi\colon \mathbb{H}_q \to \mathcal{A}_{(q,\boldsymbol{d})}$ is analytic.*

As for the proof, we do not enter in details, but just remark that it can be shown that the action of $\mathrm{Sp}(\boldsymbol{d}, \mathbb{Z})$ on $\mathbb{H}_q$ is properly discontinuous, i.e., for any pair of compact subsets K_1, K_2 of $\mathbb{H}_q$ the set

$$\Gamma(K_1, K_2) := \{N \in \mathrm{Sp}(\boldsymbol{d}, \mathbb{Z}) : N.K_1 \cap K_2 \neq \emptyset\}$$

is finite. This is not difficult to prove (for a similar argument, see the proof of Theorem 3.5.1 below).

Finally one concludes by applying a well known theorem of H. Cartan's (see [Ca]), which ensures that the quotient of a complex variety of dimension n by the action of a group which is properly discontinuous is a complex analytic space of dimension n, and that the quotient map is analytic.

2. Real structures

In this chapter we introduce the main concepts about real structures and real algebraic varieties, and we study their basic homological and cohomological properties. Many results (like Weil's Theorem 2.2.2) could be stated and proved in a more general setting. We preferred however to avoid a generality which would not be used later on in these notes.

Unless explicitely stated, the varieties we will consider are supposed to be smooth.

2.1. Definition of real structures

Let X be a complex manifold. One can define the *conjugate variety* of X, denoted by X^σ or $\overline{X}$. As a topological space X^σ is homeomorphic to X, but the coordinate charts are the complex conjugate charts of those defining X. We note that $(X^\sigma)^\sigma = X$. One denotes by $\sigma_X: X \to X^\sigma$ the homeomorphism between X and X^σ, which is called the *complex conjugation* of X. Note that $\sigma_{X^\sigma} = \sigma_X^{-1}$. One simply writes σ instead of σ_X if no confusion arises. Notice that, if X is a smooth, quasi-projective, algebraic variety over $\mathbb{C}$, then X^σ is defined by the complex conjugate equations of those defining X.

If X and Y are complex manifolds, a map $f: X \to Y$ is said to be *antiholomorphic* if $\sigma_Y \circ f$ is holomorphic. Of course the complex conjugation of a manifold is antiholomorphic. Given a map $f: X \to Y$, one can consider the *conjugate* map $f^\sigma: X^\sigma \to Y^\sigma$, with the property that $\sigma_Y \circ f = f^\sigma \circ \sigma_X$. One has $(f^\sigma)^\sigma = f$ and f is holomorphic [resp. antiholomorphic] if and only if f^σ is such. It is easy to see that by composing antiholomorphic maps [resp. a holomorphic with an antiholomorphic map], one gets an holomorphic [resp. an antiholomorphic] map. Note that, if X and Y are smooth, quasi-projective, algebraic varieties over $\mathbb{C}$, one can define in an obvious way the concept of *antiregular map*, etc.

Let X be as above and let $S: X \to X$ be an *antiinvolution*, i.e., an antiholomorphic map such that $S^2 = \mathrm{id}_X$, the identity map of X. A pair (X, S) as above is said to be a *real structure* on X, and the antiinvolution S is called the *conjugation* of the real structure. The subset X_S of X formed by the fixed points for S is a real subvariety of X, called *real part* of the real structure (X, S). If X_S is empty we will say that (X, S) has *no real points*. If X has dimension n, it is easily seen that X_S, if not empty, is a real manifold of dimension n. In what follows, we will denote by $\nu(X, S)$, or simply by $\nu(X)$, the number of connected components of X_S with respect to the euclidean topology. These components are called the *branches* of (X, S), or simply the branches of X. This is a basic invariant of a real structure, and the problem of computing it is the so-called *Harnack's problem*.

If (X, S) and (X', S') are real structures, a *morphism* f between them is a holomorphic map $f: X \to X'$ such that $S' \circ f = f \circ S$. Of course f carries X_S to $X'_{S'}$. It is obvious how the notion of *isomorphism* between two real structures is given. If two real structures are isomorphic, we will also say that they are *equivalent*. One can also define in a natural way the *product* of the two real structures (X, S) and (X', S') and consider the graph of f, etc. We omit the details.

Examples 2.1.1. (i) We can consider the obvious real structures $(\mathbb{C}^n, \sigma)$, $(\mathbb{P}^n_{\mathbb{C}}, \sigma)$, where we denote by σ the usual complex conjugation. The real parts are respectively $\mathbb{R}^n$, $\mathbb{P}^n_{\mathbb{R}}$. We will call these structures the *standard* real structures on $\mathbb{C}^n$, $\mathbb{P}^n_{\mathbb{C}}$.

(ii) An *antiprojectivity* of $\mathbb{P}^n_{\mathbb{C}}$ is, by definition, the composite map of a projectivity and of the usual complex conjugation considered in the previous example. A real structure $(\mathbb{P}^n_{\mathbb{C}}, S)$ on $\mathbb{P}^n_{\mathbb{C}}$ is clearly given by an antiprojectivity S of $\mathbb{P}^n_{\mathbb{C}}$.

We consider the *antiprojectivity* τ of $\mathbb{P}^1_{\mathbb{C}}$ defined by $\tau: z \to -1/\bar{z}$. This is an antiinvolution which has no fixed point, and determines therefore a real structure $(\mathbb{P}^1_{\mathbb{C}}, \tau)$ with no real points. This real structure is not equivalent to the standard one on $\mathbb{P}^1_{\mathbb{C}}$. It is an easy exercise, which we leave to the reader, to see that, up to equivalence, this is the only other real structure on $\mathbb{P}^1_{\mathbb{C}}$ besides the standard one.

(iii) Let $n \equiv 0 \pmod 2$. Then it is an exercise, which we leave to the reader, to see that every real structure $(\mathbb{P}^n_{\mathbb{C}}, S)$ is equivalent to the standard one. (Suggestion: take $\frac{n+2}{2}$ general pairs of points conjugate by S and send them with a projective transformation to $\frac{n+2}{2}$ general pairs of points conjugate by σ).

A *real algebraic variety* is a ringed space $(X, \mathcal{O}_X)$ which is locally isomorphic to an affine real algebraic variety, i.e., $(X, \mathcal{O}_X)$ is locally isomorphic to some algebraic set in $\mathbb{R}^n$ with the usual sheaf of regular functions. One often says that X itself is a real algebraic variety. One can consider the category of real algebraic varieties, with the obvious notions of dimension, smoothness, etc.

An algebraic variety Y defined over $\mathbb{R}$ is a reduced separated scheme of finite type over $\mathbb{R}$. We denote by $Y(\mathbb{R})$ the set of $\mathbb{R}$-rational points of Y. If $Y(\mathbb{R})$ is not empty, we will denote by $\mathcal{R}(Y)$ the ringed space $(Y(\mathbb{R}), \mathcal{O}_{Y|Y(\mathbb{R})})$. Then we have the following result: *if $Y(\mathbb{R})$ is Zariski dense in Y, then $\mathcal{R}(Y)$ is a real algebraic variety* (see [H], Proposition 1.1.8.

Let X be an algebraic variety over $\mathbb{C}$ endowed with a real structure (X, S). We will often assume, without further specification, that $S: X \to X$ is an antiregular involution. Clearly X may not be defined over $\mathbb{R}$, but X_S is always in a natural way a real algebraic variety, perhaps the empty set.

2.2. Real models

Let X be an algebraic variety over $\mathbb{C}$. By a *real model* of X we mean a pair (X_0, f), where X_0 is an algebraic variety defined over $\mathbb{R}$ and $f: (X_0)_{\mathbb{C}} \to X$ is an isomorphism of algebraic varieties over $\mathbb{C}$, where $(X_0)_{\mathbb{C}}$ denotes the fibre product $X_0 \times_{\mathrm{Spec}\,\mathbb{R}} \mathrm{Spec}\,\mathbb{C}$. If no confusion arises, one simply says that X_0 is a real model of X.

If (X_0, f) is a real model of X, there is a natural action of the Galois group $G := Gal(\mathbb{C}, \mathbb{R}) = \{1, \sigma\} \simeq \mathbb{Z}_2 := \mathbb{Z}/2\mathbb{Z}$ on $(X_0)_{\mathbb{C}}$ which induces, via f, a G-action on X. In particular, σ determines an antiregular involution $S: X \to X$. In conclusion the datum of the real model (X_0, f) gives rise to a real structure (X, S) on X, which we will say to be *determined* by the real model (X_0, f). Now we want to indicate that also the converse holds.

We notice that, if X is a complex algebraic variety defined over $\mathbb{R}$, then clearly X^σ is the same as X. Therefore, if we are given a real model (X_0, f) of X, and if $\sigma: X \to X^\sigma$ is the conjugation, then the holomorphic map $\phi := f^\sigma \circ f^{-1}: X \to X^\sigma$

is an isomorphism such that

$$\sigma^{-1} \circ \phi = \phi^{-1} \circ \sigma \tag{2.2.1}$$

which, in local coordinates $(z_1, \ldots, z_n)$, means that $\overline{\phi(z_1, \ldots, z_n)} = \phi^{-1}(\overline{z_1}, \ldots, \overline{z_n})$.

Given a complex variety X, an isomorphism $\phi \colon X \to X^\sigma$ enjoying property (2.2.1) will be called a *descent datum* for X. As we saw, the datum of a real model (X_0, f) of X gives rise to a descent datum for X, which is said to be *determined* by the real model. This result can be inverted by means of the following theorem.

Theorem 2.2.2 (A. Weil). *The following data on a quasi-projective, complex variety X are equivalent:*

(i) *a real model (X_0, f);*

(ii) *a real structure (X, S) with S antiregular (determined by (X_0, f));*

(iii) *a descent datum $\phi \colon X \to X^\sigma$ for X, with ϕ an isomorphism of complex algebraic varieties.*

Proof. As we saw, the datum (i) determines data (ii) and (iii). On the other hand, once the real structure (X, S) is given, we have that $\phi := \sigma_X \circ S$ is a descent datum. Finally a descent datum determines a real model, as in [W1]. □

Two equivalent data as in the previous theorem will be said to be *related*. We have the following.

Corollary 2.2.3. *Two real structures (X, S) and (X, S') on a smooth complex quasi-projective variety X, with S and S' antiregular, are equivalent if and only if the related real models are isomorphic.*

We leave the easy proof as an exercise for the reader.

Examples 2.2.4. (i) The two different real structures on $\mathbb{P}^1_{\mathbb{C}}$ introduced in Example 2.1.1, (ii), correspond to the two real models given by

$$\mathrm{Proj}(\mathbb{C}[x, y]), \quad \mathrm{Proj}(\mathbb{C}[x, y, t]/(x^2 + y^2 + t^2)).$$

In other words, the former corresponds to a real projective line, the latter to a conic with no real points.

(ii) Now we give another example of a complex variety with two non equivalent real structures. Let us consider the two cubic curves X_0, X_0' in the complex projective plane $\mathbb{P}^2_{\mathbb{C}}$ respectively given by the equations

$$y^2 z = x^3 - xz^2, \ y^2 z + x^3 + xz^2 = 0.$$

The projective transformation

$$x' = ix, \ y' = \zeta y, \ z' = z$$

where $\zeta^2 = i$ and, as usual, $i^2 = -1$, carries X_0 to X'_0, which are therefore both isomorphic to a given compact Riemann surface X. Since X_0, X'_0 are both defined over $\mathbb{R}$, they provide real models for X, and therefore real structures on X. We claim that these real structures are not equivalent. In fact it suffices to remark that while $X'_0(\mathbb{R})$ is connected, $X_0(\mathbb{R})$ is not connected, as one easily sees by looking at the affine equations

$$y^2 = x^3 - x, \; y^2 + x^3 + x = 0$$

of the two curves.

2.3. The action of conjugation on functions and forms

Let (X, S) and (X', S') be real structures, and let $\mathcal{F}(X, X')$ be the set of holomorphic maps from X to X'. There is a natural bijective map

$$\varphi := \varphi_{S',S}\colon f \in \mathcal{F}(X, X') \to S' \circ f \circ S \in \mathcal{F}(X, X')$$

such that $\varphi^2 = \text{id}$. We notice that if x' is a point in X' and $f_{x'}$ is the constant map with value x', then $\varphi(f_{x'}) = f_{S'(x')}$.

This can be applied, in particular, when (X', S') is either $\mathbb{C}$ or $\mathbb{P}^1_{\mathbb{C}}$ with the standard structure. In that case we have that $\varphi := \varphi_{\sigma,S}$ is an automorphism of $\mathcal{O}(X)$, or of $\mathcal{M}(X)$, where $\mathcal{O}(X)$ [resp. $\mathcal{M}(X)$] is the ring [resp. the field] of holomorphic [resp. meromorphic] functions on X.

We notice the following proposition.

Proposition 2.3.1. *A continuous involution $S\colon X \to X$ on a complex variety X is an antiinvolution if and only if for every non empty open subset U of X there is the map*

$$\varphi := \varphi_{U,S}\colon f \in \mathcal{O}(U) \to \sigma \circ f \circ S \in \mathcal{O}(S(U))$$

which is a ring isomorphism.

The easy proof can be left as an exercise for the reader.

In a more general setting, let us consider a real structure (X, S) on X and a vector bundle $\pi\colon F \to X$ on X. We can define a new vector bundle $\pi^S\colon F^S \to X$ in the following way. Suppose $\pi\colon F \to X$ is determined by a cocycle (U_{ij}, g_{ij}), i.e., by an open covering (U_i) and transition maps g_{ij} on $U_{ij} = U_i \cap U_j$. Then one defines $\pi^S\colon F^S \to X$ by the cocycle $(S(U_{ij}), \varphi(g_{ij}))$. It is not difficult to see that the definition is independent of the particular cocycle, and therefore it is a good definition. It is also clear that there is an antiholomorphic map of vector bundles, denoted by $S_F\colon F^S \to F$, which is fiberwise antilinear and such that $\pi \circ S_F = S \circ \pi^S$. If no confusion arises, we will denote S_F simply by S. For every non empty open subset U of X there is an antilinear map

$$\varphi := \varphi_{U,S}\colon f \in \Gamma(U, F) \to S \circ f \circ S \in \Gamma(S(U), F^S).$$

We will often write f^S instead of $\varphi(f)$. We record the following facts:

(i) $F \simeq G$ if and only if $F^S \simeq G^S$. Therefore F is trivial if and only if F^S is trivial. In particular if F is trivial of rank 1 the maps $\varphi_{U,S}$ are exactly those defined in the statement of Proposition 2.3.1.

(ii) $\mathrm{rk}(F^S) = \mathrm{rk}(F)$.

(iii) $(F^S)^* = (F^*)^S$ and $(F^S)^S = F$.

(iv) $F^S \oplus G^S = (F \oplus G)^S$, $F^S \otimes G^S = (F \otimes G)^S$, etc. In particular the real structure (X, S) determines an action of the conjugation S on the *Picard group* of X, i.e., an involutory group homomorphism

$$\varphi := \varphi_S : L \in \mathrm{Pic}(X) \to L^S \in \mathrm{Pic}(X)$$

on which we will return later on.

(v) If Θ_X is the tangent bundle of X, then $\Theta_X^S \simeq \Theta_X$. Indeed, if (U_i) is a system of local charts for X with coordinates $(z_1^i, \dots, z_n^i)$, also $(S(U_i))$ is a system of local charts with coordinates $(\overline{\xi}_1^i, \dots, \overline{\xi}_n^i)$, where we have set $\xi_j^i = z_j^i \circ S$ for all $j = 1, \dots, n$. The vector bundle Θ_X is defined by the cocycle (U_{ij}, θ_{ij}), where

$$\theta_{ij} = \frac{\partial(z_1^i, \dots, z_n^i)}{\partial(z_1^j, \dots, z_n^j)}.$$

On the other hand, according to our definition, Θ_X^S is defined by the cocycle $(S(U_{ij}), \vartheta_{ij})$, where

$$\vartheta_{ij} = \overline{\theta_{ij} \circ S} = \overline{\frac{\partial(\xi_1^i, \dots, \xi_n^i)}{\partial(\xi_1^j, \dots, \xi_n^j)}} = \frac{\partial(\overline{\xi}_1^i, \dots, \overline{\xi}_n^i)}{\partial(\overline{\xi}_1^j, \dots, \overline{\xi}_n^j)}$$

whence our assertion follows.

Let $\Omega_X^p = \wedge^p \Theta_X^*$ be the bundle of p-forms on X which are also such that $(\Omega_X^p)^S \simeq \Omega_X^p$. Then, for every non empty open subset U in X, we have an antilinear map

$$\varphi := \varphi_S : \omega \in \Gamma(U, \Omega_X^p) \to S \circ \omega \circ S \in \Gamma(S(U), \Omega_X^p)$$

which we can describe in terms of local coordinates. Indeed, if on the open subset U of X we have coordinates $(z_1, \dots, z_n)$, and if

$$\omega = \sum_{i_1, \dots, i_p} a_{i_1, \dots, i_p}(z_1, \dots, z_n) dz_{i_1} \wedge \dots \wedge dz_{i_p}$$

on U, then we have

$$\omega^S = \sum_{i_1, \dots, i_p} \overline{a}_{i_1, \dots, i_p}(\overline{\xi_1}, \dots, \overline{\xi_n}) d\,\overline{\xi}_{i_1} \wedge \dots \wedge d\,\overline{\xi}_{i_p}$$

on $S(U)$ with local coordinates $(\overline{\xi}_1, \dots, \overline{\xi}_n)$.

With this description in mind it is quite easy to see that, if ω is a holomorphic p-form on X and if γ is a regular p-cycle, then one has

$$\int_{\gamma} \omega = \overline{\int_{S(\gamma)} \omega^S}. \tag{2.3.2}$$

In particular we can consider the antilinear involution ϕ on global sections

$$\varphi \colon H^0(X, \Omega_X^p) \to H^0(X, \Omega_X^p)$$

which is an $\mathbb{R}$-linear involution. Then we have a decomposition

$$H^0(X, \Omega_X^p) = H^0(X, \Omega_X^p)^+ \oplus H^0(X, \Omega_X^p)^-$$

in ± 1-eigenspaces as an $\mathbb{R}$-vector space. We will say that the forms in $H^0(X, \Omega_X^p)^+$ [resp. in $H^0(X, \Omega_X^p)^-$] are *real* forms [resp. *purely imaginary* forms]. Indeed a given form $\omega \in H^0(X, \Omega_X^p)$ can be uniquely written as

$$\omega = \omega^+ + \omega^-, \quad \omega^\pm \in H^0(X, \Omega_X^p)^\pm$$

where

$$\omega^\pm = \frac{\omega \pm \omega^S}{2}.$$

Furthermore $iH^0(X, \Omega_X^p)^\pm = H^0(X, \Omega_X^p)^\mp$. This implies that the two spaces $H^0(X, \Omega_X^p)^\pm$ are isomorphic as real vector spaces, and both generate $H^0(X, \Omega_X^p)$ as a complex vector space. In other words $H^0(X, \Omega_X^p)$ is the complexification of $H^0(X, \Omega_X^p)^+$, and summing up we can state the following result.

Proposition 2.3.3. *There are bases of $H^0(X, \Omega_X^p)$ formed by real [resp. purely imaginary] forms.*

Remark 2.3.4. Given a differentiable vector bundle $\pi \colon F \to X$ one can consider the vector bundle $S^*(F)$ on X which is the pull-back of F via S. If F is defined by the cocycle (U_{ij}, g_{ij}), then $S^*(F)$ is defined by the cocycle $(S(U_{ij}), g_{ij} \circ S)$. The maps $\varphi_{U,S}$ defined in Proposition 2.3.1 should not be confused with the maps

$$S^* := S_U^* \colon f \in \mathcal{C}^\infty(U, F) \to f \circ S \in \mathcal{C}^\infty(S(U), S^*(F))$$

where $\mathcal{C}^\infty(U, F)$ is the space of differentiable sections of F. For instance S^* sends a holomorphic p-form to a form of type $(0, p)$ on X, and, more generally, a form ω of type (p, q) to a form of type (q, p). If in local coordinates, with the usual meaning of the notation, we have

$$\omega = \sum_{|I|=p, |J|=q} a_{I,J}(z, \bar{z})\, dz_I \wedge d\bar{z}_J$$

then

$$S^*(\omega) = \sum_{|I|=p,|J|=q} a_{I,J}(\overline{z}, z) d\overline{z}_I \wedge dz_J.$$

It is also easy to see that

$$S^* \circ \partial = \overline{\partial} \circ S^*, S^* \circ \overline{\partial} = \partial \circ S^*$$

where ∂ and $\overline{\partial}$ are the usual operators (see [GH], p. 24). Therefore

$$S^*(H^{p,q}(X)) = H^{q,p}(X)$$

(see again [GH] for the basics of Hodge theory).

2.4. The action of conjugation on the cohomology

Let (X, S) be a real structure on the variety X, which we will assume now to be smooth, projective and irreducible. There is then a natural action of $G = \mathrm{Gal}(\mathbb{C}, \mathbb{R}) = \{1, S\} \simeq \mathbb{Z}_2 := \mathbb{Z}/2\mathbb{Z}$ on X due to the existence of the continuous involution S. We set $Y := X/G$ and we denote by $\pi: X \to Y$ the natural projection. We will regard Y as a topological space, with the natural quotient topology. Hence π is continuous. We will set $B_i = \dim_{\mathbb{Q}} H^i(X, \mathbb{Q})$, $b_i = \dim_{\mathbb{Q}} H^i(Y, \mathbb{Q})$, for the *Betti numbers* of X and Y. We note that the G-action on X determines a G-action on the cohomology of X. It is then clear that we have the following decomposition of $\mathbb{Q}$-vector spaces

$$H^i(X, \mathbb{Q}) = H^i(X, \mathbb{Q})^+ \oplus H^i(X, \mathbb{Q})^-$$

where $H^i(X, \mathbb{Q})^{\pm}$ is the ± 1-eigenspace of $H^i(X, \mathbb{Q})$ for the G-action. Of course one has the natural isomorphism

$$H^i(Y, \mathbb{Q}) \simeq H^i(X, \mathbb{Q})^+$$

hence

$$b_i = \dim_{\mathbb{Q}} H^i(X, \mathbb{Q})^+.$$

We have the following:

Theorem 2.4.1. *The following formula for the Euler–Poincaré characteristic $\chi(X_S)$ of X_S holds:*

$$\chi(X_S) = \sum_{i \equiv 0 \pmod 2} (2b_i - B_i).$$

Proof. We can triangulate X in such a way that X_S is a subcomplex of the triangulation and the triangulation is fixed by the G-action. So the triangulation descends to a triangulation of Y. Using this triangulation to compute the Euler–Poincaré characteristic, one easily finds

$$\chi(X) = 2\chi(Y) - \chi(X_S)$$

whence

$$\chi(X_S) = \sum_i (2b_i - B_i).$$

The conclusion now follows from the fact that for odd values of i one has $B_i = 2b_i$, as immediately follows from Remark 2.3.4. □

Now we go deeper into the study of the G-action on the cohomology of X. We consider the constant sheaf $\mathbb{Z}_2$ on X and the sheaf $\mathcal{F} := \pi_*(\mathbb{Z}_2)$ on Y. Note that X_S can be naturally considered both as a subspace of X and as a subspace of Y. Hence we have the restriction map $r: \mathcal{F} \to \mathcal{F}_{|X_S}$. The G-action on X extends to an obvious G-action on $\mathcal{F}$.

Theorem 2.4.2. *The following exact sequence of sheaves on Y*

$$0 \to (1+S)\mathcal{F} \to \mathcal{F} \overset{(1+S)\oplus r}{\longrightarrow} (1+S)\mathcal{F} \oplus \mathcal{F}_{|X_S} \to 0 \tag{2.4.3}$$

is exact.

Proof. It suffices to check exactness on the stalks. If $x \in X_S$, then $(1+S)\mathcal{F}_x = 0$ and the sequence is clearly exact. If $x \notin X_S$, then $\mathcal{F}_x \simeq \mathbb{Z}_2 \oplus \mathbb{Z}_2$, and

$$(1+S)_x : (a,b) \in \mathcal{F}_x \simeq \mathbb{Z}_2 \oplus \mathbb{Z}_2 \to (a+b, a+b) \in \mathcal{F}_x \simeq \mathbb{Z}_2 \oplus \mathbb{Z}_2.$$

It is then clear that $\ker(1+S)_x \simeq (1+S)_x \mathcal{F}_x$. □

Now we can take the long exact cohomology sequence of (2.4.3), getting in this way the so called *Smith's sequence*:

$$\cdots \to H^m(Y, (1+S)\mathcal{F}) \overset{\beta_m}{\to} H^m(Y, \mathcal{F}) \overset{\alpha_m}{\to} H^m(Y, (1+S)\mathcal{F}) \oplus H^m(X_S, \mathcal{F}_{|X_S}) \to \ldots$$

We set

$$a_m := \dim_{\mathbb{Z}_2}(\ker \alpha_m)$$

and we prove the following lemma.

Lemma 2.4.4. *One has*

$$\sum_m \dim_{\mathbb{Z}_2} H^m(X_S, \mathbb{Z}_2) = \sum_m (\dim_{\mathbb{Z}_2} H^m(X, \mathbb{Z}_2) - 2a_m).$$

Proof. By Leray's theorem, we have

$$H^m(Y, \mathcal{F}) \simeq H^m(X, \mathbb{Z}_2)$$

and moreover one clearly has

$$H^m(X_S, \mathcal{F}_{|X_S}) \simeq H^m(X_S, \mathbb{Z}_2).$$

Hence from Smith's sequence we deduce the exact sequence

$$0 \to \operatorname{Im}\alpha_{m-1} \to H^{m-1}(Y,(1+S)\mathcal{F}) \oplus H^{m-1}(X_S,\mathbb{Z}_2) \\ \to H^m(Y,(1+S)\mathcal{F}) \to \ker\alpha_m \to 0$$

which yields

$$\dim_{\mathbb{Z}_2} H^m(Y,(1+S)\mathcal{F}) - \dim_{\mathbb{Z}_2} H^{m-1}(Y,(1+S)\mathcal{F}) - \dim_{\mathbb{Z}_2} H^{m-1}(X_S,\mathbb{Z}_2) \\ = a_m - \dim_{\mathbb{Z}_2} H^{m-1}(X,\mathbb{Z}_2) + a_{m-1}.$$

The assertion then follows by adding up the above relations on m. □

We now prove the following basic lemma.

Lemma 2.4.5. *One has*

$$\ker\alpha_m \supseteq (1+S)H^m(Y,\mathcal{F}) = (1+S)H^m(X,\mathbb{Z}_2).$$

Proof. The assertion follows from the following facts

$$(1+S)^2 = 0, \quad r_{|(1+S)H^m(Y,\mathcal{F})} = 0$$

which are easily proved by observing that we are working with $\mathbb{Z}_2$-vector spaces. □

Now we set

$$\lambda_m := \lambda_m(X,S) := \dim_{\mathbb{Z}_2}(1+S)H^m(X,\mathbb{Z}_2),$$

and we will call this number the *m-th Comessatti character* of the real structure (X, S), or simply of X, if no confusion arises. In view of Lemma 2.4.5, we have the basic inequality

$$\lambda_m \le a_m. \tag{2.4.6}$$

By plugging (2.4.6) into the equality of Lemma 2.4.4, we find:

Theorem 2.4.7 (Harnack–Thom–Krasnov). *If (X, S) is a real structure, one has*

$$\sum_m \dim_{\mathbb{Z}_2} H^m(X_S,\mathbb{Z}_2) \le \sum_m (\dim_{\mathbb{Z}_2} H^m(X,\mathbb{Z}_2) - 2\lambda_m).$$

A real structure (X, S), or simply a variety X, is said to be *Galois maximal* (GM) if the equality holds in (2.4.6) for all $m \ge 0$, i.e., if the Harnack–Thom–Krasnov (HTK) inequality turns into an equality. We will give later on examples of GM-varieties, but we warn the reader: there also exist varieties which are not GM-varieties, as indicated in [S1], chapt. V, §4, and [K].

Remarks 2.4.8. (i) The reason why one uses the HTK-inequality instead of the equality provided by Lemma 2.4.4 is that the Comessatti characters of real structures are more computable than the invariants a_m, as we will indicate in the next section.

(ii) Suppose $X_S \neq \emptyset$. Of course $a_m = 0$ yields $\lambda_m = 0$. Since

$$r\colon H^0(Y, \mathcal{F}) \simeq H^0(X, \mathbb{Z}_2) \simeq \mathbb{Z}_2 \to H^0(X_S, \mathcal{F}_{|X_S}) \simeq H^0(X_S, \mathbb{Z}_2) \simeq \mathbb{Z}_2{}^{\nu(X)}$$

is clearly injective, we have $a_0 = \lambda_0 = 0$. Also, if we let $n := \dim X$ and we assume $n \geq 1$, we have $a_m = \lambda_m = 0$ for all $m \geq 2n+1$. But we claim that also $a_{2n} = \lambda_{2n} = 0$. Indeed we have $H^{2n}(X_S, \mathcal{F}_{|X_S}) \simeq H^{2n}(X_S, \mathbb{Z}_2) = 0$ and $H^{2n}(Y, \mathcal{F}) \simeq H^{2n}(X, \mathbb{Z}_2) \simeq \mathbb{Z}_2$. From the exact sequence

$$\cdots \to H^{2n}(Y, (1+S)\mathcal{F}) \to H^{2n}(Y, \mathcal{F}) \xrightarrow{\alpha_{2n}} H^{2n}(Y, (1+S)\mathcal{F}) \to 0$$

we deduce that $H^{2n}(Y, (1+S)\mathcal{F}) \simeq \mathbb{Z}_2$ and that α_{2n} is an isomorphism.

(iii) In the case X is a Riemann surface, i.e., $n = 1$ with the above notation, the only relevant Comessatti character is λ_1, which we shall briefly denote by λ. We will return later on on the meaning of this important invariant. For the time being we notice that in this case we have

$$\dim_{\mathbb{Z}_2} H^0(X_S, \mathbb{Z}_2) = \dim_{\mathbb{Z}_2} H^1(X_S, \mathbb{Z}_2) = \nu(X)$$

and $\dim_{\mathbb{Z}_2} H^1(X, \mathbb{Z}_2) = 2g$, where $g := g(X)$ is the *genus* of the Riemann surface X. Then the HTK-inequality reads

$$\nu(X) \leq g + 1 - \lambda, \tag{2.4.9}$$

whence we deduce

$$\nu(X) \leq g + 1, \tag{2.4.10}$$

which is the classical *theorem of Harnack*.

2.5. A theorem of Comessatti

The purpose of this section is to study, in a certain generality, the situation which we already came accross in the previous paragraph, of an involution acting on the cohomology of a variety. We will prove a basic result, due to Comessatti [C1], which enables us to find a *normal form* for such an involution. This will turn out to be crucial in what follows.

As announced it is convenient to put ourselves in a rather general setting. Indeed we will consider a pair (M, S) where M is a $\mathbb{Z}$-module (i.e., an abelian group) and S is an involutory automorphism of M. Such a pair will be called a *Comessatti pair*. If M is free of rank m, the Comessatti pair itself will be said to be *free*. For the rest of this section we will assume to be in this situation. We introduce the following notation: for every positive integer k, we set $M_k := M/kM = M \otimes_{\mathbb{Z}} (\mathbb{Z}/k\mathbb{Z}) \simeq \mathbb{Z}_k^m$.

Of course, S induces involutions on M_k for all k's, which, by abuse of notations, will be still denoted by S.

We consider the two submodules

$$M^+ = \{x \in M : S(x) = x\}, \quad M^- = \{x \in M : S(x) = -x\},$$

and we set

$$\text{rk}(M^\pm) := s^\pm.$$

Of course we have $s^+ + s^- = m$. Indeed, $s^\pm$ is the dimension of the eigenspace corresponding to the eigenvalue ± 1 of the extension of S to the $\mathbb{C}$-vector space $M \otimes_{\mathbb{Z}} \mathbb{C}$. We will call s^+ the *symmetry index* of the Comessatti pair (M, S). We also define $N := M^+ \oplus M^-$. This is a free submodule of rank m of M. Hence M/N is a torsion module, which we want to study.

First of all we notice that $N \supseteq 2M$. Indeed, if $x \in M$ we have $2x = (x + S(x)) + (x - S(x))$ and $x \pm S(x) \in M^\pm$. Therefore there is a surjective homomorphism $M_2 \to M/N$, hence M/N, as well as M_2, is a $\mathbb{Z}_2$-vector space. We set $\lambda(M) := \dim_{\mathbb{Z}_2} M/N$ and we call $\lambda := \lambda(M)$ the *Comessatti character* of the pair (M, S). We let μ be the maximum number of elements of N which belong to some basis of M, and we notice that

$$\mu \leq m - \lambda.$$

In fact, if $(x_1, \ldots, x_m)$ is a basis of M whose first t elements belong to N, then the images of $x_{t+1}, \ldots, x_m$ in M/N have to span M/N, and therefore one has $m - t \geq \lambda$.

Now we prove the following result.

Lemma 2.5.1. *One has*

$$x \in N \Leftrightarrow x + S(x) \in 2M \Leftrightarrow x - S(x) \in 2M$$

and

$$\lambda = \dim_{\mathbb{Z}_2}(1 + S)M_2 = \dim_{\mathbb{Z}_2}(1 - S)M_2.$$

Proof. The second assertion easily follows from the first. The last equivalence in the first assertion is clear. The first implication is easy to check, and therefore we leave it to the reader. As to the converse implication, we observe that if $2y = x + S(x)$, then clearly $y \in M^+$. Therefore we have $x + S(x) = 2y = y + S(y)$, hence $x - y \in M^-$, whence the assertion. □

Remarks 2.5.2. (i) Let (X, S) be a real structure where X is a projective variety such that its cohomology has no torsion. Then for every positive integer i we have the free Comessatti pair $(H^i(X, \mathbb{Z}), S)$ where we again denote by S the action of S on the cohomology. Then the i-th Comessatti character λ_i of the real structure, as defined in the previous section is nothing but the Comessatti character of the pair $(H^i(X, \mathbb{Z}), S)$. We will denote by s_i^+ the symmetry index of $(H^i(X, \mathbb{Z}), S)$ and we will call it the *i-th symmetry index* of (X, S).

(ii) By the very definition of the Comessatti character of the pair (M, S), we have that $\lambda \leq m$. But we can actually say more. Indeed one has

$$\lambda \leq \min\{s^+, s^-\}.$$

In fact it is easy to see that $M^\pm \cap 2M = 2M^\pm$. On the other hand, we clearly have an injection

$$(1 \pm S)M_2 \to (M^\pm + 2M)/2M \simeq M^\pm/M^\pm \cap 2M = (M^\pm)_2$$

and the assertion follows from the fact that $\dim_{\mathbb{Z}_2} M^\pm = s^\pm$.

Now we are ready to prove the main result of this section.

Theorem 2.5.3 (A. Comessatti). *Let (M, S) be a free Comessatti pair. Then there are elements $\sigma_1, \ldots, \sigma_{s^+} \in M^+$, $\epsilon_1, \ldots, \epsilon_{s^-} \in M^-$, $\tau_1, \ldots, \tau_\lambda \in M$ such that:*

(i) $(1 + S)(\tau_i) = \sigma_i$, $(1 - S)(\tau_i) = \epsilon_i$, *for all* $i = 1, \ldots, \lambda$;

(ii) *the two m-tuples*

(I) $(\sigma_1, \ldots, \sigma_{s^+}, \tau_1, \ldots, \tau_\lambda, \epsilon_{\lambda+1}, \ldots, \epsilon_{s^-})$

(II) $(\tau_1, \ldots, \tau_\lambda, \sigma_{\lambda+1}, \ldots, \sigma_{s^+}, \epsilon_1, \ldots, \epsilon_{s^-})$

are both bases of M.

Proof. The proof is by induction on λ. If $\lambda = 0$, then $\dim_{\mathbb{Z}_2} M/N = 0$, i.e., $M = N = M^+ \oplus M^-$. Then it suffices to take for (I) a basis $(\sigma_1, \ldots, \sigma_{s^+})$ of M^+ and for (II) a basis $(\epsilon_1, \ldots, \epsilon_{s^-})$ of M^-.

If $\lambda > 0$, then N is a proper submodule of M. Then we can choose an element $\tau \in M - N$ such that $(1 + S)(\tau)$ and $(1 - S)(\tau)$ are both non-zero. Let σ_1 and ϵ_1 be *primitive* elements of M (i.e., $\sigma_1 \neq 0$ and $\sigma_1 = ky$, with k an integer and $y \in M - \{0\}$, implies $k = \pm 1$, and similarly for ϵ_1), such that

$$(1 + S)(\tau) = k\sigma_1, \; (1 - S)(\tau) = l\tau$$

where $k, l \in \mathbb{Z}$. Since $\tau \in M - N$, the integers k and l are odd by Lemma 2.5.2. Let

$$\tau_1 = \tau - \frac{k-1}{2}\sigma_1 - \frac{l-1}{2}\epsilon_1.$$

Then

$$(1+S)(\tau_1) = k\sigma_1 - \frac{k-1}{2}(\sigma_1 + S(\sigma_1)) + \frac{l-1}{2}(\epsilon_1 + S(\epsilon_1)) = k\sigma_1 - (k-1)\sigma_1 = \sigma_1$$

and similarly

$$(1 - S)(\tau_1) = \epsilon_1.$$

Let P be the submodule of M generated by σ_1 and τ_1. We claim that $M' := M/P$ is torsion free. Assume for the moment that this is true, and let us see how to conclude the proof. Let $p: M \to M'$ be the canonical epimorphism. Since $S(P) \subseteq P$, we have that S induces an involution S' on M' such that $S'(p(x)) = p(S(x))$ for any

$x \in M$. So we have a new free Comessatti pair (M', S'), and it is clear that $\lambda' := \lambda(M') < \lambda$ and $\mathrm{rk}(M')^{\pm} = s^{\pm} - 1$. Moreover we notice that the canonical maps $p^{\pm}: M^{\pm} \to (M')^{\pm}$ induced by p are both surjective. In fact, if $m \in M$ is such that $S(m) - m = a\sigma_1 + b\tau_1 \in P$, then $a\sigma_1 + b\tau_1 \in M^-$ and therefore

$$0 = S(a\sigma_1 + b\tau_1) + (a\sigma_1 + b\tau_1) = (a+b)\sigma_1 - b\tau_1 + (a\sigma_1 + b\tau_1) = (2a+b)\sigma_1$$

which implies $2a + b = 0$. But then we have

$$S(m - a\tau_1) = S(m) - aS(\tau_1) = m + a\sigma_1 + b\tau_1 + a(\tau_1 - \sigma_1) = m + (b+a)\tau_1 = m - a\tau_1$$

proving that the map p^+ is surjective. A similar proof works for p^-.

By applying the induction to (M', S') and using the surjectivity of $p^{\pm}$ one sees that there are elements $\sigma_2, \ldots, \sigma_{s^+} \in M^+$, $\epsilon_2, \ldots, \epsilon_{s^-} \in M^-$, $\tau_2, \ldots, \tau_\lambda \in M$ such that:

(i) $(1 + S)(\tau_i) = \sigma_i$, $(1 - S)(\tau_i) = \epsilon_i$, for all $i = 2, \ldots, \lambda$;

(ii) the two $m - 2$-tuples

$$(p(\sigma_2), \ldots, p(\sigma_{s^+}), p(\tau_2), \ldots, p(\tau_\lambda), p(\epsilon_{\lambda+1}), \ldots, p(\epsilon_{s^-}))$$

$$(p(\tau_2), \ldots, p(\tau_\lambda), p(\sigma_{\lambda+1}), \ldots, p(\sigma_{s^+}), p(\epsilon_2), \ldots, p(\epsilon_{s^-}))$$

are both bases of M'. Then the elements $\sigma_1, \ldots, \sigma_{s^+} \in M^+$, $\epsilon_1, \ldots, \epsilon_{s^-} \in M^-$, $\tau_1, \ldots, \tau_\lambda \in M$ clearly satisfy the statement of the theorem.

Finally we prove that $M' = M/P$ is torsion free. Since σ_1 is primitive, then $M/\mathbb{Z} \cdot \sigma_1$ is torsion free. Since $M' \simeq (M/\mathbb{Z} \cdot \sigma_1)/(P/\mathbb{Z} \cdot \sigma_1)$, and $P/\mathbb{Z} \cdot \sigma_1$ is generated by $\tau_1 + \mathbb{Z} \cdot \sigma_1$, it suffices to prove that $\tau_1 + \mathbb{Z} \cdot \sigma_1$ is primitive in $M/\mathbb{Z} \cdot \sigma_1$. Let $x \in M$ and $n \in \mathbb{Z}$ be such that $nx \equiv \tau_1 \pmod{\mathbb{Z} \cdot \sigma_1}$. Then $n(1 - S)(x) = (1 - S)(nx) = (1 - S)(\tau_1) = \epsilon_1$, and since ϵ_1 is also primitive, one has $n = \pm 1$, whence the assertion. □

Remarks 2.5.4. (i) The bases of type (I) [resp. (II)] in the statement of Theorem 2.5.3 are called *bases of the first type* [resp. *bases of the second type*] of the Comessatti pair (M, S). We notice that, by exchanging (M, S) with $(M, -S)$, the bases of the two types are also exchanged.

(ii) We want to write down the matrix of S in a basis of the first [resp. second] type. In order to do this, we introduce the following notation. We denote, as usual, by I_n the unitary matrix of order n, and by $0_{a,b}$ the 0-matrix of type $a \times b$. Furthermore we denote by $I_{a,b,c}$ the matrix of type $a \times b$ having the first c entries of the main diagonal equal to 1, and all other elements equal to 0. Then the matrix of S in a basis of the first type is clearly of the form

$$\begin{pmatrix} I_{s^+} & I_{s^+,s^-,\lambda} \\ 0_{s^-,s^+} & -I_{s^-} \end{pmatrix}$$

whereas the matrix of S in a basis of the second type is of the form

$$\begin{pmatrix} I_{s^+} & 0_{s^+,s^-} \\ -I_{s^-,s^+,\lambda} & -I_{s^-} \end{pmatrix}.$$

2.6. Group cohomology

In this section we give, with no aim to completeness, a brief account on the basic facts concerning cohomology of groups, which will be used later. For further information on this subject, see for example [Se1], [R] or [HS].

Let G be a multiplicative group. We will consider the *group algebra* $\mathbb{Z}(G)$, which by definition is the set of all finite sums $\sum_i m_i g_i$, where $m_i \in \mathbb{Z}$ and $g_i \in G$. This set has a natural structure of algebra, with the obvious sum and the multiplication which naturally extends the one in G, i.e.,

$$\sum_i m_i g_i \cdot \sum_j m'_j g'_j := \sum_{i,j} m_i m'_j g_i g'_j.$$

We note that $\mathbb{Z}(G)$ is commutative if and only if G is abelian. Let M be a right $\mathbb{Z}(G)$-module. We remark that this is the same as giving a $\mathbb{Z}$-module M with a right G-action, i.e., an abelian group with a representation of G as a group of endomorphisms of M. If $g \in G$ and $x \in M$, we denote by x^g the result of the action of g on x. We will systematically use in the rest of this section this *exponential notation* to denote maps, actions, etc.

For all non negative integers n we consider the set $C^n(G, M)$ of all maps

$$\phi\colon (g_1, \dots, g_n) \in G^n \to (g_1, \dots, g_n)^\phi \in M$$

such that $(g_1, \dots, g_n)^\phi = 0$ as soon as $g_i = 1$ for some $i = 1, \dots, n$. The elements of $C^n(G, M)$ are called *n-cochains* of G with values in M. Of course $C^n(G, M)$ itself is a right $\mathbb{Z}(G)$-module with the obvious operations: if $\phi, \psi \in C^n(G, M)$ and if $m_1 g_1 + \dots + m_k g_k \in \mathbb{Z}(G)$ then

$$(g_1, \dots, g_n)^{\phi+\psi} := (g_1, \dots, g_n)^\phi + (g_1, \dots, g_n)^\psi$$

$$(g_1, \dots, g_n)^{\phi\cdot(m_1 g_1 + \dots + m_k g_k)} := ((g_1, \dots, g_n)^\phi)^{(m_1 g_1 + \dots + m_k g_k)}.$$

Now we define a *coboundary operator*

$$\delta := \delta^n\colon \phi \in C^n(G, M) \to \phi^\delta \in C^{n+1}(G, M)$$

in the following way:

$$(g_1, \dots, g_{n+1})^{\phi^\delta} := (g_2, \dots, g_{n+1})^\phi$$
$$+ \sum_{i=1}^{n} (-1)^i (g_1, \dots, g_{i-1}, g_i g_{i+1}, g_{i+2}, \dots, g_{n+1})^\phi + (-1)^{n+1}((g_1, \dots, g_n)^\phi)^{g_{n+1}}.$$

It is quite easy to check that $\phi^\delta \in C^{n+1}(G, M)$ and that $\delta^{n+1} \circ \delta^n = 0$. We set $Z^n(G, M) := \ker \delta^n$ and $B^n(G, M) := \operatorname{Im} \delta^{n-1}$. The elements of $Z^n(G, M)$ [resp. of $B^n(G, M)$] are called *n-cocycles* [resp. *n-coboundaries*] of G with coefficients in M. Of course we have $Z^n(G, M) \supseteq B^n(G, M)$, and we set $H^n(G, M) := Z^n(G, M)/B^n(G, M)$ and call this $\mathbb{Z}(G)$-module the *m-th cohomology module* of G with coefficients in M.

Examples 2.6.1. (i) Let $n = 0$. By definition we have $C^0(G, M) = M$. If $x \in M$, we have $g^{x^\delta} = x - x^g$. Therefore $Z^0(G, M) = H^0(G, M)$ is the submodule of fixed points by the action of G on M, which we will denote by M^G.

(ii) Let $n = 1$ and let $\phi \in C^1(G, M)$. We have

$$(g_1, g_2)^{\phi^\delta} = g_2{}^\phi - (g_1g_2)^\phi + (g_1{}^\phi)^{g_2}.$$

Thus we have

$$Z^1(G, M) = \{\phi: G \to M : (g_1g_2)^\phi = g_2{}^\phi + (g_1{}^\phi)^{g_2}\}$$
$$B^1(G, M) = \{\phi_x: G \to M : g^{\phi_x} = x - x^g, x \in M\}$$

and this enables us to compute $H^1(G, M)$.

(iii) We consider now the particular case in which $G = \{1, S\} \simeq \mathbb{Z}_2$. Then to give a $\mathbb{Z}(G)$-module is equivalent to give a Comessatti pair (M, S). We note that, for every non negative integer n, we have an isomorphism of $\mathbb{Z}(G)$-modules

$$\phi \in C^n(G, M) \to (S, \ldots, S)^\phi \in M$$

and from now on we will identify $C^n(G, M)$ with M via this isomorphism. Then the n-th coboundary operator can be regarded as a map $\delta^n: M \to M$, which can be easily seen to be nothing else than the map $1 + (-1)^{n+1}S$, where, by abusing notation, we denote by 1 the identity map of M. Hence for $n \geq 1$ we have

$$H^n(G, M) = \frac{\ker(1 + (-1)^{n+1}S)}{(1 + (-1)^n S)M}.$$

In particular, if the Comessatti pair (M, S) is free, we have

$$\operatorname{rk} H^0(G, M) = \operatorname{rk} M^G = s^+$$

whereas $H^i(G, M)$ is, for $i \geq 1$, a $\mathbb{Z}_2$-vector space, and more precisely, for $i \geq 1$ we have

$$H^i(G, M) = \begin{cases} \frac{M^-}{(1-S)M} & i \equiv 1 \pmod 2, \\ \frac{M^+}{(1+S)M} & i \equiv 0 \pmod 2. \end{cases}$$

Then by Lemma 2.5.1 we have

$$\dim_{\mathbb{Z}_2} H^i(G, M) = \begin{cases} s^- - \lambda & i \equiv 1 \pmod 2, \\ s^+ - \lambda & i \equiv 0 \pmod 2. \end{cases}$$

To the other extreme, we can suppose M to be a $\mathbb{Z}_2$-vector space of dimension m. Again we set

$$\lambda := \dim_{\mathbb{Z}_2}(1+S)M.$$

Then from the exact sequence

$$0 \to M^G \to M \xrightarrow{1+S} (1+S)M \to 0$$

we get

$$\dim_{\mathbb{Z}_2} M^G = m - \lambda.$$

Then, according to the previous discussion, we have

$$H^i(G, M) = \begin{cases} M^G & i = 0, \\ \frac{M^G}{(1+S)M} & i \geq 1, \end{cases}$$

and therefore

$$\dim_{\mathbb{Z}_2} H^i(G, M) = \begin{cases} m - \lambda & i = 0, \\ m - 2\lambda & i \geq 1. \end{cases}$$

One of the main features of group cohomology is the existence of the *long exact sequence* of cohomology. We state in fact the following theorem, which we are not going to prove, referring the reader to [HS] for its proof.

Theorem 2.6.2. *Let*

$$0 \to M' \to M \to M'' \to 0$$

be a short exact sequence of $\mathbb{Z}(G)$*-modules. Then there is a long exact sequence*

$$\cdots \to H^i(G, M') \xrightarrow{a_i} H^i(G, M) \xrightarrow{b_i} H^i(G, M'') \xrightarrow{d_i} H^{i+1}(G, M') \to \ldots$$

where a_i, b_i *are the obvious maps and* d_i *are suitable connecting homomorphisms.*

There is also a notion of *non abelian cohomology* which is useful to mention in this context. Let G be a *finite* group and let M be a multiplicative group with a right G-action which is compatible with the group structure, i.e., we are given a map $(M, G) \to M$ such that

$$(mn)^g = m^g n^g$$

for all m, n in M and g in G. One defines

$$H^0(G, M) := M^G,$$

which is a subgroup of M. Then one defines the *first cohomology set* $H^1(G, M)$ in the following way. Let $Z^1(G, M)$ be the set of *cocycles* of G with values in M. An element of $Z^1(G, M)$ is, by definition, a map $\phi: g \in G \to g^\phi \in M$, such that

$$(g_1 g_2)^\phi = ((g_1)^\phi)^{g_2} g_2^\phi.$$

In $Z^1(G, M)$ one defines an equivalence relation $\sim$ by declaring $\phi \sim \phi'$ if and only if there is an element $x \in M$ such that for every $g \in G$ one has

$$g^{\phi'} = x^{-1} g^{\phi} x^g.$$

Then one defines

$$H^1(G, M) := Z^1(G, M)/\sim .$$

Notice that $H^1(G, M)$ is a *pointed set*, i.e., there is a distinguished element in $H^1(G, M)$, denoted by 1 and called the *neutral cohomology class*, which corresponds to the constant cocycle whose value is $1 \in M$.

It is interesting to notice that at least a part of the long exact sequence of cohomology related to a short exact sequence exists also for *non abelian cohomology*. First of all one notes that the definition of non abelian cohomology is functorial, i.e., if M, N are groups with a G-action, and if $f: M \to N$ is a homomorphism compatible with the G-action (in the obvious sense), which we will call a *G-homomorphism*, then we have induced maps in cohomology $H^i(G, M) \to H^i(G, N)$, $i = 0, 1$, which are homomorphisms only for $i = 0$, and maps of pointed sets for $i = 1$. Furthermore one has the following result, for the proof of which we refer to [Se ,§5].

Theorem 2.6.3. *Let*

$$0 \to M' \to M \to M'' \to 0$$

be a short exact sequence of groups with G-action and G-homomorphisms. Then there is a long exact sequence (of pointed sets)

$$0 \to H^0(G, M') \to H^0(G, M) \to H^0(G, M'') \xrightarrow{\delta} H^1(G, M') \to H^1(G, M).$$

It is interesting to notice a connection between real structures on a complex manifold X and non abelian cohomology with value in the group $\mathrm{Aut}(X)$ of its complex analytic automorphisms. First of all let us denote by $\mathcal{S}(X)$ the set of all real structures on X, and by $\mathcal{T}(X)$ the set $\mathcal{S}(X)$ modulo the *equivalence* relation. Let (X, S) be a fixed real structure on X, and let, as usual, $G = \{1, S\} = \mathbb{Z}_2$. Then there is a G-action on $\mathrm{Aut}(X)$ defined by

$$f^S = S \circ f \circ S$$

if $f \in \mathrm{Aut}(X)$. One has the following theorem.

Theorem 2.6.4. *Let (X, S) be a real structure. Then there is a natural bijection*

$$f: \mathcal{T}(X) \to H^1(G, \mathrm{Aut}(X))$$

sending (X, S) to the neutral cohomology class.

Proof. Let (X, S') be a real structure on X and let us consider the map

$$\phi_{S'}: g \in G \to \begin{cases} \mathrm{id}_X \in \mathrm{Aut}(X) & \text{if } \quad g = 1, \\ S \circ S' \in \mathrm{Aut}(X) & \text{if } \quad g = S. \end{cases}$$

It is easy to see that $\phi_{S'}$ is a cocycle. Thus we have a map

$$F: (X, S') \in \mathcal{T}(X) \to \phi_{S'} \in Z^1(G, \mathrm{Aut}(X)).$$

To finish the proof it is enough to show that (X, S') is equivalent to (X, S'') if and only if $\phi_{S'} \sim \phi_{S''}$. This is an easy exercise which can be left to the reader. □

Corollary 2.6.5. *Let X be an abelian variety. Then there are only finitely many non-equivalent real structures on X.*

Proof. By Theorem 2.6.4 it suffices to show that the group $H^1(G, \mathrm{Aut}(X))$ is finite. This follows from (6.1) of [BS]. □

A different, direct proof of this corollary will be given later on (see Theorem 3.2.2).

2.7. The action of conjugation on the Albanese variety and the Picard group

Let (X, S) be, as usual, a real structure on a smooth, complex, projective variety X. We will use the standard notation

$$q := \dim_{\mathbb{C}} H^1(X, \mathcal{O}_X) = \dim_{\mathbb{C}} H^0(X, \Omega^1_X)$$

to denote the *irregularity* of X. Let us recall the exact sequence defining the *Albanese variety*:

$$0 \to H_1(X, \mathbb{Z})_0 \xrightarrow{f} H^0(X, \Omega^1_X)^* \to \mathrm{Alb}(X) \to 0 \tag{2.7.1}$$

(see §1.4). The action of G on $H_1(X, \mathbb{Z})_0$ and on $H^0(X, \Omega^1_X)^*$ is compatible with f, i.e., f is a map of $\mathbb{Z}(G)$-modules. Therefore G also acts on $\mathrm{Alb}(X)$, namely $\mathrm{Alb}(X)$ inherits an induced antiinvolution which we still denote by S, and we can consider the real structure $(\mathrm{Alb}(X), S)$. Moreover (2.7.1) turns out to be a short exact sequence of $\mathbb{Z}(G)$-modules.

Theorem 2.7.2. *Let (X, S) be a real structure and suppose that $H_1(X, \mathbb{Z})$ is torsion free. Then for the real part of* $(\mathrm{Alb}(X), S)$ *one has the following expression as a real compact Lie group:*

$$\mathrm{Alb}(X)_S \simeq (\mathbb{R}/\mathbb{Z})^q \times \mathbb{Z}_2{}^{q-\lambda_1}$$

where λ_1 is the first Comessatti character of (X, S). Furthermore the first symmetry index of (X, S) is $s_1^+ = q$, and therefore also $s_1^- = q$.

Proof. We can consider the long exact sequence of group cohomology associated to (2.7.1). By the results in §2.3 we have that

$$H^0(G, H^0(X, \Omega_X^1)^*) \simeq (H^0(X, \Omega_X^1)^G)^* \simeq \mathbb{R}^q.$$

By Remark 2.6.1, (iii), we have also

$$H^1(G, H^0(X, \Omega_X^1)^*) \simeq \frac{\ker(1+S)}{\mathrm{Im}(1-S)} = 0.$$

Therefore we have

$$0 \to H_1(X,\mathbb{Z})^G \to \mathbb{R}^q \to \mathrm{Alb}(X)_S \to H^1(G, H_1(X,\mathbb{Z})) \to 0.$$

By Remark 2.6.1, (iii), we have

$$H^1(G, H_1(X,\mathbb{Z})) \simeq \mathbb{Z}_2{}^{q-\lambda_1}.$$

On the other hand $\mathrm{Alb}(X)_S$ is compact and therefore $\mathbb{R}^q/H_1(X,\mathbb{Z})^G$ has to be a subgroup of $\mathrm{Alb}(X)$ of finite index. Therefore $H_1(X,\mathbb{Z})^G \simeq \mathbb{Z}^q$. The assertion then follows. □

As a direct consequence we have the following:

Corollary 2.7.3. *With the same hypotheses of Theorem* 2.7.2, *one has*

$$\nu(\mathrm{Alb}(X)) = 2^{q-\lambda_1}.$$

One can dually consider the G-action on the Picard group of X. First we present some generalities.

Let D be a divisor on X, defined, as usual, by a collection (U_i, f_i), where (U_i) is a covering of X and $f_i \in \mathcal{O}(U_i)$ such that $\frac{f_i}{f_j} \in \mathcal{O}^*(U_i \cap U_j)$, whenever $U_i \cap U_j \neq \emptyset$. The *conjugate divisor* D^S is defined by the collection $(S(U_i), \sigma \circ f_i \circ S)$. If $\mathcal{O}(D)$ is the line bundle associated to D, i.e., the line bundle defined by the cocycle $(U_{ij}, \frac{f_i}{f_j})$, then of course $\mathcal{O}(D^S) \simeq \mathcal{O}(D)^S$.

If D is S-invariant, or *real*, i.e., if $D^S = D$, then S acts on $H^0(X, \mathcal{O}(D))$ as an antilinear involution. Hence, as in §2.3 one sees that there is a splitting

$$H^0(X, \mathcal{O}(D)) = H^0(X, \mathcal{O}(D))^+ \oplus H^0(X, \mathcal{O}(D))^-$$

as a real vector space, and indeed $H^0(X, \mathcal{O}(D))$ is the complexification of $H^0(X, \mathcal{O}(D))^+$, hence

$$\dim_{\mathbb{C}} H^0(X, \mathcal{O}(D)) = \dim_{\mathbb{R}} H^0(X, \mathcal{O}(D))^+,$$

and therefore there is a basis of $H^0(X, \mathcal{O}(D))$ formed by S-invariant, i.e., *real,* sections.

We will denote by $\mathrm{Pic}(X)^S$ the subgroup of equivalence classes of line bundles $\mathcal{L}$ on X such that $\mathcal{L}^S \simeq \mathcal{L}$. Furtermore we will denote by $\mathrm{Pic}(X)^+$ the subgroup of equivalence classes of line bundles on X of the form $\mathcal{O}(D)$ such that D is real. Of course $\mathrm{Pic}(X)^+ \subseteq \mathrm{Pic}(X)^S$. We want to point out the following result.

Proposition 2.7.4. *Let (X, S) be a real structure and let X be smooth and projective. Then there is some ample real divisor D on X. Furthermore, if X is such that $X_S \neq \emptyset$ and if D is a divisor on X such that $\mathcal{O}(D^S) \simeq \mathcal{O}(D)$, then there is some real divisor D' such that $\mathcal{O}(D') \simeq \mathcal{O}(D)$. In other words, if $X_S \neq \emptyset$ then* $\mathrm{Pic}(X)^+ = \mathrm{Pic}(X)^S$.

Proof. As for the first assertion, we note that if D is ample, D^S is also ample. Hence $D + D^S$ is ample and real.

Let us prove the second part of the proposition. It is easy to see that it suffices to prove it for D effective. By the hypothesis, the linear system $|D|$ contains D^S. If $D = D^S$ there is nothing to prove. Otherwise $|D|$ is a projective space on which S induces an antiprojective involution. In particular the line L in $|D|$ joining D and D^S is fixed by S. If x is a general point of X_S, there is a unique divisor D' in L containing x. This is clearly fixed by S, proving the assertion. □

Now we consider the exact sequence (1.4.2). The group G acts on each of the sheaves of the sequence, but this is not a sequence of G-sheaves, i.e., maps are not compatible with the G-action. Indeed, the action on $\mathbb{Z}$ is trivial, whereas it is not trivial on $\mathcal{O}_X$ and we embed $\mathbb{Z}$ into $\mathcal{O}_X$ by multiplication by i. In order to make the above an exact sequence of G-sheaves, we have to modify the G-action on $\mathbb{Z}$, and let S act on $\mathbb{Z}$ by sending the integer n to $-n$. We call $\mathbb{Z}(-1)$ this new group (or sheaf). We note that in general, given a $\mathbb{Z}(G)$-module A, one can consider the *twisted* module $A(-1)$ by letting S act on $x \in A$ by sending x to $-x^S$ instead of x^S.

Now we have the same exact sequence as above

$$0 \to \mathbb{Z}(-1) \xrightarrow{\times i} \mathcal{O}_X \to \mathcal{O}_X^* \to 0 \tag{2.7.5}$$

which is a sequence of G-sheaves. From (2.7.5) we deduce an exact sequence of $\mathbb{Z}(G)$-modules

$$\begin{aligned} 0 \to H^1(X, \mathbb{Z}(-1)) \to H^1(X, \mathcal{O}_X) \to H^1(X, \mathcal{O}_X^*) \\ \simeq \mathrm{Pic}(X) \to NS(X) \subseteq H^2(X, \mathbb{Z}(-1)) \end{aligned} \tag{2.7.6}$$

where the G-action on $\mathrm{Pic}(X)$ is clearly the one described in §2.3.

Remark 2.7.7. We note here that the *Néron–Severi group* $NS(X)$ naturally appears as a subgroup of $H^2(X, \mathbb{Z}(-1))$. Therefore, if c_1 is the first Chern class map as defined in (1.4.3) and if D is a divisor, one has

$$c_1(D)^S = -c_1(D^S).$$

In particular if D is *algebraically equivalent* to D^S, then

$$c_1(D)^S = -c_1(D).$$

The sequence (2.7.6) tells us how G acts on $\mathrm{Pic}^0(X)$. Indeed we have the following exact sequence of $\mathbb{Z}(G)$-modules

$$0 \to H^1(X, \mathbb{Z}(-1)) \to H^1(X, \mathcal{O}_X) \to \mathrm{Pic}^0(X) \to 0. \tag{2.7.8}$$

In particular $\mathrm{Pic}^0(X)$ inherits a G-action, hence a real structure which we still denote by $(\mathrm{Pic}^0(X), S)$. We then have the following theorem, which may be thought as a dual statement to Theorem 2.7.2, since the Albanese variety is *dual* to the Picard variety (see [GH]).

Theorem 2.7.9. *Let (X, S) be a real structure and suppose that $H_1(X, \mathbb{Z})$ is torsion free. Then the real part of $(\mathrm{Pic}^0(X), S)$ has the following structure as a compact Lie group:*

$$\mathrm{Pic}^0(X)_S \simeq (\mathbb{R}/\mathbb{Z})^q \times \mathbb{Z}_2{}^{q-\lambda_1}$$

where λ_1 is the first Comessatti character of (X, S). Furthermore

$$\nu(\mathrm{Pic}^0(X)) = 2^{q-\lambda_1}.$$

Proof. We consider the long exact cohomology sequence related to (2.7.8). The isomorphism

$$H^1(X, \mathcal{O}_X) \simeq H^0(X, \Omega^1_X)$$

is clearly compatible with the G-action (see the proof of Dolbeault's theorem in [GH], p. 45). Hence, by the results in §2.3 and by Example 2.6.1, (iii), we have

$$H^0(G, H^1(X, \mathcal{O}_X)) \simeq H^0(X, \Omega^1_X)^G \simeq \mathbb{R}^q,$$

$$H^1(G, H^1(X, \mathcal{O}_X)) = 0.$$

Therefore taking the G-action on (2.7.8) we get the long exact cohomology sequence:

$$0 \to H^1(X, \mathbb{Z}(-1))^G \to H^1(X, \mathcal{O}_X)^G \simeq \mathbb{R}^q \to \mathrm{Pic}^0(X)^G$$
$$\to H^1(G, H^1(X, \mathbb{Z}(-1))) \to 0$$

Since clearly $H^1(X, \mathbb{Z}(-1)) \simeq H^1(X, \mathbb{Z})(-1)$, arguing as in the proof of Theorem 2.7.2 one sees that

$$H^1(X, \mathbb{Z}(-1))^G \simeq \mathbb{Z}^q, \quad H^1(G, H^1(X, \mathbb{Z}(-1))) \simeq \mathbb{Z}_2^{q-\lambda_1},$$

whence the conclusion follows. □

Before closing this section, we make an application of these results to the case of curves. Let (X, S) be a real structure, where X is a compact Riemann surface of genus g. We introduce an invariant of (X, S), which is defined as

$$a(X) = \begin{cases} 0 & \text{if } X - X_S \text{ is disconnected,} \\ 1 & \text{if } X - X_S \text{ is connected.} \end{cases}$$

We can prove the following:

Proposition 2.7.10. (i) *If* $\nu(X) = g + 1$ *then* $a(X) = 0$.

(ii) *If* $a(X) = 0$ *then the number of connected components of* $X - X_S$ *is two.*

(iii) *If* $a(X) = 0$ *then* $\nu(X) \equiv g + 1 \pmod 2$.

Proof. Let γ_i, $(i = 1, \dots, g + 1)$, be the connected components of X_S. Of course γ_i, with suitable orientations can be seen as elements of $H_1(X, \mathbb{Z})$. Since clearly $\gamma_i^S = \gamma_i$, the γ_i's are in fact elements of $H_1(X, \mathbb{Z})^G$, which, by Theorem 2.7.2 is free of rank g. Therefore the γ_i's are dependent, i.e., a suitable linear combination of them is homologous to zero, which clearly yields the assertion (i).

As for (ii), we observe that two adjacent connected components A, B of $X - X_S$ are homeomorphic, since they are exchanged by the conjugation. This homeomorphism has to fix the common boundary of $\overline{A}$ and $\overline{B}$, hence $\overline{A \cup B}$ is a compact subvariety of X, therefore it coincides with X.

Let us finally prove (iii). By (ii) we know that $X - X_S$ is the union of two homeomorphic components A, B. Let us put $Y := \overline{A}$, which is a compact surface with boundary $\partial Y = X_S$. By triangulating X in such a way that X_S is a subcomplex, it is easy to see that

$$2 - 2g = \chi(X) = 2\chi(Y) - \chi(X_S) = 2\chi(Y)$$

since every connected component γ of X_S is a real analytic compact curve, hence it is homeomorphic to a circle, and therefore $\chi(X_S) = 0$. By squeezing the $\nu(X)$ components of X_S to points, Y becomes a compact surface Z without boundary, of genus say g'. Hence $\chi(Y) = \chi(Z) - \nu(X)$, and therefore

$$1 - g = \chi(Y) = \chi(Z) - \nu(X) = 2 - 2g' - \nu(X)$$

whence the assertion. □

2.8. Period matrices in pseudonormal form and the Albanese map

Let us go on keeping the notation of the previous section, and let us recall that a *period matrix* of X is a matrix of type $q \times 2q$ of the type

$$\Omega = (u_{ij})_{i=1,\dots,q;\, j=1\dots,2q}$$

with

$$u_{ij} = \int_{\gamma_j} \omega_i$$

where $\{\gamma_j\}_{j=1,\dots,2q}$ is a basis of $H_1(X, \mathbb{Z})_0$ and $\{\omega_i\}_{i=1,\dots,q}$ is a basis of $H^0(X, \Omega_X^1)$ (see §1.4). We know how such a matrix changes by a change of the two bases: if $A \in \mathrm{GL}(q, \mathbb{C})$ is the matrix of the base change in $H^0(X, \Omega_X^1)$ and if $B \in \mathrm{GL}(2q, \mathbb{Z})$

is the matrix of the base change in $H_1(X, \mathbb{Z})_0$, then the new period matrix Ω' is

$$\Omega' = A \cdot \Omega \cdot B. \tag{2.8.1}$$

Furthermore

$$\mathrm{Alb}(X) \simeq \mathbb{C}^q/\Lambda \tag{2.8.2}$$

where Λ is the lattice of rank $2q$ of $\mathbb{C}^q$ generated by the columns of Ω, which is a period matrix of $\mathrm{Alb}(X)$ as well.

Let us now see how the real structure (X, S) on X helps in finding suitable bases of $H_1(X, \mathbb{Z})_0$ and of $H^0(X, \Omega^1_X)$ with respect to which the period matrix takes a particularly simple form.

First of all we notice that (2.3.2) implies that $\overline{\Omega}$ is, as well as Ω, a period matrix of X. This corresponds to the fact that the real structure gives an isomorphism between X and the conjugate variety X^σ (see Theorem 2.2.2).

If we take $\{\omega_i\}_{i=1,\dots,q}$ to be a basis of $H^0(X, \Omega^1_X)$ formed by real forms (see Proposition 2.3.3), again (2.3.2) tell us that

$$\overline{\int_{\gamma_j} \omega_i} = \int_{S(\gamma_j)} \omega_i \tag{2.8.3}$$

for all $i = 1, \dots, q$ and $j = 1, \dots, 2q$.

Now, in relation with the Comessatti pair $(H_1(X, \mathbb{Z})_0, S)$, we can find a basis of the first type $(\sigma_1, \dots, \sigma_{s^+}, \tau_1, \dots, \tau_\lambda, \epsilon_{\lambda+1}, \dots, \epsilon_{s^-})$ of $H_1(X, \mathbb{Z})_0$, where λ is the Comessatti character of the pair (see Theorem 2.5.3). We already know that $s^+ = s^- = q$ (see Theorem 2.7.2), but we can give another proof of this fact in the present setting. Indeed, the period matrix is now of the form

$$\Omega = (\int_{\sigma_j} \omega_i, \int_{\tau_h} \omega_i, \int_{\epsilon_k} \omega_i)_{i=1,\dots,q;\, j=1,\dots,s^+, h=1,\dots,\lambda; k=\lambda+1,\dots,s^-}$$

From (2.8.3) we have that

$$\int_{\sigma_j} \omega_i \in \mathbb{R}, \qquad \int_{\epsilon_k} \omega_i \in i\mathbb{R}.$$

Then it is clear from (1.1.1) and (1.1.2) that $s^+ \leq q$ and $s^- \leq q$. But $s^+ + s^- = 2q$, whence $s^+ = s^- = q$.

Now it is an easy consequence of (1.1.1) and (1.1.2) that the real matrix

$$(\int_{\sigma_j} \omega_i)_{i=1,\dots,q;\, j=1,\dots,q}$$

has maximal rank. This implies that we can choose a basis of real forms $\{\omega_i\}_{i=1,\dots,q}$ of $H^0(X, \Omega^1_X)$ in such a way that

$$\int_{\sigma_j} \omega_i = \delta_{ij}.$$

But then by (2.8.3) we have

$$\overline{\int_{\tau_h} \omega_i} = \int_{S(\tau_h)} \omega_i = \int_{\sigma_h - \tau_h} \omega_i = \delta_{ih} - \int_{\tau_h} \omega_i$$

for all $i = 1, \dots, q$ and $h = 1, \dots, \lambda$, whence

$$2\,\mathrm{Re} \int_{\tau_h} \omega_i = \delta_{ih}.$$

In conclusion we have that the corresponding period matrix is of the form

$$\Omega = (I_q \quad \frac{1}{2} I_{q,q,\lambda} + iT) \tag{2.8.4}$$

where T is a real $q \times q$ matrix (see Remark 2.5.4, (ii)). Such a matrix will be called a period matrix in *pseudonormal form*. Now we can state the next result.

Theorem 2.8.5. *Let* (X, S) *be a real structure. Then there are period matrices of* X *in pseudonormal form* (2.8.4), *where* T *is in* $\mathrm{GL}(q, \mathbb{R})$ *and* λ *is the Comessatti invariant of the pair* $(H_1(X, \mathbb{Z})_0, S)$. *The number* $\delta := |T|^2$, *as well as* λ, *is an invariant of the real structure.*

Proof. We only have to prove that δ is non-zero and it is invariant, since the rest of the theorem has already been proved before. The first assertion easily follows from (1.1.1). Indeed we have

$$0 \neq \det \begin{pmatrix} \Omega \\ \overline{\Omega} \end{pmatrix} = \det \begin{pmatrix} I_q & \frac{1}{2} I_{q,q,\lambda} + iT \\ I_q & \frac{1}{2} I_{q,q,\lambda} - iT \end{pmatrix} = \det \begin{pmatrix} 0_q & 2iT \\ 2I_q & I_{q,q,\lambda} \end{pmatrix} \tag{2.8.6}$$

whence it follows that T has maximal rank.

Now suppose we have another period matrix in pseudonormal form

$$\Omega' = (I_q \quad \frac{1}{2} I_{q,q,\lambda} + iT')$$

related to Ω by the formula (2.8.1), where we can write

$$B = \begin{pmatrix} B_1 & B_2 \\ B_3 & B_4 \end{pmatrix}$$

and where the B_i's are $q \times q$ matrices over $\mathbb{Z}$, $i = 1, 2, 3, 4$. Note that, since A is the matrix of a base change between two bases of real forms of $H^0(X, \Omega^1_X)$, one has $A \in \mathrm{GL}(q, \mathbb{R})$. Then

$$\begin{aligned} &(I_q \quad \frac{1}{2} I_{q,q,\lambda} + iT') \\ &\quad = (A \cdot (B_1 + (\frac{1}{2} I_{q,q,\lambda} + iT) \cdot B_3) \quad A \cdot (B_2 + (\frac{1}{2} I_{q,q,\lambda} + iT) \cdot B_4)). \end{aligned}$$

But then it is clear that $B_3 = 0_q$ and that $A = B_1^{-1}$. Since

$$\pm 1 = |B| = |B_1| \cdot |B_4|$$

we have

$$|A| = \pm 1.$$

Furthermore

$$\det\left(\frac{\Omega'}{\overline{\Omega'}}\right) = \det\left(\frac{\Omega}{\overline{\Omega}}\right) \cdot |A|^2 \cdot |B| = \pm \det\left(\frac{\Omega}{\overline{\Omega}}\right)$$

(see the proof of Theorem 1.3.5). The conclusion now easily follows by (2.8.6). □

Remark 2.8.7. This is the right place to make a comment on the *Albanese map* of a variety with a real structure (see §1.4). Let (X, S) be a real structure on a compact Kähler manifold, such that $X_S \neq \emptyset$. Then we can chose $\{\omega_i\}_{i=1,\dots,q}$ to be a basis of $H^0(X, \Omega^1_X)$ formed by real forms and we can chose p to be a point in X_S. Then the Albanese map α_p considered in §1.4 can be written as

$$\alpha_p : x \in X \to \left(\int_p^x \omega_1, \dots, \int_p^x \omega_q\right) + \Lambda \in \mathrm{Alb}(X)$$

where, with the notation of §1.4, $\Lambda := f(H_1(X, \mathbb{Z})_0)$. Then it is clear that α_p is a morphism of real structures between (X, S) and the induced real structure $(\mathrm{Alb}(X), S)$.

3. Real abelian varieties

In this chapter we will study real structures on complex tori. First we will classify them and then we will study their relations with line bundles on complex tori, specializing to real abelian varieties. Finally we will introduce moduli spaces of real abelian varieties and consider real theta functions.

3.1. Real structures on complex tori

Let $X = V/\Lambda$ be a complex torus of dimension q, where V is a complex vector space of dimension q and Λ is a maximal lattice contained in V (see §1.1). We will denote by $p: V \to X$ the natural projection. We will use this notation all over this chapter. We will say that a real structure (X, S) on X is *proper* if $S: X \to X$ is a group isomorphism. Of course this happens if and only if S is an antiholomorphic involution and a group isomorphism of X as a compact real Lie group. Note that if (X, S) is proper then $X_S \neq \emptyset$, since $0 \in X_S$.

Proposition 3.1.1. (i) *Let* (X, S) *be a real structure on a complex torus* $X = V/\Lambda$. *Then there is a pair* $(V, \tilde{S})$ *where* $\tilde{S}: V \to V$ *is an antiaffine map (i.e., a map which is affine if we put the usual complex structure on the domain* V *and the conjugate structure on the range* V *of* $\tilde{S}$*), such that* $S \circ p = p \circ \tilde{S}$ *and that* $\tilde{S}^2$ *is the translation by an element of* Λ.

(ii) *Let* (X, S) *be a real structure on the complex torus* $X = V/\Lambda$. *Then:*

(a) *there is a proper real structure* (X, S') *such that*

$$S(x) = S'(x) + S(0)$$

for all $x \in X$;

(b) (X, S) *is equivalent to a proper real structure if and only if* $X_S \neq \emptyset$.

Proof. The proof of (i) is similar to the proof of Proposition 1.3.1. We just sketch it. By lifting S to the universal cover of X we find the antiaffine map $\tilde{S}: V \to V$. It is an exercise to prove that $\tilde{S}$ enjoys the properties of the statement.

Let us prove (ii). Let $\tilde{S}: V \to V$ be an antiaffine map corresponding to (X, S). We have $\tilde{S} = \tau_a \circ \tilde{S}'$ where $\tilde{S}': V \to V$ is antilinear (i.e., antiaffine and fixing the origin) and τ_a is the translation by $a = \tilde{S}(0)$. Since $\tilde{S}(x + \lambda) - \tilde{S}(x) \in \Lambda$ for all $x \in V$ and all $\lambda \in \Lambda$, it is clear that $\tilde{S}'(\Lambda) \subset \Lambda$. This implies that $\tilde{S}'$ descends to an antiholomorphic group isomorphism $S': X \to X$ such that $S' \circ p = p \circ \tilde{S}'$ and $S(x) = S'(x) + S(0)$. Let us prove that S' is a real structure, i.e., that $S^2 = \mathrm{id}_X$. For all $x \in X$ we have

$$\begin{aligned} x = S(S(x)) &= S(S'(x) + S(0)) = S'(S'(x) + S(0)) + S(0) \\ &= S'(S'(x)) + S'(S(0)) + S(0) \end{aligned}$$

but

$$0 = S(S(0)) = S'(S(0)) + S(0)$$

hence $S'(S'(x)) = x$. This proves (a).

Let us prove (b). One implication is clear: if (X, S) is equivalent to a proper structure, then $X_S \neq \emptyset$.

Conversely, let us assume that $X_S \neq \emptyset$. Let $p \in X_S$ and let $\tau_p: X \to X$ be the translation by p. The map $S'' = \tau_p^{-1} \circ S \circ \tau_p$ is again an antiinvolution on X and (X, S'') is equivalent to (X, S). Moreover, let (X, S') be the proper real structure existing by (a). Then we have

$$S''(x) = S(x + p) - p = S'(x + p) + S(0) - p = S'(x) + S(0) + S'(p) - p$$

for all $x \in X$. But we have

$$p = S(p) = S'(p) + S(0)$$

proving that $S' = S''$. Therefore (X, S) is equivalent to the proper structure (X, S'). □

Now we can describe all real structures on a complex torus (an implicit description is encoded in Theorem 2.6.4). First of all we let (X, S) be a real structure on a complex torus X. We let $\mathrm{End}(X)^*$ be the group of invertible endomorphisms of X, i.e., the elements of $\mathrm{End}(X)^*$ are the complex Lie group automorphisms of X. Then we consider the set

$$E_S = \{\pi \in \mathrm{End}(X)^* : (S \circ \pi)^2 = \mathrm{id}_X\}.$$

For all $\pi \in E_S$ we consider the set

$$X_{(S,\pi)} := \{a \in X : \pi(a) + S(a) = 0\}.$$

Then, given $\pi \in E_S$ and $a \in X_{(S,\pi)}$, we define

$$S_{(\pi,a)} : x \in X \to S(\pi(x)) + a \in X$$

which is clearly an antiholomorphic map. The following theorem holds.

Theorem 3.1.2. *Let (X, S) be a proper real structure on a complex torus X. Then for all $\pi \in E_S$ and for all $a \in X_{(S,\pi)}$, the pair $(X, S_{(\pi,a)})$ is a real structure on X and all real structures on X are obtained in this way.*

Proof. It is easy to verify, and we leave it to the reader, that for all $\pi \in E_S$ and for all $a \in X_{(S,\pi)}$, the map $S_{(\pi,a)}$ is an antiinvolution. Let us prove the converse. Let (X, S') be a real structure on X. Then $S \circ S'$ is an automorphism of X, hence there is a $\pi \in \mathrm{End}(X)^*$ and there is a $b \in X$ such that for all $x \in X$ one has

$$S(S'(x)) = \pi(x) + b$$

(see Corollary 1.3.2, (ii)), whence

$$S'(x) = S(\pi(x)) + a$$

where we put $a := S(b)$. Then

$$x = S^2(x) = (S \circ \pi)^2(x) + S(\pi(a)) + a$$

for all $x \in X$. If we put $x = 0$ we find $S(\pi(a)) + a = 0$ and therefore $\pi \in E_S$ and $a \in X_{(S,\pi)}$. □

Example 3.1.3. For any complex torus X, the ring $\mathbb{Z}$ is contained in $\mathrm{End}(X)$. In fact, for all $n \in \mathbb{Z}$, one can consider the endomorphism $n_X : x \in X \to nx \in X$, and the map $n \in \mathbb{Z} \to n_X \in \mathrm{End}(X)$ is injective. In particular $\pm 1 \in \mathrm{End}(X)^*$.

For any given real structure (X, S) on X clearly $\pm 1 \in E_S$. Accordingly we have

$$X_{(S,\pm 1)} = \{a \in X : S(a) \pm a = 0\} = X_{\mp S}$$

and we can consider the real structures $(X, S_a^{\pm})$, where $S_a^{\pm} = S_{(\pm 1,a)}$, i.e.,

$$S_a^{\pm}(x) = \pm S(x) + a$$

for all $a \in X_{\mp S}$ and for all $x \in X$. This construction yields particular real structures, corresponding to the $+$ [resp. the $-$] sign, which, according to Comessatti, we call real structures of the *first type* [resp. of the *second type*] with respect to the given (X, S). Notice that all these structures are clearly distinct, even if some of them could be equivalent. Among them the only proper ones are (X, S) and $(X, -S)$, and the latter is called *complementary structure* to (X, S).

We will go deeper into the study of real structures of complex tori in the next section. Meanwhile we conclude the present section by applying to tori some of the results from §§2.7, 2.8.

Let (X, S) be a real structure on a complex torus X, such that $X_S \neq \emptyset$. By Proposition 3.1.1, (X, S) is equivalent to a proper real structure, hence we may assume that it is proper. By keeping the notation of §2.8, we have an induced real structure $(\mathrm{Alb}(X), S)$ on the Albanese variety of X. On the other hand, by Remark 2.8.7 and by §1.4 we have that (X, S) is isomorphic to $(\mathrm{Alb}(X), S)$. But now, by Theorem 2.7.2 and Corollary 2.7.3 we can derive the next result.

Theorem 3.1.4. *Let (X, S) be a real structure on a complex torus X, such that $X_S \neq \emptyset$. Then we have the following isomorphism of real compact Lie goups*

$$X_S \simeq (\mathbb{R}/\mathbb{Z})^q \times \mathbb{Z}_2^{q-\lambda}$$

where λ is the first Comessatti character of (X, S). Furthermore, one has

$$\nu(X) = 2^{q-\lambda}.$$

In conclusion, we point out the following theorem, which is a sort of reformulation of Proposition 3.1.1.

Theorem 3.1.5. *Let X be a complex torus of dimension q. Then:*

(i) *there is a real structure (X, S) on X if and only if X has a period matrix Ω in pseudonormal form* (2.8.4)*, where λ is the first Comessatti character of X;*

(ii) *if (X, S) is a real structure on X such that $X_S \neq \emptyset$ and with first Comessatti character λ, then (X, S) is isomorphic to $(\mathbb{C}^q/\Lambda, \sigma)$, where Λ is the lattice generated by the columns of a period matrix Ω in pseudonormal form* (2.8.4)*, and σ is the antiinvolution induced on $\mathbb{C}^q/\Lambda$ by the usual conjugation of $\mathbb{C}^q$;*

(iii) *if (X, S) is a real structure on X such that $X_S = \emptyset$ and with first Comessatti character λ, then (X, S) is isomorphic to $(\mathbb{C}^q/\Lambda, \sigma')$, where Λ is the lattice generated by the columns of a period matrix Ω in pseudonormal form* (2.8.4)*, and $\sigma' = \tau_a \circ \sigma$, where τ_a is the translation by a suitable $a \in \mathbb{C}^q$ such that $a + \overline{a} \in \Lambda$.*

Proof. One implication of (i) follows by Theorem 2.8.5. Conversely, if X has a period matrix Ω in pseudonormal form, then $X \simeq \mathbb{C}/\Lambda$, where Λ is the lattice generated by the columns of the matrix Ω. We notice that the complex conjugation σ of $\mathbb{C}^q$ fixes Λ, and therefore induces a proper real structure $(\mathbb{C}^q/\Lambda, \sigma)$ on $\mathbb{C}^q/\Lambda$, and therefore on X. This finishes the proof of (i)

As for (ii), we keep the above notation and notice that the discussion of §2.8 shows that $(\mathrm{Alb}(X), S)$ is isomorphic to $(\mathbb{C}^q/\Lambda, \sigma)$. Then (ii) follows by Remark 2.8.7, since in this case (X, S) is isomorphic to $(\mathrm{Alb}(X), S)$.

The proof of (iii) is completely analogous and therefore omitted. □

Given a real structure (X, S) on a complex torus, we can consider, together with the first Comessatti character λ of (X, S) also the invariant δ introduced in Theorem 2.8.5 via the consideration of a period matrix in pseudonormal form. The pair (λ, δ) will be called the *pair of Comessatti invariants* of (X, S). We conclude this section by proving the following proposition.

Proposition 3.1.6. *Let (X, S) be a real structure on a complex torus X with the pair of Comessatti invariants given by (λ, δ). Let $(X, S_a^{\pm})$ be any real structure of the first or second type with respect to (X, S), and let $(\lambda^{\pm}, \delta^{\pm})$ be its pair of Comessatti invariants. Then*

$$\lambda^{\pm} = \lambda, \quad \delta^{+} = \delta, \quad 4^{\lambda}\sqrt{\delta^{+}\delta^{-}} = 1.$$

Proof. Only the identity $4^{\lambda}\sqrt{\delta^{+}\delta^{-}} = 1$ needs some comment. We denote by

$$\Omega^{+} = (I_q \quad \frac{1}{2} I_{q,q,\lambda} + iT^{+})$$

a matrix in pseudonormal form relative to the real structures (X, S^{+}). We claim that

$$\Omega^{-} := A \cdot \Omega^{+} \cdot B,$$

where

$$A = -i(T^{+} \cdot (I_{q,q,\lambda} + I_q))^{-1}$$
$$B = \begin{pmatrix} -I_{q,q,\lambda} & I_q - I_{q,q,\lambda} \\ I_q + I_{q,q,\lambda} & I_{q,q,\lambda} \end{pmatrix}$$

is a period matrix in pseudonormal form relative to a real structure (X, S^{-}) of the second type. This is easily proved by observing that

$$\Omega^{-} = (I_q \quad \frac{1}{2} I_{q,q,\lambda} - i(I_q + I_{q,q,\lambda}^{-1} \cdot (T^{+})^{-1} \cdot (I_q - \frac{1}{2} I_{q,q,\lambda}))$$

and moreover that B is the matrix of the base change from a basis of $H_1(X, \mathbb{Z})$ of the first type $(\sigma_1, \ldots, \sigma_q, \tau_1, \ldots, \tau_\lambda, \epsilon_{\lambda+1}, \ldots, \epsilon_q)$ to one of the second type

$(\epsilon_1, \ldots, \epsilon_q, \tau_1, \ldots, \tau_\lambda, \sigma_{\lambda+1}, \ldots, \sigma_q)$. From the above expression of Ω^- one gets

$$\delta^- = (2^{-\lambda} \det(T^+)^{-1} (\frac{1}{2})^\lambda)^2 = 2^{-4\lambda} (\delta^+)^{-1},$$

hence the assertion. □

An important consequence of the results of this section is the following theorem, whose proof we omit referring to [K].

Theorem 3.1.7. *Let* (X, S) *be a real structure on a complex torus* X*, such that* $X_S \neq \emptyset$*. Then* (X, S) *is GM.*

3.2. Equivalence classes for real structures on complex tori

In this section we go back to the study of real structures on a complex torus X of dimension q. Specifically we want to study the set $\mathcal{T}(X)$ of real structures on X up to equivalence, as defined in §2.6. In particular we recall Theorem 2.6.4 which states that $\mathcal{T}(X)$ is bijective to $H^1(G, \mathrm{Aut}(X))$. Unfortunately $H^1(G, \mathrm{Aut}(X))$ is in general rather difficult to compute, and, as far as we know, it is still an open question to determine $H^1(G, \mathrm{Aut}(X))$ for any complex torus X. However Theorem 3.1.2 provides us with a rather efficient tool for making explicit computations. Using it we will be able to give a couple of interesting general result, the first of which is the following.

Theorem 3.2.1. *Let* X *be a complex torus of dimension* q *such that* $\mathrm{End}(X)^* \simeq \{\pm 1\}$*. If* $\mathcal{S}(X)$ *is non empty, then all real structures on* X *have the same first Comessatti character* λ *and* $\mathcal{T}(X)$ *has order* $2^{q+1-\lambda}$*, and only 2 elements of* $\mathcal{T}(X)$ *are classes of proper real structures.*

Proof. First of all we consider a proper real structure (X, S) on X. So we have the structures $(X, S_a^\pm)$ of the first and second type with respect to (X, S). By Theorem 3.1.2 a real structure of X is either of the first or of the second type with respect to S. This, together with Theorem 3.1.6, proves the first assertion.

Now we have:

Claim 1: A real structure of X of the first type is never equivalent to a real structure of the second type.

Proof of the claim. A real structure (X, S') is equivalent to $(X, S_a^\pm)$ if and only there is a $\tau \in \mathrm{Aut}(X)$ such that $S' = \tau \circ S_a^\pm \circ \tau^{-1}$. Since $\mathrm{End}(X)^* \simeq \{\pm 1\}$, then τ is of the type

$$\tau : x \in X \to \pm x + b \in X$$

where b is a suitable element of X. But, for all $x \in X$ we have

$$\begin{aligned} S'(x) &= \tau(S_a^+(\tau^{-1}(x))) = \tau(S_a^+(\pm x - b)) \\ &= \tau(\pm S(x) - S(b) + a)S(x) \mp S(b) \pm a + b = S^+_{\mp S(b) \pm a + b}(x) \\ S'(x) &= \tau(S_a^-(\tau^{-1}(x))) = \tau(S_a^-(\pm x - b)) \\ &= \tau(\mp S(x) + S(b) + a) = -S(x) \pm S(b) \pm a + b = S^-_{\pm S(b) \pm a + b}(x) \end{aligned}$$

which proves the assertion.

Now we consider the antiholomorphic maps

$$\mathrm{id}_X \pm S \colon X \to X$$

which are homomorphisms of real Lie groups. We set $\Gamma^{\pm} := \mathrm{Im}(\mathrm{id}_X \pm S)$.

Claim 2: $\Gamma^{\pm}$ is the connected component $(X_{\pm S})_0$ of the origin of $X_{\pm S}$.

Proof of the claim. We prove the assertion only for the structures of the first type, i.e., for the $+$ sign, since the proof in the other case is similar. First of all we remark that the inclusion $\Gamma^+ \subseteq X_S$ is trivial. Since X is connected, then $\Gamma^+ \subseteq (X_S)_0$. Since X is compact as well as $(X_S)_0$ and since the real dimension of X is $2q$ and the dimension of X_S is q, in order to conclude it suffices to prove that $\ker(\mathrm{id}_X + S)$ has dimension q. But this is an easy consequence of the fact that

$$\ker(\mathrm{id}_X + S) = \frac{(H^0(X, \Omega_X^1)^*)^-}{H_1(X, \mathbb{Z})^-}$$

since by Proposition 2.3.3 we have $\dim_{\mathbb{R}} H^0(X, \Omega_X^1)^- = q$, and by Theorem 2.7.2, $\mathrm{rk}\, H_1(X, \mathbb{Z})^- = q$.

Next we show the following statement.

Claim 3: Two real structures of the same type $(X, S_a^{\pm})$, $(X, S_{a'}^{\pm})$ are equivalent if and only if $a - a' \in \Gamma^{\mp}$.

Proof of the claim. We prove the assertion for structures of the first type, the proof being analogous in the other case. In the proof of claim 1 above, we saw that S_a^+ is equivalent to $S_{a'}^+$ if and only if there is a $b \in X$ such that $a' = \pm a \mp S(b) + b$. If $a - a' \in \Gamma^-$, then there is a $b \in X$ such that $a' - a = b - S(b)$, which proves that S_a^+ is equivalent to $S_{a'}^+$. Conversely if this happens, then $a' = \pm a \mp S(b) + b$. If the upper sign holds, then $a - a' \in \Gamma^-$. If the lower sign holds, then S is equivalent to $S_{a+a'}$. Since S is proper, there is an $x \in X$ such that $x = S_{a+a'}(x) = S(x) + a + a'$, i.e., $a + a' = x - S(x) \in \Gamma^-$. On the other hand X_{-S}/Γ^- is a finite group isomorphic to $H^1(G, H_1(X, \mathbb{Z})) \simeq \mathbb{Z}_2^{q-\lambda}$, hence $2a \in \Gamma^-$ and therefore $a - a' \in \Gamma^-$.

Now we can easily compute the order of $\mathcal{T}(x)$. Indeed the equivalence classes of the structures of the first type are bijective to X_{-S}/Γ^-, which, as we saw, has order $2^{q-\lambda}$. Similarly the structures of the second type are bijective to X_S/Γ^+, whose order is also $2^{q-\lambda}$. Whence the assertion about the order of $\mathcal{T}(x)$. The last assertion easily follows by (ii) of Proposition 3.1.1, and by Claims 1 and 3. □

Our next result is a new proof of Corollary 2.6.5. This theorem is more general than Theorem 3.2.1, since we do not assume $\mathrm{End}(X)^* \simeq \{\pm 1\}$, but it is unfortunately less precise.

Theorem 3.2.2. *Let X be a complex torus. Then $\mathcal{T}(X)$ is finite.*

Proof. Let (X, S) be a real structure on S. First of all one remarks that the set $E_S \subseteq \mathrm{End}(X)^*$ is finite. In fact for all $\pi \in \mathrm{End}(X)^*$ we have an induced rational representation $\rho_r(\pi)$ (see §1.3). Similarly we have the rational representation $\rho_r(S)$ of S. The condition $\pi \in E_S$ yields that $\det(\rho_r(\pi)) \cdot \det(\rho_r(S)) = \pm 1$. Since $\mathrm{End}(X)$ is a subalgebra of $\mathrm{End}(H_1(X, \mathbb{Z}))$ (see §1.3 again), this clearly yields that E_S is finite. On the other hand one sees, with an argument similar to the one used in the proof of Theorem 3.2.1, that, once $\pi \in E_S$ has been fixed, the equivalence classes of the structures of the type $(X, S_{(\pi,a)})$, with $a \in X_{(S,\pi)}$, is a finite set. □

In conclusion we want to point out that the hypothesis $\mathrm{End}(X)^* \simeq \{\pm 1\}$ of Theorem 3.2.1, although restrictive, holds for the *general* complex torus of given dimension (see [Ci]).

We conclude this section by recording two results in the same circle of ideas but somehow going in *opposite directions*.

Let X be a simple complex torus of dimension q (see Proposition 1.3.7). Then $\mathrm{End}_0(X) = \mathrm{End}_{\mathbb{Q}}(X)$ is a semisimple algebra of finite dimension over $\mathbb{Q}$. Moreover it is well known that, if X is an abelian variety, then $\mathrm{End}_0(X)$ is of one of the following types (see [M] for a more precise statement, involving also the so called *Rosati involution* on $\mathrm{End}_0(X)$):

(I) $\mathrm{End}_0(X)$ is isomorphic to a totally real number field K;

(II) $\mathrm{End}_0(X)$ is isomorphic to a quaternion algebra over a totally real number field and $\mathrm{End}(X) \otimes \mathbb{R}$ is isomorphic to a matrix algebra $M_2(\mathbb{R})$;

(III) $\mathrm{End}_0(X)$ is isomorphic to a quaternion algebra over a totally real number field and $\mathrm{End}(X) \otimes \mathbb{R}$ is isomorphic to a product of copies of H, where H is the field of ordinary quaternions;

(IV) $\mathrm{End}_0(X)$ is isomorphic to a central simple algebra over a purely imaginary quadratic extension K of a totally real number field K_0.

In all cases if we denote by Z the center of $\mathrm{End}_0(X)$, by e the degree of Z over $\mathbb{Q}$ and by d^2 the degree of $\mathrm{End}_0(X)$ over C then $d^2 \cdot e$ divides $2q$. We will say that a simple complex torus X has *real multiplication* if $\mathrm{End}_0(X)$ is of type (I). In this case, if (X, S) is a real structure on X, we denote by $r_S(X)$ the degree of the subfield $\mathrm{End}_0(X)^S$ of $\mathrm{End}_0(X)$ fixed by S over $\mathbb{Q}$. On the other hand X is said to have *complex multiplication* if $\mathrm{End}_0(X)$ is an imaginary quadratic extension of a totally real field K_0

of degree q over $\mathbb{Q}$, i.e., if $\mathrm{End}_0(X)$ is of type (IV) with $d = 1$. With these definitions we can state the following result.

Theorem 3.2.3 (R. Silhol). (i) *If X has complex multiplication, then X has at most two proper non equivalent real structures.*

(ii) *If X has real multiplication and if it has a real structure (X, S), then the order of $T(X)$ does not exceed $2^{rs(X)+1}$.*

Proof. See [S2]. □

It is worth remarking that the bounds in the above theorem are sharp. For instance Comessatti [C2] (see §7, no. 30) produced infinitely many families of real abelian surfaces X with real multiplication and with $\mathrm{End}_0(X) = \mathrm{End}_0(X)^S$ of degree 2 over $\mathbb{Q}$, having 8 non equivalent real structures. More recent results on the subject are contained in [H].

3.3. Line bundles on complex tori with a real structure

Let (X, S) be a real structure on a complex torus X of dimension q. We will make one of the following equivalent assumptions:

(i) $X_S \neq \emptyset$;

(ii) (X, S) is equivalent to a proper structure;

(iii) (X, S) is isomorphic to $(\mathbb{C}^q/\Lambda, \sigma)$, where Λ is a maximal lattice in $\mathbb{C}^q$ fixed by the complex conjugation σ, and one still denotes by σ the induced antiinvolution on $\mathbb{C}^q/\Lambda$ (see Proposition 3.1.1 and Theorem 3.1.5).

We will actually identify (X, S) with $(\mathbb{C}^q/\Lambda, \sigma)$. We will use the exponential notation for the action of the conjugation.

Now we want to describe the action of the conjugation on $\mathrm{Pic}(X)$. In order to do that, we will use the explicit description of line bundles on $X = \mathbb{C}^q/\Lambda$ provided by the Appèll–Humbert Theorem (see Theorem 1.5.3). We recall that any line bundle on X is of the type $L(H, \alpha)$, where H is a hermitian form on $\mathbb{C}^q$ whose imaginary part E takes integral values on Λ, and α is a quasi-character for H. We have the following lemma.

Lemma 3.3.1. *For every line bundle $L(H, \alpha)$ on X one has*

$$L(H, \alpha)^\sigma = L(\alpha^\sigma, H^\sigma),$$

where α^σ and H^σ are defined as follows:

$$\alpha^\sigma(\lambda) := \overline{\alpha(\lambda^\sigma)} \quad \textit{for all } \lambda \in \Lambda,$$

$$H^\sigma(x, y) := \overline{H(x^\sigma, y^\sigma)} \quad \textit{for all } (x, y) \in \mathbb{C}^q \times \mathbb{C}^q.$$

Proof. Recall the action of Λ on $\mathbb{C}^q \times \mathbb{C}$ given by

$$\lambda.(x, z) = (x + \lambda, e_\lambda(x) \cdot z)$$

for all $\lambda \in \Lambda$, $x \in \mathbb{C}^q$, $z \in \mathbb{C}$, where $e_\lambda := e_\lambda(H, \alpha)$ is the automorphy factor related to the Appèll–Humbert datum (H, α) and to λ. Then

$$(\lambda.(x, z))^\sigma = (x + \lambda, e_\lambda(x) \cdot z)^\sigma = (x^\sigma + \lambda^\sigma, \overline{e_\lambda(x) \cdot z}).$$

But

$$\begin{aligned}\overline{e_\lambda(x)} &= \overline{\alpha(\lambda) \cdot \exp(\pi H(x, \lambda) + \frac{\pi}{2} H(\lambda, \lambda))} \\ &= \alpha^\sigma(\lambda^\sigma) \exp(\pi H^\sigma(x^\sigma, \lambda^\sigma) + \frac{\pi}{2} H(\lambda^\sigma, \lambda^\sigma)) = e_{\lambda^\sigma}(x^\sigma)\end{aligned}$$

where $e_{\lambda^\sigma} := e_{\lambda^\sigma}(\alpha^\sigma, H^\sigma)$ is the automorphy factor related to $(H^\sigma, \alpha^\sigma)$ and to λ^σ. Therefore

$$(\lambda.(x, z))^\sigma = (x^\sigma + \lambda^\sigma, e_{\lambda^\sigma}(x^\sigma) \cdot \overline{z}) = \lambda^\sigma(x, z)^\sigma$$

which proves the assertion. □

Remarks 3.3.2. (i) In view of the hypothesis $X_S \neq \emptyset$, if X is projective, then we have $\mathrm{Pic}(X)^G = \mathrm{Pic}(X)^+$ (see Proposition 2.7.4). We stress that the line bundles in $\mathrm{Pic}(X)^G$ are those of the type $L(H, \alpha)$ with $H = H^\sigma, \alpha = \alpha^\sigma$. Note that the equality $H = H^\sigma$ is equivalent to

$$E(x^\sigma, y^\sigma) = -E(x, y), \quad E := \mathrm{Im}\, H$$

for all $(x, y) \in \mathbb{C}^q \times \mathbb{C}^q$. Indeed, this follows by the identity

$$H(x, y) = E(x, iy) + iE(x, y).$$

Also we remark that, once H is given such that $H = H^\sigma$, there are certainly quasi-characters α for H such that $\alpha = \alpha^\sigma$. Indeed, in view of the relation (1.5.2), it suffices to define the quasi-character α on the elements of a basis of Λ (see §1.5). Furthermore, if $\alpha(\lambda^\sigma) = \overline{\alpha(\lambda)}$ and $\alpha(\mu^\sigma) = \overline{\alpha(\mu)}$ for two elements $\lambda, \mu \in \Lambda$, then one has

$$\begin{aligned}\alpha(\lambda^\sigma + \mu^\sigma) &= (-1)^{E(\lambda^\sigma, \mu^\sigma)} \cdot \alpha(\lambda^\sigma) \cdot \alpha(\mu^\sigma) \\ &= (-1)^{-E(\lambda, \mu)} \cdot \overline{\alpha(\lambda)} \cdot \overline{\alpha(\mu)} \\ &= (-1)^{E(\lambda, \mu) - E(\lambda, \mu)} \overline{\alpha(\lambda + \mu)} = \overline{\alpha(\lambda + \mu)}.\end{aligned}$$

Therefore it suffices to define α on a basis of Λ in such a way that it verifies

$$\alpha(\lambda^\sigma) = \overline{\alpha(\lambda)} \tag{3.3.3}$$

for all the elements λ of this basis. If we choose, as we can, a pseudonormal basis $(\sigma_1, \ldots, \sigma_q, \tau_1, \ldots, \tau_\lambda, \epsilon_{\lambda+1}, \ldots, \epsilon_q)$ of the first type for $H_1(X, \mathbb{Z})$, the conditions

(3.3.3) become

$$\alpha(\sigma_i) = \overline{\alpha(\sigma_i)} \Longleftrightarrow \alpha(\sigma_i) = \pm 1, \quad i = 1, \ldots, q,$$
$$\alpha(-\epsilon_j) = \alpha(\epsilon_j)^{-1} = \overline{\alpha(\epsilon_j)}, \quad j = \lambda + 1, \ldots, q,$$

$$\begin{aligned}(-1)^{-E(\sigma_i,\tau_i)} \cdot \alpha(\sigma_i) \cdot \alpha(\tau_i)^{-1} &= (-1)^{E(\sigma_i,-\tau_i)} \cdot \alpha(\sigma_i) \cdot \alpha(-\tau_i) \\ &= \alpha(\sigma_i - \tau_i) = \overline{\alpha(\tau_i)}, \quad i = 1, \ldots, \lambda.\end{aligned}$$

The first set of relations can certainly be satisfied with an appropriate choice of the values of α. The second set of relations imposes no condition at all on the values of α and the last relation is equivalent to

$$|\alpha(\tau_i)|^2 = (-1)^{E(\sigma_i,\tau_i)} \cdot \alpha(\sigma_i), \quad i = 1, \ldots, \lambda$$

which imposes conditions on the values of α which can be certainly satisfied. This proves our assertion.

(ii) Let $L = L(H, \alpha)$ be a line bundle on X. Then the map

$$\phi_L : x \in X \to t_x^* L \otimes L^* \in \mathrm{Pic}^0(X)$$

is a homomorphism, which is an isogeny if and only if L is ample, i.e., if and only if H is non-degenerate(see [LB], §2.4), in which case the degree of the isogeny equals $\det(E)$. It is easy to check with a direct computation that

$$t_x^* L \otimes L^* = L(\alpha_x, 0)$$

where

$$\alpha_x : \lambda \in \Lambda \to \exp(2\pi i E(v, \lambda)) \in S_1$$

and $v \in \mathbb{C}^q$ is a representative of $x \in X$. Now it is easy to see that, if $L = L^\sigma$, then ϕ_L is a morphism of real structures between (X, S) and $(\mathrm{Pic}^0(X), S)$. Indeed

$$\begin{aligned}\alpha_x^\sigma(\lambda) &= \overline{\alpha_x(\lambda^\sigma)} = \exp(-2\pi i E(v, \lambda^\sigma)) \\ &= \exp(2\pi i E(v^\sigma, \lambda)) = \alpha_{x^\sigma}(\lambda)\end{aligned}$$

which proves that

$$\phi_L(x^\sigma) = \phi_L(x)^\sigma.$$

We notice that in the previous argument we implicitly used the fact that $X_S \neq \emptyset$ since we identified (X, S) with $(\mathbb{C}^q/\Lambda, \sigma)$ and used Lemma 3.3.1, which applies to this case. If $X_S = \emptyset$, then the map ϕ_L is no longer a morphism of real structures between (X, S) and $(\mathrm{Pic}^0(X), S)$.

We conclude this section by starting the discussion of the problem of classifying real abelian varieties. Namely we will consider the following:

Problem 3.3.4. *Construct a classifying space for the isomorphism classes of all triples (X, S, H), where:*

(i) *$X = V/\Lambda$ is a complex torus of dimension q;*

(ii) *H is a positive definite hermitian form on V such that its imaginary part E is integral on Λ, i.e., H, seen as an element of $NS(X)$ (see* §1.5*), is the first Chern class $c_1(L)$ of an ample line bundle L on X of a fixed type $\mathbf{d} = (d_1, \ldots, d_q)$;*

(iii) *(X, S) is a real structure such that $X_S \neq \emptyset$;*

(iv) *$H = H^S$, i.e., $H(x, y) = \overline{H(S(x), S(y))}$ for all $(x, y) \in V \times V$ or equivalently $E(S(x), S(y)) = -E(x, y)$ for all $(x, y) \in V \times V$ (see Remark* 3.3.2, (i)*).*

A triple (X, S, H) will be called a *polarized real abelian variety* of type $\mathbf{d} = (d_1, \ldots, d_q)$. An isomorphims between two such abelian varieties (X, S, H), (X', S', H') is an isomorphism $f: X \to X'$ of (X, S) to (X', S') which carries H, seen as an element of $NS(X) \subset H^2(X, \mathbb{Z})$ to $H' \in NS(X') \subset H^2(X', \mathbb{Z})$.

At this point a remark is in order.

Remark 3.3.5. We know of the existence of the moduli space $\mathcal{A}_{(q,\mathbf{d})}$ (see §1.7). This variety has a natural real structure determined by the antiinvolution σ mapping X to its conjugate variety X^σ (see §2.2), which is still an element of $\mathcal{A}_{(q,\mathbf{d})}$. In order to solve Problem 3.3.4, one could be tempted to take $(\mathcal{A}_{(q,\mathbf{d})})^\sigma$ as the right classifying space. This is completely wrong! Here we give two reasons why this is so:

(i) there are varieties $X \in (\mathcal{A}_{(q,\mathbf{d})})^\sigma$ with different real structures (see Example 2.2.4, (ii));

(ii) if X is in $\mathcal{A}^\sigma_{(q,\mathbf{d})}$ it is true that there is an isomorphism $\phi: X \to X^\sigma$ of polarized abelian varieties. However it is not in general the case that ϕ is a descent datum for X (see §2.2), hence it is not in general the case that ϕ determines a real structure on X according to Theorem 2.2.2. In other words, there are points of $(\mathcal{A}_{(q,\mathbf{d})})^\sigma$ which do not correspond to real abelian varieties.

In order to illustrate what we said in (ii), let us give the following example of Shimura's. Let X be the principally polarized abelian surface corresponding to the point

$$Z = \begin{pmatrix} a + ic & b \\ b & -a + ic \end{pmatrix}$$

of the Siegel upper-half-space $\mathbb{H}_2$, where $a, b, c \in \mathbb{R} - \mathbb{Q}$ and $c > 0$. Let us set

$$N = \begin{pmatrix} 0 & 1 & 0 & 0 \\ -1 & 0 & 0 & 0 \\ 0 & 0 & 0 & -1 \\ 0 & 0 & 1 & 0 \end{pmatrix}.$$

One has $N.(I_2 \quad Z) = (I_2 \quad \overline{Z})$, where the action of $N \in \mathrm{Sp}(4, \mathbb{Z})$ on $\mathbb{H}_2$ has been defined in §1.7. Then, according to (1.7.3), there is an isomorphism $\phi: X \to X^\sigma$, which is determined by the linear map

$$f\colon (v_1, v_2) \in \mathbb{C}^2 \to (-v_2, v_1) \in \mathbb{C}^2.$$

But a direct computation shows that $\sigma \circ f \circ \sigma^{-1} \neq f^{-1}$, proving that ϕ is not a descent datum.

The last remark shows that, in order to construct a suitable moduli space which solves problem (3.3.4), the existence of $\mathcal{A}_{(q,d)}$ is of no help. We will need a few ad hoc considerations which unfortunately work only in the case of principal polarizations. In particular we will need a real version, due to Comessatti, of the Riemann bilinear relations, which we will discuss in the next section.

3.4. Riemann bilinear relations for principally polarized real varieties

In order to state and prove the main result of this section, i.e., Comessatti's theorem (Theorem 3.4.3), we need some preliminaries on symmetric matrices over $\mathbb{Z}_2$. Let $\mathcal{S}(q, \mathbb{Z}_2)$ be the set of all symmetric matrices of order q on $\mathbb{Z}_2$. Of course $\mathcal{S}(q, \mathbb{Z}_2)$ can be seen as the set of all symmetric bilinear forms of a vector space V of dimension q over $\mathbb{Z}_2$, once a basis of V has been fixed. The group $\mathrm{GL}(q, \mathbb{Z}_2)$ acts on $\mathcal{S}(q, \mathbb{Z}_2)$ in the usual way

$$(A, N) \in \mathrm{GL}(q, \mathbb{Z}_2) \times \mathcal{S}(q, \mathbb{Z}_2) \to A.N := A^t \cdot N \cdot A \in \mathcal{S}(q, \mathbb{Z}_2)$$

where the action corresponds to base changes in V. The above action induces an equivalence relation $\sim$, called *conjugation*, in $\mathcal{S}(q, \mathbb{Z}_2)$ such that for $N, N' \in \mathcal{S}(q, \mathbb{Z}_2)$ one has

$$N \sim N' \iff \text{there is an } A \in \mathrm{GL}(q, \mathbb{Z}_2) \text{ such that } N' = A.N.$$

Of course the rank of a matrix is a conjugacy invariant. Let us point out another more subtle conjugacy invariant. Let $N = (n_{ij})_{i,j=1,\dots,q}$ be a symmetric matrix. We will say that N is of *diasymmetric type* [resp. of *orthosymmetric type*] if there is an index $i = 1, \dots, q$ such that $n_{ii} = 1$ [resp. if for all $i = 1, \dots, q$ one has $n_{ii} = 0$].

Lemma 3.4.1. *The diasymmetric [resp. orthosymmetric] type is a conjugacy invariant. Furthermore, if N is orthosymmetric of maximal rank, then q is even.*

Proof. Let $A = (a_{ij})_{i,j=1,\dots,q} \in \mathrm{GL}(q, \mathbb{Z}_2)$ and let $N' = N.A := (n'_{ij})_{i,j=1,\dots,q}$. One has

$$n'_{rs} = \sum_{i,j=1,\dots,q} a_{ir} a_{js} n_{ij}$$

hence

$$n'_{rr} = \sum_{i,j=1,\dots,q} a_{ir}a_{jr}n_{ij} = \sum_{i=1}^{q} a_{ir}^2 n_{ii}$$

since we work over $\mathbb{Z}_2$ and therefore the repeated terms $a_{ir}a_{js}n_{ij}$, for $i \neq j$, in the sum cancel out because $n_{ij} = n_{ji}$. Thus, if N is orthosymmetric, so is any matrix conjugate to N. This proves the first assertion. As to the second assertion, we note that an orthosymmetric matrix in $\mathcal{S}(q, \mathbb{Z}_2)$ is both symmetric and antisymmetric (i.e., it is symmetric and all the elements on the main diagonal are zero). Therefore

$$\det(N) = \sum_{\sigma \in \mathcal{S}_q : \sigma = \sigma^{-1}} n_{1,\sigma(1)} \dots n_{q,\sigma(q)}$$

because N is symmetric and its coefficients are in $\mathbb{Z}_2$. If q is odd every permutation σ in the symmetric group $\mathcal{S}_q$ on q objects such that $\sigma = \sigma^{-1}$ has a fixed point, and this implies that $\det(N) = 0$ because N is antisymmetric. □

Now we introduce for each possible rank $\lambda > 0$ of a symmetric matrix over $\mathbb{Z}_2$, two typical matrices, one of diasymmetric type, the other of orthosymmetric type. The two matrices are $I_{q,\lambda} := I_{q,q,\lambda}$, having the first λ elements of the main diagonal equal to 1, all the other entries equal to 0 (see §2.5), and $J_{q,\lambda}$, having the (i, j)-entries, with $i + j = \lambda + 1$, equal to 1, all the other entries equal to 0. Now we can prove the following proposition due to Comessatti (see [C3,4]).

Proposition 3.4.2. *A matrix N of rank $\lambda > 0$ in $\mathcal{S}(q, \mathbb{Z}_2)$ is either conjugate to $I_{q,\lambda}$ or to $J_{q,\lambda}$ according to whether N is diasymmetric or orthosymmetric.*

Proof. Let us interpret N as non-degenerate, bilinear symmetric form on the $\mathbb{Z}_2$-vector space V of dimension q. Then N is an orthogonal sum of a zero form, i.e., its radical, and of a symmetric form of rank λ. Accordingly there is a matrix N' conjugate to N of the form

$$N' = \begin{pmatrix} N'_1 & 0_{\lambda,q-\lambda} \\ 0_{q-\lambda,\lambda} & 0_{q-\lambda,q-\lambda} \end{pmatrix}$$

where N'_1 is symmetric of order λ and has maximal rank λ. This clearly shows that we can restrict our attention to the case that N has maximal rank.

Let us now treat the diasymmetric case first. Let $N = (n_{ij})_{i,j=1,\dots,q}$ be a diasymmetric matrix. We claim that in the conjugacy class of N there is a matrix $N' = (n'_{ij})_{i,j=1,\dots,q}$ such that $n_{ii} = 1$ for all $i = 1, \dots, q$. To prove this we may assume that $n_{ii} = 1$ for $i = 1, \dots, h$, whereas $n_{ii} = 0$ for $i = h + 1, \dots, q$. Then we pass to the conjugate matrix $N' = A^t \cdot N \cdot A$, where $A = (a_{ij})_{i,j=1,\dots,q}$ is such that its first h rows are

$$\begin{pmatrix} I_h & \Delta \end{pmatrix}$$

where Δ is the (h, q)-matrix whose first row consists of 1's and all the other entries are 0. It is clear that such a matrix can be a submatrix of a matrix in $GL(q, \mathbb{Z}_2)$ since it has maximal rank. Now we have

$$n'_{rr} = \sum_{i=1}^{q} a_{ir}^2 n_{ii} = \sum_{i=1}^{h} a_{ir}^2 n_{ii} = 1$$

for all $r = 1, \dots, q$, which proves our claim.

So let $N = (n_{ij})_{i,j=1,\dots,q}$ be a diasymmetric matrix of maximal rank q such that $n_{ii} = 1$ for all $i = 1, \dots, q$. We may assume that $n_{12} = 0$, since not all the entries of N can be equal to 1. By summing the first column of N to the columns having the first entry equal to 1, and correspondingly doing the symmetric operation on the lines, we see that N is equivalent to a matrix of the type

$$\begin{pmatrix} 1 & 0 & \dots & 0 \\ 0 & & \dots & \\ 0 & & N' & \\ 0 & & \dots & \end{pmatrix}$$

where $N' = (n'_{ij})_{i,j=2,\dots,q}$ is symmetric of type $(q-1, q-1)$ of maximal rank. By the assumption on n_{12} we have that $n'_{22} = n_{22} = 1$. Hence N' is diasymmetric and therefore we can argue by induction, proving the theorem.

Now we turn to the orthosymmetric case. We remark that in any line of $N = (n_{ij})_{i,j=1,\dots,q}$ there is an entry, not lying on the main diagonal, which is equal to 1, since we are assuming N to be of maximal rank. Then we may suppose, up to reordering the lines of N, which does not change the conjugacy class, and perhaps up to summming the first two rows and columns to the others, that N is of the form

$$\begin{pmatrix} J_{2,2} & 0_{2,q-2} \\ 0_{q-2,2} & N' \end{pmatrix}$$

where N' is clearly orthosymmetric again. Then one concludes by induction. □

Now we are ready to state the main result of this section.

Theorem 3.4.3 (A. Comessatti). *Let (X, H) be a principally polarized abelian variety of dimension q. The following are equivalent:*

(i) *there is a real structure (X, S) on X with Comessatti invariant λ, such that $X_S \neq \emptyset$ and there is a line bundle $L = L(H, \alpha)$ such that $L^S = L$;*

(ii) *X has a period matrix corresponding to the given polarization H (see §1.7), which is either of the form*

$$\Omega = (I_q \quad \frac{1}{2} I_{q,\lambda} + iT) \tag{3.4.4}$$

or of the form

$$\Omega = (I_q \quad \frac{1}{2} J_{q,\lambda} + iT) \tag{3.4.5}$$

where T *is real symmetric and positive definite.*

Proof. Let us first prove that (ii) implies (i). Let us assume that X has a period matrix of the form Ω given in (3.4.4) or (3.4.5). Then $X \simeq \mathbb{C}^q/\Lambda$ where Λ is generated by the columns of Ω. It is clear that Λ is stable by the conjugation σ of $\mathbb{C}^q$. Hence there is a real structure (X, S) such that $X_S \neq \emptyset$ (see the first part of the proof of Theorem 3.1.5). Since $Z = \frac{1}{2}M + iT$, where $M = I_{q,\lambda}$ or $M = J_{q,\lambda}$, is symmetric with positive definite imaginary part, then there is a principal polarization H on X, which has to be seen as a hermitian form on $\mathbb{C}^q$ whose imaginary part E is integral on Λ and has the columns of Ω as a symplectic basis. Let $(\lambda_1, \dots, \lambda_q, \mu_1, \dots, \mu_q)$ be this basis. The action of S on it coincides with the action of σ, hence it has matrix of the form

$$\begin{pmatrix} I_q & M \\ 0_{q,q} & -I_q \end{pmatrix}.$$

But now we have

$$\begin{aligned} &E(\lambda_i^\sigma, \lambda_j^\sigma) = E(\lambda_i, \lambda_j) = 0, \quad i, j = 1, \dots, q, \\ &E(\lambda_i^\sigma, \mu_j^\sigma) = E(\lambda_i, -\mu_j), \quad i, j = 1, \dots, q, \\ &E(\mu_i^\sigma, \mu_j^\sigma) = E(\sum_{h=1}^q m_{ih}\lambda_h - \mu_i, \sum_{k=1}^q m_{jk}\lambda_k - \mu_j) = -m_{ij} + m_{ji} = 0, \end{aligned}$$

where $M = (m_{ij})_{i,j=1,\dots,q}$. Whence we deduce that $H = H^\sigma$ and that there is a line bundle $L = L(H, \alpha)$ such that $L^S = L$ (see Remark 3.3.2, (i)).

Let us now prove that (i) implies (ii). Let (X, S) be a real structure on X such that $X_S \neq \emptyset$ and with first Comessatti invariant λ. Then, according to Theorem 3.1.5, (ii), we may assume that (X, S) coicides with $(\mathbb{C}^q/\Lambda, \sigma)$, where Λ is the lattice generated by the columns of a period matrix Ω in pseudonormal form (2.8.4), and σ is the antiinvolution induced on $\mathbb{C}^q/\Lambda$ by the usual conjugation σ on $\mathbb{C}^q$. Let $L = L(H, \alpha)$ be a line bundle such that $L^S = L$. This means in particular that $H^\sigma = H$. Let $(\sigma_1, \dots, \sigma_q, \tau_1, \dots, \tau_\lambda, \epsilon_{\lambda+1}, \dots, \epsilon_q)$ be a pseudonormal basis of Λ of the first type. We have

$$\sigma_i^\sigma = \sigma_i, \quad i = 1, \dots, q$$

hence

$$-E(\sigma_i, \sigma_j) = E(\sigma_i^\sigma, \sigma_j^\sigma) = E(\sigma_i, \sigma_j), \quad i, j = 1, \dots, q$$

and therefore the sublattice Λ^+ spanned by $(\sigma_1, \dots, \sigma_q)$ is isotropic for E. Thus the matrix of E with respect to the pseudonormal basis $(\sigma_1, \dots, \sigma_q, \tau_1, \dots, \tau_\lambda,$

$\epsilon_{\lambda+1}, \dots, \epsilon_q)$ is of the form

$$\begin{pmatrix} 0_{q,q} & A^t \\ -A & B \end{pmatrix}$$

where A and B are matrices of type (q, q) over the integers, such that $|A| = \pm 1$ and B is antisymmetric. A base change

$$(\sigma_1, \dots, \sigma_q, \tau_1, \dots, \tau_\lambda, \epsilon_{\lambda+1}, \dots, \epsilon_q) \to (\sigma_1, \dots, \sigma_q, \mu_1, \dots, \mu_q)$$

whose matrix is

$$\begin{pmatrix} I_q & 0_{q,q} \\ 0_{q,q} & (A^t)^{-1} \end{pmatrix}$$

changes the matrix of E to the matrix

$$\begin{pmatrix} 0_{q,q} & I_q \\ -I_q & D \end{pmatrix}$$

where D is still antisymmetric. Now we claim that there is a matrix C such that $D = C - C^t$. For instance one may take C to be the *upper triangular part* of D. Then, since D is antisymmetric, we have $D = C - C^t$.

Now we make a base change of the type

$$(\sigma_1, \dots, \sigma_q, \mu_1, \dots, \mu_q) \to (\sigma_1, \dots, \sigma_q, \nu_1, \dots, \nu_q)$$

corresponding to the matrix

$$\begin{pmatrix} I_q & C \\ 0_{q,q} & I_q \end{pmatrix}.$$

Accordingly the matrix of E in the new basis changes to

$$\begin{pmatrix} 0_{q,q} & I_q \\ -I_q & 0_{q,q} \end{pmatrix},$$

i.e., the basis $(\sigma_1, \dots, \sigma_q, \nu_1, \dots, \nu_q)$ is symplectic. The matrix of the antiinvolution σ in this basis has the form

$$\begin{pmatrix} I_q & N \\ 0_{q,q} & -I_q \end{pmatrix}.$$

We want to investigate the nature of the integral matrix $N = (n_{ij})_{i,j=1,\dots,q}$, and eventually reduce it to a *normal form* by a sequence of suitable base changes. To this aim, let $(\omega_1, \dots, \omega_q)$ be a basis of real forms of $H^0(X, \Omega^1_X)$, such that

$$\int_{\sigma_i} \omega_j = \delta_{ij}, \quad i, j = 1, \dots, q.$$

The corresponding period matrix is therefore of the form

$$\Omega = (I_q \quad Z)$$

where $Z \in \mathbb{H}_q$, since $(\sigma_1, \ldots, \sigma_q, \nu_1, \ldots, \nu_q)$ is symplectic (see §1.7). Remember that

$$z_{ij} = \int_{\nu_j} \omega_i .$$

Therefore we have

$$\begin{aligned}\overline{z_{ij}} = \overline{\int_{\nu_j} \omega_i} = \int_{\nu_j^\sigma} \omega_i \\ = n_{ij} - \int_{\nu_j} \omega_i = -z_{ij} + n_{ij}\end{aligned}$$

whence

$$n_{ij} = 2\,\mathrm{Re}(z_{ij}).$$

Therefore N is symmetric and we have

$$\Omega = (I_q \quad \frac{1}{2}N + iT)$$

where T is real, symmetric and positive definite. Now we are going to make *symplectic* base changes with matrices of the form

$$\begin{pmatrix} (a^t)^{-1} & b \\ 0_{q,q} & a \end{pmatrix}$$

where a and b are $q \times q$ integral matrices such that $a^t \cdot b$ is symmetric and $a \in \mathrm{GL}(q, \mathbb{Z})$. We leave to the reader the easy proof of the fact that such matrices belong to $\mathrm{Sp}(2q, \mathbb{Z})$. The corresponding action on $\mathbb{H}_q$ sends Z to

$$\begin{aligned} a^t \cdot (b + Z \cdot a) &= a^t \cdot (\frac{1}{2}N + iT) \cdot a + a^t \cdot b \\ &= \frac{1}{2}a^t \cdot N \cdot a + ia^t \cdot T \cdot a + a^t \cdot b \\ &= \frac{1}{2}(a^t \cdot N \cdot a + 2 \cdot a^t \cdot b) + ia^t \cdot T \cdot a. \end{aligned}$$

Now we remark that by taking transformations of the above type with $a = I_q$, we can reduce N modulo 2. Then, by taking $a \in \mathrm{GL}(q, \mathbb{Z}_2)$ and $b = 0_{q,q}$, and applying Proposition 3.4.2 we finally conclude the proof of the theorem. □

A principally polarized real abelian variety (X, S, H) having a period matrix of the form (3.4.4) [resp. of type (3.4.5)] is said to be of type (λ, d) [resp. of type (λ, o)]. We note that, by the proof of Theorem 3.4.3 and according to Lemma 3.4.1, a variety is of type (λ, o) only if λ is even. Note that for $\lambda = 0$ the two types coincide. Nonetheless we consider this case as orthosymmetric, hence it will be denoted by $(0, o)$. Note also that clearly the type of a principally polarized real abelian variety is invariant under isomorphisms.

Before concluding this section, we prove another basic result for the construction of the moduli spaces.

Theorem 3.4.6. *Let* (X, S, H) *and* (X', S', H') *be principally polarized real abelian varieties of dimension* q *with corresponding matrices of the form*

$$\omega = (I_q \quad \frac{1}{2}M + iT), \quad \omega' = (I_q \quad \frac{1}{2}M' + iT')$$

as dictated by Theorem 3.4.3. *Then the following are equivalent:*

(i) (X, S, H) *and* (X', S', H') *are isomorphic;*

(ii) *there is a matrix* $a \in \mathrm{GL}(q, \mathbb{Z})$ *such that*

$$a^t \cdot M \cdot a \equiv M' \pmod 2, \quad a^t \cdot T \cdot a = T'.$$

Proof. The implication $(ii) \Rightarrow (i)$ has already been proved during the proof of Theorem 3.4.3. Let us prove the converse implication.

Let $f: X \to X'$ be the isomorphism between the two principally polarized real abelian varieties. Corresponding to f there is a symplectic matrix

$$N = \begin{pmatrix} a & b \\ c & d \end{pmatrix}$$

where a, b, c, d are $q \times q$ integral matrices such that

$$\frac{1}{2}M' + iT' = (a + (\frac{1}{2}M + iT) \cdot c)^{-1} \cdot (b + (\frac{1}{2}M + iT) \cdot d),$$

and the isomorphism f is induced by the map

$$\phi: \boldsymbol{x} \in \mathbb{C}^q \to (a + (\frac{1}{2}M + iT) \cdot c)^{-1} \cdot \boldsymbol{x} \in \mathbb{C}^q$$

where $\boldsymbol{x}$ is a column vector, $X \simeq \mathbb{C}^q/\Lambda$ [resp. $X' \simeq \mathbb{C}^q/\Lambda'$] and Λ [resp. Λ'] are generated by the columns of Ω [resp. Ω']. Since f is an isomorphism between the real structures (X, S) and (X', S'), then ϕ has to commute with the usual conjugation σ of $\mathbb{C}^q$. This implies that $T \cdot c = 0_{q,q}$. Since T has maximal rank, this yields $c = 0_{q,q}$. But then the conditions that N is symplectic read

$$d = (a^t)^{-1}, \quad d^t \cdot b = (d^t \cdot b)^t$$

and the assertion follows from the proof of Theorem 3.4.3. □

3.5. Moduli spaces of principally polarized real abelian varieties

We denote by $\mathcal{A}_{q,(\lambda,\epsilon)}$ the set of all isomorphism classes of principally polarized real abelian varieties of type (λ, ϵ), where either $\epsilon = d$ or $\epsilon = o$ (the latter case only for even λ). The triple determined by q and by the pair (λ, ϵ) which identifies the type of a principally polarized real abelian variety is equally well determined by assigning the matrix M, where $M = I_{q,\lambda}$ if $\epsilon = d$ and $M = J_{q,\lambda}$ if $\epsilon = o$ (if $\lambda = 0$ the matrix M is 0_q). So we may write $\mathcal{A}_M$ istead of $\mathcal{A}_{q,(\lambda,\epsilon)}$.

Now we define the group

$$\Gamma_M := \Gamma_{q,(\lambda,\epsilon)} = \{a \in \mathrm{GL}(q, \mathbb{Z}) : a^t \cdot M \cdot a \equiv M \pmod 2\}.$$

This group acts on

$$\mathbb{H}(\mathbb{R})_q := \{T \in \mathrm{GL}(q, \mathbb{R}) : T = T^t,\ T > 0\}$$

in the following way:

$$(a, T) \in \Gamma_M \times \mathbb{H}(\mathbb{R})_q \to T.a := a^t \cdot T \cdot a \in \mathbb{H}(\mathbb{R})_q.$$

By Theorem 3.4.6 there is an obvious bijection

$$\phi_M \colon \mathcal{A}_M \to \mathbb{H}(\mathbb{R})_q / \Gamma_M.$$

Next we will see that there is a natural structure of real analytic space of dimension $\frac{q(q+1)}{2}$ on $\mathbb{H}(\mathbb{R})_q/\Gamma_M$. This structure is inherited by $\mathcal{A}_M$ via ϕ_M, and therefore $\mathcal{A}_M$ itself can be considered in a natural way as a $\frac{q(q+1)}{2}$-dimensional real analytic space. The main step in this direction is the following result.

Theorem 3.5.1. *The action of Γ_M on $\mathbb{H}(\mathbb{R})_q$ is properly discontinuous, i.e., for any pair of compact subsets K_1, K_2 of $\mathbb{H}(\mathbb{R})_q$ the set*

$$\Gamma_M(K_1, K_2) := \{a \in \Gamma_M : K_1 \cap K_2.a \neq \emptyset\}$$

is finite.

Proof. We note that the action of Γ_M on $\mathbb{H}(\mathbb{R})_q$ comes by restriction from the $\mathrm{GL}(q, \mathbb{R})$-action

$$(a, T) \in \mathrm{GL}(q, \mathbb{R}) \times \mathbb{H}(\mathbb{R})_q \to T.a := a^t \cdot T \cdot a \in \mathbb{H}(\mathbb{R})_q$$

Indeed this action is well known to be transitive, so that $\mathbb{H}(\mathbb{R})_q$ is the orbit of any of its elements, for example of I_q. So the map

$$h \colon a \in \mathrm{GL}(q, \mathbb{R}) \to a^t \cdot a \in \mathbb{H}(\mathbb{R})_q$$

is surjective. We claim that it is, in addition, proper. In fact it is well known that one has the following homeomorphism of topological spaces:

$$\mathrm{GL}(q, \mathbb{R}) \simeq O(q, \mathbb{R}) \times \mathbb{H}(\mathbb{R})_q$$

since every non-degenerate real matrix can be uniquely written as the product of an orthogonal matrix and of a symmetric positive definite one. If $a = u \cdot b$, where $u \in O(q, \mathbb{R})$ and $b \in \mathbb{H}(\mathbb{R})_q$, then $h(a) = b^t \cdot b = b^2$. Since the map $b \in \mathbb{H}(\mathbb{R})_q \to b^2 \in \mathbb{H}(\mathbb{R})_q$ is a homeomorphism, the map h coincides, up to homeomorphisms, with the projection of $O(q,\mathbb{R}) \times \mathbb{H}(\mathbb{R})_q$ to the second factor, and as such it is proper, since $O(q,\mathbb{R})$ is compact.

Furthermore we claim that any discrete subgroup G of $\mathrm{GL}(q,\mathbb{R})$ acts in a properly discontinuous way on $\mathbb{H}(\mathbb{R})_q$. Indeed, let K_1, K_2 be compact subsets of $\mathbb{H}(\mathbb{R})_q$. The set

$$G(K_1, K_2) := \{a \in G : K_1 \cap K_2.a \neq \emptyset\}$$

clearly coincides with the set

$$h^{-1}(K_1)^{-1} \cdot h^{-1}(K_2) := \{x^{-1}y, x \in h^{-1}(K_1), y \in h^{-1}(K_2)\}.$$

Since K_1, K_2 are compact, h is proper and the product map is continuous, we have that $G(K_1, K_2)$ is compact. Since G is discrete, it follows that $G(K_1, K_2)$ is in fact finite. The last claim applied to Γ_M proves the assertion. □

Now the theorem of E. Cartan's (see [Ca]) says that there is a unique structure of real anaytic space on $\mathbb{H}(\mathbb{R})_q/\Gamma_M$ such that the natural projection map $\mathbb{H}(\mathbb{R})_q \to \mathbb{H}(\mathbb{R})_q/\Gamma_M$ is an analytic submersion, and $\mathbb{H}(\mathbb{R})_q/\Gamma_M$ turns out to have dimension $\frac{q(q+1)}{2}$. As we said above, this structure can be transported to $\mathcal{A}_M$ via ϕ_M. Let $\mathcal{A}(\mathbb{R})_q$ be the disjoint union of $\mathcal{A}_M$ for all possible matrices $M = I_{q,\lambda}$ or $M = J_{q,\lambda}$. Notice that $\mathcal{A}(\mathbb{R})_q$ is the set of all isomorphism classes of principally polarized abelian varieties of dimension q, i.e., is the *moduli space* of such varieties. In conclusion we have the following theorem.

Theorem 3.5.2. *The moduli space $\mathcal{A}(\mathbb{R})_q$ is the disjoint union of $q+1+[\frac{q}{2}]$ connected real analytic spaces $\mathcal{A}_M$ each of dimension $\frac{q(q+1)}{2}$.*

Proof. The only unproved assertion concerns the number $q+1+[\frac{q}{2}]$ of the components. Indeed the number of connected components $\mathcal{A}_{q,(\lambda,d)}$ is q, one for each value of $\lambda = 1, \dots, q$. The components $\mathcal{A}_{q,(\lambda,o)}$ are only present for even values of λ, included 0, hence there are $[\frac{q}{2}] + 1$ of them. □

Example 3.5.3 (The case of elliptic curves, i.e., $q = 1$). For $q = 1$ we have two one-dimensional components of $\mathcal{A}(\mathbb{R})_q$, namely $\mathcal{A}_{1,(0,o)}$ and $\mathcal{A}_{1,(1,d)}$. In both cases we have $\Gamma_M = \mathrm{GL}(1,\mathbb{Z}) = \{1, -1\}$, and the action on $\mathbb{H}(\mathbb{R})_1 = \mathbb{R}^+$ is trivial. Hence we have

$$\mathcal{A}_{1,(0,o)} \simeq \mathcal{A}_{1,(1,d)} \simeq \mathbb{R}^+$$

and precisely, by looking at period matrices, we have

$$\mathcal{A}_{1,(0,o)} = \{(1, it), t \in \mathbb{R}^+\},$$

$$\mathcal{A}_{1,(1,d)} = \{(1, \frac{1}{2} + it), t \in \mathbb{R}^+\}.$$

Let us now remark that in general there is an analytic involution $\gamma_{q,(\lambda,\epsilon)}$ in each moduli space $\mathcal{A}_{q,(\lambda,\epsilon)}$, which sends the isomorphism class of any variety to the class of the complementary structure. Clearly two complementary structures belong to $\mathbb{C}$-isomorphic varieties, which can very well be non isomorphic as real principally polarized abelian varieties.

In the present case, by Proposition 3.1.6 we have

$$\gamma_{1,(0,o)} : (1, it) \in \mathcal{A}_{1,(0,o)} \to (1, \frac{i}{t}) \in \mathcal{A}_{1,(0,o)},$$

$$\gamma_{1,(1,d)} : (1, \frac{1}{2} + it) \in \mathcal{A}_{1,(1,d)} \to (1, \frac{1}{2} + \frac{i}{4t}) \in \mathcal{A}_{1,(1,d)}.$$

Furthermore the elliptic curves corresponding to period matrices of the type

$$(1, \frac{1}{2} + it), \quad \frac{1}{2} \le t \le \frac{\sqrt{3}}{2},$$

are $\mathbb{C}$-isomorphic to those corresponding to period matrices of the type

$$(1, \cos\theta + i \sin\theta), \quad \frac{\pi}{3} \le \theta \le \frac{\pi}{2}.$$

The appropriate transformation is

$$z' = \frac{z-1}{z}.$$

As we know from Theorem 3.2.1, the elliptic curves X having $\mathrm{End}(X)^* \simeq \{\pm 1\}$, with $\mathcal{S}(X)$ non empty and with $\lambda = 1$ have exactly two non equivalent real structures with non empty real part, namely the two complementary structures. Let us now examine the case of elliptic curves X with $\mathrm{End}(X)^* \not\simeq \{\pm 1\}$, with $\mathcal{S}(X)$ non empty and with $\lambda = 1$. We have only two cases (see [Hr], p. 321):

(i) the curve X_η corresponding to the period matrix $(1, \eta)$, where $\eta = \frac{1+i\sqrt{3}}{2}$;

(ii) the curve $X_{\eta'}$ corresponding to the period matrix $(1, \eta')$, where $\eta' = \frac{1+i}{2}$.

The curve $X_\eta \simeq \mathbb{C}/\langle 1, \eta\rangle$ has *multiplication* by η, namely the multiplication by $\overline{\eta}$ in $\mathbb{C}$ fixes the lattice $\langle 1, \eta\rangle$ and therefore descends to an automorphism π of X_η, which generates the group $\mathrm{End}(X_\eta)^*$. If S is the antiinvolution on X_η, one immediately sees that $(S \circ \pi)^2 = \mathrm{id}_{X_\eta}$. On the other hand one also sees that $(S \circ \pi) \circ \pi = \pi \circ S^-$, hence $S \circ \pi$ gives no new real structures on X. Furthermore (X_η, S) and (X_η, S^-) are not equivalent, since clearly $(1, \eta)$ is not a fixed point by the involution $\gamma_{1,(1,d)}$. A similar argument applied to the other elements of $\mathrm{End}(X)^*$ shows that, although $\mathrm{End}(X_\eta)^* \not\simeq \{\pm 1\}$, the curve X_η behaves exactly like all other curves in its component of the moduli space as far as the number of non equivalent real structures on it is concerned.

By contrast, the curve $X_{\eta'}$ has a different behaviour. As we saw, this curve is also $\mathbb{C}$-isomorphic to the curve X_i corresponding to the period matrix $(1, i)$. Therefore the curve has a multiplication π' by i, in the same sense as above. Here we have (compare with Example 2.2.4), (ii)):

(a) A real structure (X_i, σ), where σ is induced by the usual conjugation on $\mathbb{C}$, with $\lambda = 0$. This structure is isomorphic to the complementary one, indeed the corresponding point on $\mathcal{A}_{1,(0,o)}$ is fixed by $\gamma_{1,(0,o)}$.

(b) Two real structures with $\lambda = 0$ and with no real points. In fact, by Theorem 3.1.2, a real structure corresponding to π' is determined by an element $a \in \mathbb{C}$ such that $a + \overline{a} \in \Lambda$ and $ia + \overline{a} \in \Lambda$. Therefore a is of the form $a = x + iy$ where $2x \in \mathbb{Z}$ and $x - y \in \mathbb{Z}$. An easy computation shows that the real structures corresponding to $a = \frac{1}{2}$ and to $a = \frac{1+i}{2}$ are the only non equivalent ones, i.e., the structures in question are $(X_i, \tau_{\frac{1}{2}} \circ \sigma)$ and $(X_i, \tau_{\frac{1+i}{2}} \circ \sigma)$.

(c) A real structure $(X_{\eta'}, \sigma)$, where again σ is induced by the usual conjugation on $\mathbb{C}$, with $\lambda = 1$. Also this structure is isomorphic to the complementary one, as one sees from the description of the moduli space and of the map $\gamma_{1,(1,d)}$.

For all real elliptic curves X with $\lambda = 0$ non isomorphic to X_i we have $\mathrm{End}(X)^* = \{\pm 1\}$ (see [Hr]). Hence by Theorem 3.2.1 they have four real, pairwise complementary, structures, two with real points (appearing in the moduli space $\mathcal{A}(\mathbb{R})_1$) and two without real points.

Remark 3.5.4. For the results of this paragraph we refer to Comessatti's papers quoted among the references, and to the paper of Shimura [Sm]. Unfortunately not too many results are available for real abelian varieties with non principal polarization. The problem of their classification has been classically treated by Cherubino [Ch1,2,3]. A small account on the subject is also contained in [S1], p. 94. Certainly there is still open space for further investigations.

3.6. Real theta functions

Let (X, S, H) be a principally polarized abelian variety, having period matrix $\Omega = (I_q \quad Z)$ of the form (3.4.4) or (3.4.5), where $Z = \frac{1}{2}M + iT$. As in the proof of Theorem 3.4.3, we have that $X \simeq \mathbb{C}^q/\Lambda$, where Λ is the lattice generated by the columns $(\lambda_1, \ldots, \lambda_q, \mu_1, \ldots, \mu_q)$ of Ω, which is a symplectic basis for the imaginary part E of the hermitian form H determining the principal polarization. Furthermore the antiinvolution S on X is induced by the conjugation σ on $\mathbb{C}^q$.

We let α_0 be the quasi-character for H which is uniquely defined by the assumption that $\alpha_0(\lambda_i) = \alpha_0(\mu_i) = 1$, $i = 1, \ldots, q$ (see §1.5). Let us consider the line bundle $L_0 := L(\alpha_0, H)$. Since H is a principal polarization, the Theorem 1.6.4 of Riemann–Roch says that $h^0(X, L_0) = 1$. Actually the proof of the Theorem 1.6.4 shows that the unique section of L_0 arises from the even theta function

$$\begin{aligned}\theta(x) &= \sum_{\lambda\in\mathbb{Z}^q} \exp(2\pi i\lambda^t\cdot x + \frac{\pi}{2}H(x,x) + \pi i\lambda^t\cdot Z\cdot\lambda)\\ &= \exp(\frac{\pi}{2}H(x,x))\cdot\sum_{\lambda\in\mathbb{Z}^q}\exp(2\pi i\lambda^t\cdot x + \frac{\pi i}{2}\lambda^t\cdot M\cdot\lambda - \pi\lambda^t\cdot T\cdot\lambda),\end{aligned}$$

called the *Riemann theta function.* Let us set

$$\vartheta(x) := \sum_{\lambda\in\mathbb{Z}^q}\exp(2\pi i\lambda^t\cdot x + \frac{\pi i}{2}\lambda^t\cdot M\cdot\lambda - \pi\lambda^t\cdot T\cdot\lambda)$$

so that

$$\theta(x) = \exp(\frac{\pi}{2}H(x,x))\cdot\vartheta(x)$$

and therefore the zero-locus of θ is the same as the zero locus of ϑ.

We know that $H^\sigma = H$ but not that $\alpha_0^\sigma = \alpha_0$. Hence the conjugation sends L_0 to L_0^σ, that can very well be different from L_0. Accordingly the theta function above is changed by σ into the conjugate function

$$\begin{aligned}\vartheta^\sigma(x) &= \sum_{\lambda\in\mathbb{Z}^q}\exp(-2\pi i\lambda^t\cdot x - \pi\lambda^t\cdot\overline{Z}\cdot\lambda)\\ &= \sum_{\lambda\in\mathbb{Z}^q}\exp(-2\pi i\lambda^t\cdot x - \frac{\pi i}{2}\lambda^t\cdot M\cdot\lambda - \pi\lambda^t\cdot T\cdot\lambda)\\ &= \sum_{\lambda\in\mathbb{Z}^q}\exp(-2\pi i\lambda^t\cdot x + \frac{\pi i}{2}\lambda^t\cdot M\cdot\lambda - \pi\lambda^t\cdot T\cdot\lambda - \pi i\lambda^t\cdot M\cdot\lambda).\end{aligned}$$

Comparing with the expression of θ and taking into account that for all integers λ, one has $\lambda^2 \equiv \lambda \pmod 2$, in the diasymmetric case one has

$$\begin{aligned}\vartheta^\sigma(x) &= \sum_{\lambda\in\mathbb{Z}^q}\exp(-2\pi i\lambda^t\cdot(x+\epsilon) + \pi i\lambda^t\cdot Z\cdot\lambda)\\ &= \vartheta(-(x+\epsilon)) = \vartheta(x+\epsilon)\end{aligned}$$

where ϵ is the column vector whose first λ entries are $\frac{1}{2}$, the others are 0.

In general the theta function $\vartheta(x+a)$ is changed by the conjugation into the function $\vartheta(x+a^\sigma+\epsilon)$. Now we abuse notation and denote points in $\mathbb{C}^q$ and corresponding points in X with the same letter. The above discussion yields that for all $a\in X$ one has $(\tau_a^*L_0)^\sigma \simeq \tau_{a^\sigma+\epsilon}^*L_0$. Now we note that, since H is a principal polarization, the map ϕ_{L_0} considered in Remark 3.3.2, (ii), is an isomorphism, in particular it is injective (see [LB], p. 38). Therefore $\tau_a^*L_0$ is a real line bundle if and only if $a = a^\sigma+\epsilon$ in X. On the other hand we know that real line bundles do exist (see Theorem 3.4.3). This means that $\epsilon\in(X_{-S})_0$, and that all real line bundles are of the form $\tau_a^*L_0$ with a in the q-dimensional real variety which is the fibre of $\mathrm{id}_X - S$ over ϵ (see the proof of Theorem 3.2.1, Claim 2).

In the orthosymmetric case the term $\lambda^t \cdot M \cdot \lambda$ is clearly even, and therefore we have

$$\vartheta^\sigma(x) = \vartheta(-x) = \vartheta(x)$$

and more generally $\vartheta(x + a)$ is changed by conjugation into the function $\vartheta(x + a^\sigma)$. The real line bundles $\tau_a^* L_0$ correspond therefore to the real points $a \in X_S$. In particular the original Riemann theta function is real.

Example 3.6.1. For $q = 1$ and $M = 0$ (orthosymmetric elliptic case), $z = it$ and $x \in \mathbb{R}$ one has, up to the multiplicative factor $\exp(\frac{\pi}{2}x^2)$,

$$\theta(x) = \sum_{n \in \mathbb{Z}} \exp(-n^2 \pi t)(\cos(2\pi n x) + i \sin(2\pi n x)).$$

But it is clear that the $\sin(2\pi n x)$ cancels out with $\sin(-2\pi n x)$ for all integers n, and one is left with the real Fourier series

$$\theta(x) = \sum_{n \in \mathbb{Z}} \exp(-n^2 \pi t) \cos(2\pi n x)$$

which is the typical real ortosymmetric elliptic theta function.

4. Applications to real curves

In this chapter we apply Comessatti's results on real abelian varieties to Jacobians of real curves. We will relate algebraic invariants of a real structure (X, S) on a complex compact Riemann surface X with topological invariants of X_S. We prove the real version of Torelli's theorem and discuss the construction of moduli spaces for real curves. In the last section we will show, by using étale cohomology, how one can extend classical results on smooth projective curves to any real, even affine or singular, curve.

4.1. The Jacobian of a real curve

Let X be a smooth projective curve of genus $g := g(X)$ over $\mathbb{C}$ with a real structure (X, S). We will briefly say that (X, S) is a *real curve* of genus g. First of all we need the following result, first proved by Witt (see [Wi]) using Artin's proof of Hilbert's 17th problem (see [Ar]). We give here a sketch of the proof, essentially following Witt's ideas. Another proof is given in [GrHa].

Lemma 4.1.1. *Let (X, S) be an irreducible real variety without real points and let X_0 be a real model of X. Let K be the field of rational functions on X_0. Then there exist* *elements* $u_1, \dots, u_s \in K$ *such that*

$$-1 = u_1^2 + \cdots + u_s^2.$$

If X *is a curve then there exist a rational function* f *on* X *such that* $f \cdot f^S = -1$.

Proof. Without any loss of generality, we may assume X_0 is affine and contained in $\mathbb{A}^n_{\mathbb{C}}$. Let $(f_1, \dots, f_q)$ be the real polynomials defining X_0, and let I be the prime ideal generated by them in $\mathbb{R}[x_1, \dots, x_n]$. Since $X_S = \emptyset$, there is no point $p \in \mathbb{A}^n_{\mathbb{R}}$ such that $f_1(p) = \cdots = f_q(p) = 0$. By the so-called *Positivstellensatz* (see [BCR], thm. 4.4.2), there exist polynomials h_i, $i = 1, \dots, m$, g in $\mathbb{R}[x_1, \dots, x_n]$ with $g \notin I$ and a polynomial $f \in I$ such that

$$h_1^2 + \cdots + h_m^2 + g^2 + f = 0$$

If we consider the above relation modulo I, we get that in K we have

$$\overline{h}_1^2 + \cdots + \overline{h}_m^2 + \overline{g}^2 = 0$$

where $\overline{g} \neq 0$. Therefore the field K cannot be ordered, and this is equivalent, by a well known property of ordered fields, to our first statement.

If X is a curve, then K has degree of transcendence one over $\mathbb{R}$. Therefore every element in K which is a sum of squares, is actually a sum of two squares (see [BCR], thm. 6.3.15). This implies that there are rational functions $u, v \in \mathbb{R}(X))^*$ such that $u^2+v^2 = -1$. Therefore we have a rational map $\psi\colon X_0 \to \mathrm{Spec}([u, v]/(u^2+v^2+1))$. The rational function $f := u + iv \in \mathbb{C}(X)^*$ is such that $f \cdot f^S = -1$. □

Now we need to go back to Proposition 2.7.4, from which we keep the notation, in order to make it more precise in the present case.

Proposition 4.1.2. *Let* (X, S) *be a real curve of genus* g.

(i) *If* $X_S \neq \emptyset$ *then* $\mathrm{Pic}(X)^+ = \mathrm{Pic}(X)^S$.

(ii) *If* $X_S = \emptyset$ *then* $\mathrm{Pic}(X)^+ \neq \mathrm{Pic}(X)^S$ *and* $\mathrm{Pic}(X)^S / \mathrm{Pic}(X)^+ \simeq \mathbb{Z}_2$.

(iii) *Let* $\mathcal{L}$ *be a line bundle in* $\mathrm{Pic}(X)^S$ *whose image in* $\mathrm{Pic}(X)^S / \mathrm{Pic}(X)^+$ *is a generator; then*

$$h^0(X, \mathcal{L}) = h^1(X, \mathcal{L}) = 0 \pmod 2$$

and

$$\deg(\mathcal{L}) = g - 1 \pmod 2.$$

Proof. Part (i) has been proved in 2.7.4.

(ii) Let X_0 be a real model of X (see §2.2). Notice that X_0 is a scheme over $\mathbb{R}$, hence $\mathrm{Pic}(X_0)$ is the group of isomorphism classes of the line bundles on X_0 defined over $\mathbb{R}$. Therefore $\operatorname{Pic}(X_0) \simeq \operatorname{Pic}(X)^+$ and there is an exact sequence:

$$0 \to \mathbb{R}^* \to \mathbb{R}(X)^* \xrightarrow{\text{div}} \operatorname{Div}(X_0) \to \operatorname{Pic}(X)^+ \to 0 \tag{4.1.3}$$

where $\operatorname{Div}(X_0)$ is the group of divisors on X_0, i.e., the group $\oplus_x \mathbb{Z}$, where x runs over the closed points of X_0, and $\mathbb{R}(X)$ denotes the field of rational functions on X_0. Similarly there is the exact sequence involving $\operatorname{Pic}(X)$

$$0 \to \mathbb{C}^* \to \mathbb{C}(X)^* \xrightarrow{\text{div}} \operatorname{Div}(X) \to \operatorname{Pic}(X) \to 0$$

where $\operatorname{Div}(X)$ is the group of divisors on X, i.e., the group $\oplus_x \mathbb{Z}$, where x runs over the closed points of X, and $\mathbb{C}(X)$ is the field of rational functions on X.

Let $M := \operatorname{Im}(\text{div}) \subset \operatorname{Div}(X)$. By taking the G-cohomology of the above exact sequence where $G = \{1, S\}$ and using the fact that $H^1(G, \mathbb{C}^*) = 0$ (see Example 2.6.1, (ii)), we have $M^G \simeq \mathbb{R}(X)^*/\mathbb{R}^*$, and we get the exact sequence

$$0 \to M^G \to \operatorname{Div}(X)^G \to \operatorname{Pic}(X)^S \to H^1(G, M) \to H^1(G, \operatorname{Div}(X)) \to \dots \tag{4.1.4}$$

The right end group in the above sequence is 0. Indeed $H^1(G, \operatorname{Div}(X)) \simeq \frac{\ker(1+S)}{\operatorname{Im}(1-S)}$ (see Example 2.6.1, (ii)), and it is easy to verify that $\ker(1+S) \simeq \operatorname{Im}(1-S)$ is the set of all divisors D on X such that $D = -D^S$.

Finally we can see that $H^1(G, M) \simeq \mathbb{Z}_2$. In fact $H^1(G, M) \simeq \frac{\ker(1+S)}{(1-S)M}$. But $\ker(1+S) \simeq \{f \in \mathbb{C}(X)^* : f \cdot f^S \in \mathbb{R}^*\}/\mathbb{R}^*$ and $(1-S)M \simeq \{\frac{g}{\bar{g}}, g \in \mathbb{C}(X)^*\}/\mathbb{R}^*$. Thus we have a well defined homomorphism $\sigma \colon H^1(G, M) \to \mathbb{Z}_2 = \{\pm 1\}$, which assigns to the representative f of a cohomology class the signum of $f \cdot f^S$. By Lemma 4.1.1, the map σ is surjective. We claim that it is also injective. Indeed, let $f = u + iv$ be such that $f \cdot f^S = 1$, i.e., $u^2 + v^2 = 1$. Then $g := v + i(1-u)$ is such that $f = \frac{g}{\bar{g}}$, whence the assertion.

Now, by comparing (4.1.3) and (4.1.4), one gets the following exact sequence which proves (ii):

$$0 \to \operatorname{Pic}(X)^+ \to \operatorname{Pic}(X)^S \xrightarrow{d} \mathbb{Z}_2 \to 0. \tag{4.1.5}$$

(iii) The linear system $|\mathcal{L}|$ has no fixed points under the action of G. Hence

$$\dim |\mathcal{L}| \equiv 1 \pmod 2$$

(see Example 2.1.1, (iii)). Let K_X be the canonical line bundle on X. Then we have $K_X \in \operatorname{Pic}(X)^+$, hence $K_X \otimes \mathcal{L}^{-1}$ is also a generator of $\operatorname{Pic}(X)^S / \operatorname{Pic}(X)^+$. Therefore $h^1(X, \mathcal{L}) = h^0(X, K_X \otimes \mathcal{L}^{-1})$ is even as well as $h^0(X, \mathcal{L})$. The result then follows by the theorem of Riemann–Roch, which says that $h^0(X, \mathcal{L}) - h^1(X, \mathcal{L}) = \deg \mathcal{L} - g + 1$. □

Remarks 4.1.6. (i) The short exact sequence (4.1.5) is valid for any real curve, not necessarily smooth neither projective, without real points. It may be deduced from a Leray spectral sequence as we will see in Remark 4.5.2.

(ii) Every line bundle in $\mathrm{Pic}(X)^+$ has clearly even degree. Then, if g is even, the map d in the exact sequence (4.1.5) is just the degree map modulo 2, and the sequence (4.1.5) does not split. Otherwise we would have a line bundle $\mathcal{L} \in \mathrm{Pic}(X)^S$, of odd degree such that $\mathcal{L}^{\otimes 2}$ is trivial, a contradiction.

By contrast, the sequence (4.1.5) splits when g is odd. To see this, it suffices to see that there is a line bundle $\mathcal{L} \in \mathrm{Pic}(X)^S$ such that $\mathcal{L} \notin \mathrm{Pic}(X)^+$ and such that $\mathcal{L}^{\otimes 2}$ is trivial. The existence of such an $\mathcal{L}$ (in fact of 2^g such line bundles) is a consequence of the results of this and the next section (see Remark 4.2.7).

We now let $J(X) := \mathrm{Pic}^0(X)$ be the *Jacobian* of X. Notice that $J(X)$ is the group of the line bundles of degree 0 on X. Indeed the Chern map $c_1: \mathrm{Pic}(X) \to H^2(X, \mathbb{Z}) \simeq \mathbb{Z}$ is nothing but the map which associates to a line bundle its degree (see §1.4 and [GH], p. 144). As usual, we will denote by S the real structure induced on $J(X)$ by the real structure of X (see §2.7). We will denote by $\mathrm{Pic}^d(X)$ the set of all line bundles of degree d on X, for all integers d, and we notice that $\mathrm{Pic}^d(X)$ is a $\mathrm{Pic}^0(X)$-homogeneous space. Hence $\mathrm{Pic}^d(X)$ is a variety of dimension g as well as $J(X)$, and furthermore one has

$$\mathrm{Pic}(X) = \oplus_{d \in \mathbb{Z}} \mathrm{Pic}^d(X).$$

Let $X(d)$, where $d \geq 0$, be the *d-fold symmetric product* of X, i.e., $X(d) = X^d/S_d$ where S_d is the symmetric group over d objects. Of course $X(d)$ is the variety parametrizing all effective divisors of degree d on X, and it is a projective variety. One can define natural maps

$$\phi_d: D \in X(d) \to \mathcal{O}_X(D) \in \mathrm{Pic}^d(X)$$

which, by the theorem of Riemann–Roch, are surjective as soon as $d \geq g$ (see [ACGH, 1.3]). We denote by $W_d := W_d(X)$ the image of $X(d)$ via the map ϕ_d.

Notice that the antiinvolution S on X induces an obvious antiinvolution, which we still denote by S, on $X(d)$ for all $d \geq 1$, hence we have an induced real structure $(X(d), S)$. The maps ϕ_d are maps of real varieties hence there are induced maps

$$\phi_d^+: X(d)_S \to \mathrm{Pic}^d(X)^+$$

which are surjective by the definition of $\mathrm{Pic}(X)^+$. With the above notation, we prove the following theorem due to Comessatti (see [C2],I).

Theorem 4.1.7 (A. Comessatti). *Let $\nu := \nu(X)$ denote the number of connected components of X_S. Then:*

(i) *if $\nu > 0$ then the number of branches of $X(d)$ is given by the following espression*

$$\nu(X(d)) = \sum_{i=0}^{[d/2]} \binom{\nu}{d-2i} = \sum_{i=0}^{d} \binom{\nu-1}{i};$$

if $\nu = 0$ then $\nu(X(d))$ is 1 if d is even, and is 0 if d is odd;

(ii) *if* $\nu > 0$ *we have:*

$$\nu(W_d) = \nu(X(d)) \quad \textit{for all } d \in \mathbb{Z}$$

and

$$\nu(J(X)) = \nu(\mathrm{Pic}^d(X)) = 2^{\nu-1} \quad \textit{for all } d \in \mathbb{Z};$$

(iii) *if* $\nu = 0$ *then*

$$g \equiv 0 \pmod 2 \Rightarrow \nu(\mathrm{Pic}^d(X)) = 1 \quad \textit{for all } d \in \mathbb{Z}$$

and

$$g \equiv 1 \pmod 2 \Rightarrow \nu(\mathrm{Pic}^{2d+1}(X)) = 0 \quad \textit{and} \quad \nu(J(X)) = \nu(\mathrm{Pic}^{2d}(X)) = 2.$$

Proof. (i) Let C be a branch of $X(d)$ (see §2.1) and let $D := P_1 + \cdots + P_d$ be a general point of C. We claim that D contains at most 1 point on each connected component of X. If not, D would contain two such points, and we could let them come together, and then move them into a pair of complex conjugate points, hence getting a *more general* point D' in the same branch of $X(d)$.

So, if we let $(X_1, \dots, X_\nu)$ be the ν-tuple of the connected components of X_S given with a certain ordering, we can define a function which assigns to any component C of $X(d)$ an n-tuple $(t_1, \dots, t_\nu)$ where

$$t_i = \begin{cases} 0, & \text{if } D \text{ has no points on } X_i, \\ 1, & \text{if } D \text{ has 1 point on } X_i. \end{cases}$$

Then we have

$$t_1 + \cdots + t_\nu \leq d$$

and

$$t_1 + \cdots + t_\nu \equiv d \pmod 2.$$

From the above relations one easily gets the formula for $\nu(X(d))$. If $\nu = 0$ then every $D \in X(d)_S$ is the sum of pairs of conjugate points, whence the assertion.

(ii) Let $\mathcal{L} \in (W_d)_S$. Hence we may assume $\mathcal{L}$ is fixed under S. By Proposition 2.7.4 we have $\mathrm{Pic}(X)^+ = \mathrm{Pic}(X)^S$. Since $h^0(X, \mathcal{L}) \geq 1$, then $\mathcal{L}$ comes from $X(d)_S$. The fibres of the map ϕ_d^+ being real projective spaces, distinct connected components of $X(d)_S$ are mapped to distinct connected components of $(W_d)_S$. This proves the first formula in (ii).

To show that also the second formula in (ii) holds, it suffices to observe that, being $\nu \neq 0$, then $\mathrm{Pic}^d(X)$ contains real points for every d. Thus $\mathrm{Pic}^d(X)$ is isomorphic over $\mathbb{R}$ to $J(X)$. Indeed, if $\mathcal{L}$ is a real point in $\mathrm{Pic}^d(X)$, the required isomorphism is

$$\psi_{\mathcal{L}}: \mathcal{M} \in \mathrm{Pic}^d(X) \to \mathcal{L}^{-1} \otimes \mathcal{M} \in \mathrm{Pic}^0(X).$$

Finally $\nu(J(X)) = \nu(\mathrm{Pic}^d(X))$ and $\mathrm{Pic}^d(X) \simeq W_d$ for $d \gg O$. Therefore for $d \gg 0$ we have

$$\nu(J(X)) = \nu(\mathrm{Pic}^d(X)) = \nu(W_d) = \sum_{i=0}^{d} \binom{\nu - 1}{i} = 2^{\nu-1}.$$

(iii) It suffices to prove the assertion for $d \geq 0$. Let $\nu = 0$. Then by (i) we have $\nu(X(2d)) = 1$ and $\nu(X(2d+1)) = 0$. If $g \equiv 0 \pmod 2$, by Proposition 4.1.2, (iii), we get, by arguing as above,

$$\mathrm{Pic}^{2d}(X)^+ = \mathrm{Pic}^{2d}(X)^S \Rightarrow \nu(W_{2d}) = \nu(X(2d)) = 1 \quad \text{for } d \geq 0 \Rightarrow \nu(J(X)) = 1.$$

The conclusion follows by observing that $\mathrm{Pic}^{2d}(X)^S \neq 0$ for all $d \geq 0$ and $\mathrm{Pic}^{2d+1}(X)^S \neq 0$ for some d (see Proposition 4.1.2). Hence all varieties $\mathrm{Pic}^d(X)$ are isomorphic.

If $g \equiv 1 \pmod 2$ then, by Proposition 4.1.2, (iii), we have

$$\mathrm{Pic}^{2d+1}(X)^+ = \mathrm{Pic}^{2d+1}(X)^S \Rightarrow \nu(W_{2d+1}) = \nu(X(2d+1)) = 0$$

for all $d \geq 0$. Therefore we get $\nu(\mathrm{Pic}^{2d+1}(X)) = 0$ if $d \geq 0$.

From $\nu(X(2d)) = 1$ we deduce $\nu(\mathrm{Pic}^{2d}(X)^+) = 1$. From Proposition 4.1.2 it follows that, as a topological space, $\mathrm{Pic}^{2d}(X)^S$ is the disjoint union of two copies of $\mathrm{Pic}^{2d}(X)^+$. Hence

$$\nu(\mathrm{Pic}^{2d}(X)) = 2 \quad \text{for all } d \in \mathbb{Z} \quad \text{and} \quad \nu(J(X)) = 2.$$

The above equalities imply then easily

$$\nu(\mathrm{Pic}^{2d+1}(X)) = 0 \quad \text{for all } d \in \mathbb{Z}.$$

□

Corollary 4.1.8. *Let X be a smooth real projective curve, such that $\nu(X) \neq 0$. Then X is GM.*

Proof. From the equalities

$$\nu(J(X)) = 2^{\nu(X)-1} \quad \text{and} \quad \nu(J(X)) = 2^{g-\lambda}$$

where λ is the Comessatti character of X, which is the same as the first Comessatti character of $J(X)$, we deduce $\nu(X) = g + 1 - \lambda$. □

Remark 4.1.9. Let X be such that $\nu(X) = 0$ and $g \equiv 0 \pmod 2$. Then from the above results we get $\nu(J(X)) = 1$ which implies $\lambda = g$. If $\nu(X) = 0$ and $g \equiv 1 \pmod 2$, then we have $\nu(J(X)) = 2$ and $\lambda = g - 1$.

Next we apply the results in §3 to the natural principal polarization on the Jacobian $J(X)$ of a smooth projective curve X of genus g with a real structure (X, S).

First we recall without proofs a few well know facts concerning complex curves (the readers may refer either to [GH] or to [LB]). The intersection pairing E on $H_1(X,\mathbb{Z})$ extends to an antisymmetric bilinear form on $H_1(X,\mathbb{R}) \simeq H^0(X,\Omega^1_X)^*$ such that $E(ix,iy) = E(x,y)$. Let $(\lambda_1,\dots,\lambda_g,\mu_1,\dots,\mu_g)$ be a symplectic basis for $H_1(X,\mathbb{Z})$ and let $(\omega_1,\dots,\omega_g)$ be a basis for $H^0(X,\Omega^1_X)$ such that the corresponding period matrix Ω is of the form $\Omega = (I_g \quad Z)$. Then a theorem of Riemann asserts that Z is symmetric of rank g and $\operatorname{Im} Z > 0$. From §1.7 it follows that these data yield a principal polarization Θ on $\operatorname{Alb}(X)$.

It turns out that $\operatorname{Alb}(X)$ is also the Albanese variety of $X(d)$ for all positive integers d. In particular one can consider the Albanese maps $\alpha_d\colon X(d) \to \operatorname{Alb}(X)$ which depend upon the choice of a fixed divisor $D \in X(d)$. We set $V_d = \alpha_d(X(d))$. In particular V_{g-1} turns out to be a divisor on $\operatorname{Alb}(X)$ whose associated line bundle gives rise to the polarization Θ described above.

The polarization Θ gives rise to an isomorphism

$$\phi_\Theta\colon \operatorname{Alb}(X) \to \operatorname{Pic}^0(X) = J(X)$$

(see Remark 3.3.2, (ii) and §1.4). On the whole, we have the commutative diagram

$$\begin{array}{ccc} X(d) & \xrightarrow{\phi_d} & \operatorname{Pic}^d(X) \\ \alpha_d \downarrow & & \downarrow \psi_{\mathcal{L}} \\ \operatorname{Alb}(X) & \xrightarrow{\phi_\Theta} & \operatorname{Pic}^0(X) \end{array} \tag{4.1.6}$$

where $\mathcal{L}$ is the line bundle $\mathcal{O}_X(D)$ (see [LB], p. 336).

Now we return to the case in which we have a real structure (X,S) on X. The antiinvolution S is an homeomorphism of X which reverts the orientation. Hence we have

$$E(S(x),S(y)) = -E(x,y)$$

for any $x,y \in H_1(X,\mathbb{Z})$. This shows that the polarization Θ on $\operatorname{Alb}(X)$ is real (see Remark 3.3.2, (i)), and therefore the isomorphism $\phi_\Theta\colon \operatorname{Alb}(X) \to \operatorname{Pic}^0(X) = J(X)$ is real too.

If $X_S \neq \emptyset$ then we can choose a real divisor D in order to define the maps α_d and $\psi_{\mathcal{L}}$ appearing in (4.1.6), which is therefore a commutative diagram of real morphisms. If $X_S = \emptyset$ then the same holds, provided $d \equiv 0 \pmod 2$. In particular, if $g \equiv 1 \pmod 2$, then the polarization on $J(X)$ is given by the real divisor W_{g-1}.

If the polarization on $J(X)$ is real we may choose, according to the results in §3, a period matrix Ω of the form

$$\Omega = (I_g \quad 1/2M + iT)$$

where T is real, symmetric, positive definite and M is either of the diasymmetric type (λ,d) or of the orthosymmetric type (λ,o) if $\lambda \equiv 0 \pmod 2$. Thus according to the definition given in §3.4 we will say that the curve X *is of type* (λ,ϵ), where ϵ equals either o or d.

The following result, due to Comessatti, relates the topology of X_S with the character (λ, ϵ). Recall from §2.7 that $a(X)$ equals 1 if X_S does not disconnect X and $a(X) = 0$ if X_S disconnects X.

Theorem 4.1.10.

(i) *Assume* $\nu(X) \neq 0$. *Then we have:*
  (a) $\lambda = g + 1 - \nu(X)$,
  (b) $\epsilon = d \Leftrightarrow a(X) = 1$,
  (c) $\epsilon = o \Leftrightarrow a(X) = 0$.

(ii) *If* $\nu(X) = 0$ *then*
  (a′) $\lambda = g$ *if* $g \equiv 0 \pmod 2$, $\lambda = g - 1$ *if* $g \equiv 1 \pmod 2$,
  (b′) $\epsilon = o$.

Proof. (a) has already been proved in Corollary 4.1.8.

We prove (b) which is clearly equivalent to (c). Let $(\lambda_1, \dots, \lambda_g, \mu_1, \dots, \mu_g)$ be a symplectic basis for $H_1(X, \mathbb{Z})$, and let e be the reduction modulo 2 of the intersection pairing E on $H_1(X, \mathbb{Z})$, i.e.,

$$e: H_1(X, \mathbb{Z}_2) \times H_1(X, \mathbb{Z}_2) \to \mathbb{Z}_2.$$

Define the map $f: H_1(X, \mathbb{Z}_2) \to \mathbb{Z}_2$ by $f(u) = e(u, u^S)$, where S is the antiinvolution on X. Let C_X be the class of X_S in $H_1(X, \mathbb{Z}_2)$. Clearly $C_X = 0$ if and only if $a(X) = 0$. Given an element $u \in H_1(X, \mathbb{Z}_2)$ we can represent it with a cycle A such that A intersects both A^S and X_S transversally. Then we have

$$f(u) = e(u, u^S) = \mathrm{Card}\{A \cap A^S\} \pmod 2$$

and of course

$$\begin{aligned}\mathrm{Card}\{A \cap A^S\} &= \mathrm{Card}\{A \cap A^S \cap X_S\} + \mathrm{Card}\{A \cap A^S \cap (X - X_S)\}\\ &= \mathrm{Card}\{A \cap X_S\} \pmod 2\end{aligned}$$

The intersection form e being a perfect pairing is non-degenerate. Therefore

$$f \equiv 0 \Leftrightarrow C_X = 0 \Leftrightarrow a(X) = 0.$$

Representing e in the chosen symplectic basis for $H_1(X, \mathbb{Z})$ and observing that its matrix with respect to this basis is of the form

$$\begin{pmatrix} I_g & M \\ 0 & -I_g \end{pmatrix}$$

we immediately see that $f \equiv 0$ if and only if M is of orthosymmetric type, i.e., if and only if $\epsilon = o$.

(ii) Part (a′) has been already observed in Remark 4.1.9. As for (b′), it is obvious that $C_X = 0$ if $\nu(X) = 0$. Therefore $a(X) = 0$ and the same argument as in (i) applies. □

4.2. Real theta-characteristics

We briefly recall some definitions and properties of theta characteristics on a smooth projective curve X of genus g over $\mathbb{C}$, for which we refer the reader to [GH], [ACGH] and [LB].

A *theta-characteristic* on X is an element $\mathcal{L} \in \mathrm{Pic}^{g-1}(X)$ such that $\mathcal{L} \otimes \mathcal{L} \simeq K$, where $K := K_X$ is the *canonical line bundle* of X, i.e., $K_X \simeq \Omega^1_X$. If $\mathcal{L}$ and $\mathcal{M}$ are theta-characteristics, then, using the additive notation in $\mathrm{Pic}(X)$, we have $2(\mathcal{L} - \mathcal{M}) = 0$, i.e., $\mathcal{L} - \mathcal{M} \in J(X)_2$, the 2-torsion subgroup of the Jacobian of X. From the obvious isomorphism $J(X)_2 \simeq H_1(X, \mathbb{Z}_2)$ it follows that the number of theta-characteristics is 2^{2g} (see [ACGH], p. 287).

We will say that a theta characteristic $\mathcal{L}$ is *even* [resp. *odd*] if $h^0(X, \mathcal{L})$ is even [resp. odd].

Let $\mathcal{L}$ be a theta-characteristic. Then we have the morphism

$$\psi_{\mathcal{L}}^{-1}: \mathcal{M} \in J(X) \to \mathcal{M} \otimes \mathcal{L} \in \mathrm{Pic}^{g-1}(X)$$

carrying $J(X)_2$ to the set of all theta-characteristics. If X is *general enough* among all curves of genus g, then the divisor $W_{g-1} \subset \mathrm{Pic}^{g-1}(X)$ is known to contain exactly the odd theta-characteristics (see [ACGH]). On the other hand $\psi_{\mathcal{L}}^{-1}(W_{g-1})$ is a divisor which belongs to the principal polarization of $J(X)$, i.e., it is a so-called *theta-divisor* on $J(X)$. If D is such a divisor, an easy application of the theorem of Riemann–Roch, which we leave to the reader, shows that the multiplication by -1 in $J(X)$ fixes D, in symbols $(-1)^*D = D$, i.e., the divisor is *symmetric*.

Let $(\lambda_1, \dots, \lambda_g, \mu_1, \dots, \mu_g)$ be a symplectic basis for $H_1(X, \mathbb{Z})$ and let Θ be the divisor defined by the vanishing of the Riemann theta function θ (see §3.6). It is known (see [ACGH], appendix B) that points in $J(X)_2 \cap \Theta$ correspond, via the natural map

$$p: H^0(X, \Omega^1_X)^* \simeq H_1(X, \mathbb{Z}) \otimes_{\mathbb{Z}} \mathbb{R} \to J(X)$$

to the so-called *half-periods* of the form

$$1/2(\sum_{i=1}^{g} a_i \lambda_i + \sum_{i=1}^{g} b_i \mu_i)$$

where $a_i, b_i \in \{0, 1\}$, and are such that $\sum_{i=1}^{g} a_i b_i$ is odd. The sum $\sum_{i=1}^{g} a_i b_i$ is called the *characteristic* of the half-period.

It is elementary to compute the number of such half-periods, which shows that the number of odd theta-characteristics is $S^{\mathrm{odd}} = 2^{g-1}(2^g - 1)$. This in turn implies that the number of even theta-characteristics is $S^{\mathrm{even}} = 2^{g-1}(2^g + 1)$.

Now we go back to the case in which X has a real structure (X, S) and we prove the following result.

Lemma 4.2.1. *Let X have a real structure (X, S). Then X has real theta-characteristics, i.e., line bundles $\mathcal{L}$ such that $\mathcal{L} \otimes \mathcal{L} \simeq K$ and $\mathcal{L}^S \simeq \mathcal{L}$.*

Proof. We divide the proof of the lemma into two cases.

Case 1: Either $\nu(X) \neq 0$ or $\nu(X) = 0$ and $g \equiv 1 \pmod 2$. From the diagram (4.1.10), which is defined in the category of real varieties, we get maps

$$\mathrm{Pic}^{g-1}(X) \xrightarrow{\psi_{\mathcal{L}}} J(X)$$

and

$$X(g-1) \xrightarrow{\phi_{g-1}} \mathrm{Pic}^{g-1}(X).$$

A well known theorem of Riemann (see [GH], p. 338) says that

$$\psi_{\mathcal{L}}(W_{g-1}) = \Theta + \kappa$$

where $\kappa \in J(X)$ is the so-called *Riemann's constant.* If X is of the orthosymmetric type then Θ is real (see §3.6) and therefore also κ is real. Let E be an effective divisor of degree $g-1$, i.e., $E \in X(g-1)$. Then $\boldsymbol{c} = \psi_{\mathcal{L}}(\phi_{g-1}(E))$ is an element of $\Theta + \kappa$. Therefore one has $\theta(\boldsymbol{c} - \kappa) = 0$, where θ is Riemann's theta function defined in §3.6. Since θ is even we also have $\theta(\kappa - \boldsymbol{c}) = 0$, i.e., $\theta(-\boldsymbol{c} + 2\kappa - \kappa) = 0$. Since the element $-\boldsymbol{c} + 2k$ is in $\Theta + \kappa$, we get $\psi_{\mathcal{L}}^{-1}(-\boldsymbol{c} + 2\kappa) \in W_{g-1}$. Let E' be an element of $X(g-1)$ such that

$$\phi_{g-1}(E') = \psi_{\mathcal{L}}^{-1}(-\boldsymbol{c} + 2\kappa).$$

Moreover we have

$$\phi_{2g-2}(E + E') = \phi_{g-1}(E) + \phi_{g-1}(E') = \psi_{\mathcal{L}}^{-1}(-\boldsymbol{c} + 2\kappa) + \psi_{\mathcal{L}}^{-1}(\boldsymbol{c}) = \psi_{\mathcal{L}}^{-1}(2\kappa).$$

Since κ is constant, the above formula shows that $|E + E'| = |K|$. Indeed this linear series has degree $2g-2$ and there is a divisor of it containing a general divisor D of degree $g-1$ of X, hence it has dimension at least $g-1$, whence the conclusion by the theorem of Riemann–Roch. This proves that if X is of orthosymmetric type, then K is really divisible by 2, a real theta-caracteristic being κ.

If X is of diasymmetric type, then Θ is no longer real, whereas W_{g-1} is. Therefore we have

$$\Theta + \kappa = \Theta^S + \kappa^S.$$

On the other hand from §3.6 it follows that

$$\Theta + \kappa = \Theta - \epsilon + \kappa$$

where ϵ is the semiperiod $\epsilon = (1/2, \ldots, 1/2, 0, \ldots, 0)^t$, whose first λ entries are $1/2$, the others are 0. By Remark 3.3.2, (ii), the above relations yield $\kappa^S = \epsilon + \kappa$.

Now recall that ϵ belongs to the connected component of 0 in $J(X)_{-S}$ (see §3.6), which is the image of the map $\mathrm{id}_{J(X)} - S$ (see §3.2). It is then easy to see that there exist conjugate points α and α^S of order 2 such that $\epsilon = \alpha - \alpha^S$. Therefore we get $2(\kappa + \alpha) = 2\kappa$ and $\kappa + \alpha$ is real. The same argument used in the orthosymmetric case shows that for any divisor E in $X(g-1)$ there is a divisor E' in $X(g-1)$ such that

$$\phi_{2g-2}(E + E') = \psi_{\mathcal{L}}^{-1}(2\kappa) = \psi_{\mathcal{L}}^{-1}(2(\kappa + \alpha)).$$

This proves that $|E+E'|$ is the canonical series on X, which is again really divisible by 2.

Case 2: $\nu(X) = 0$ and $g \equiv 0 \pmod 2$. By Theorem 4.1.5 we have $\nu(\mathrm{Pic}^d(X)) = 1$ for all positive integers d. Consider the map

$$\rho\colon \mathrm{Pic}^{g-1}(X) \to \mathrm{Pic}^{2g-2}(X)$$

sending $\mathcal{L}$ to $\mathcal{L}^{\otimes 2}$. The map ρ clearly has maximal rank, hence it is surjective. Therefore K, being in the image of ρ, is really divisible by 2. □

Corollary 4.2.2. *Every canonical divisor has an even number of points on each branch of* X.

Next we are going to prove a result, due again to Comessatti, which gives a precise computation for the number of real theta-characteristics. Let $S_{\mathbb{R}}^{\text{even}}$ denote the number of even real theta-characteristics and let $S_{\mathbb{R}}^{\text{odd}}$ be the number of odd real theta-characteristics.

Theorem 4.2.3. *Let X be a smooth, irreducible, projective, real curve. Then we have:*

(i) *if* $\nu(X) > 0$ *then*

$$S_{\mathbb{R}}^{\text{even}} = 2^{g-1}(2^{\nu(X)-1} + 1 - a(X)),$$

$$S_{\mathbb{R}}^{\text{odd}} = 2^{g-1}(2^{\nu(X)-1} - 1 + a(X));$$

(ii) *if* $\nu(X) = 0$ *then*

$$S_{\mathbb{R}}^{\text{even}} = \begin{cases} 3 \cdot 2^{g-1}, & \text{if } g \equiv 1 \pmod 2 \\ 2^g, & \text{if } g \equiv 0 \pmod 2 \end{cases}$$

$$S_{\mathbb{R}}^{\text{odd}} = \begin{cases} 2^{g-1}, & \text{if } g \equiv 1 \pmod 2 \\ 0, & \text{if } g \equiv 0 \pmod 2. \end{cases}$$

Proof. Because of the existence of a real theta-characteristic $\mathcal{L}$, proved in Lemma 4.2.1, the map $\psi_{\mathcal{L}}\colon \mathrm{Pic}^{g-1}(X) \to J(X)$ is real. Assume that X is of the orthosymmetric type. Then, with the same notation as in the proof of Lemma 4.2.1, we have that $\Theta+\kappa$ is real and it is defined by the equation $\theta(x-\kappa) = 0$. Moreover $\theta(2\kappa - x - \kappa) = \theta(\kappa - x) = \theta(x-\kappa)$ since θ is even. Hence $\Theta + \kappa$ is stable under the transformation

$$T\colon x \in J(X) \to 2\kappa - x \in J(X).$$

The fixed points under T on $\Theta + \kappa$ are of the form $x = \sigma + \kappa$, where $\sigma \in J(X)_2$ and $\theta(\sigma) = 0$. In order to compute $S_{\mathbb{R}}^{\text{odd}}$ we have to consider points σ of order 2 with odd characteristics, such that $\sigma + \kappa$ is real. Since $\Theta + \kappa$ is real, we have that κ has to be real (see §3.6) and therefore also σ is real. If $(\lambda_1, \dots, \lambda_g, \mu_1, \dots, \mu_g)$ is a symplectic basis in which the period matrix has form (3.4.5), then

$$\sigma = 1/2\Big(\sum_{i=1}^{g} a_i\lambda_i + \sum_{i=1}^{g} b_i\mu_i\Big)$$

and the above condition is satisfied only if $b_i = 0$ for $i = 1, \dots, \lambda$, where λ is the Comessatti character of the curve. Therefore we get

$$S_{\mathbb{R}}^{\text{odd}} = 2^{g-1}(2^{g-\lambda} - 1).$$

Suppose now that X is of diasymmetric type. Then, by taking into account the proof of Lemma 4.2.1, one sees that the same argument as above yields $b_i = 1$ for $i = 1, \dots, \lambda$. Therefore we get

$$S_{\mathbb{R}}^{\text{odd}} = 2^{g-\lambda} \cdot 2^{g-1}.$$

The number of real points in $J(X)_2$ is $2^g \cdot 2^{g-\lambda} = 2^{2g-\lambda}$. Therefore the above computation give the formulae for $S_{\mathbb{R}}^{\text{odd}}$ and $S_{\mathbb{R}}^{\text{even}}$. □

We now discuss the number of components in the real locus of W_{g-1} for a curve X with $\nu(X) > 0$. We will need the following lemma.

Lemma 4.2.4. *Let X be a smooth real projective curve such that $\nu(X) > 0$ and let f be a non zero real rational function on X. Then the principal divisor (f) of f has an even number of points on each branch of X_S.*

Proof. Let Y be a branch of X_S. Then Y is homeomorphic to the real 1-sphere S_1. The function f defines a continous map $f: Y \to \mathbb{P}^1_{\mathbb{R}}$, and $\mathbb{P}^1_{\mathbb{R}}$ is also homeomorphic to S_1. Hence f induces a continuous map $\tilde{f}: S^1 \to S^1$. The result immediately follows from the definition of the topological degree of $\tilde{f}$. □

Let X be a smooth real projective curve such that $\nu(X) > 0$ and let X_i, $i = 1, \dots, \nu(X)$, be the branches of X. The previous lemma enables us to define the map

$$c: \operatorname{Pic}(X)^S \to \mathbb{Z}_2^{\nu(X)}$$

whose i-th component $\operatorname{Pic}(X)^S \to \mathbb{Z}_2$ sends the line bundle $\mathcal{O}_X(P)$ to 1 if and only if $P \in X_i$, with $0 \leq i \leq \nu(X)$. Of course we have restrictions

$$c_d: \operatorname{Pic}^d(X)^S \to \mathbb{Z}_2^{\nu(X)}$$

for any integer d. We shall denote these restrictions again by c if there is no danger of confusion.

Remark 4.2.5. Lemma 4.2.4 may be generalized as follows. Let X be a real (possibly singular and not projective) curve such that $X_S \neq \emptyset$ and let $l := l(X)$ be the number of *loops* in X_S, i.e., $l(X) = \dim_{\mathbb{Z}_2} H^1(X_S, \mathbb{Z}_2)$. Then there is an isomorphism

$$\mathrm{Pic}(X)^S \otimes \mathbb{Z}_2 \simeq \mathbb{Z}_2^l.$$

This result has been proved by Comessatti [C2] and by Witt [Wi] in the case of smooth projective real curves. An algebraic proof has been given by Knebusch [Kn] thus showing that the above isomorphism is valid over any real closed field. The general result, for any curve over $\mathbb{R}$ has been obtained by Pedrini and Weibel [PW1] (see §4.5).

In the case X is smooth and projective then $l(X) = \nu(X)$ by Poincare's duality. Then the isomorphism $\mathrm{Pic}(X)^S \otimes \mathbb{Z}_2 \simeq \mathbb{Z}_2^{\nu(X)}$ is induced by the map c defined above.

Now we concentrate on the map c_{g-1}, which in general is not surjective, since its image consists of all the $\nu(X)$-tuples $(\gamma_1, \dots, \gamma_{\nu(X)})$ such that $\sum_{i=1}^{\nu(X)} \gamma_i \equiv g-1 \pmod 2$.

If p is a point in X_S, we can consider the map

$$\mathcal{L} \in \mathrm{Pic}^{g-1}(X) \to \mathcal{L} \otimes \mathcal{O}_X(2p) \in \mathrm{Pic}^{g+1}(X)$$

which is an isomorphism of real varieties. By using this isomorphism and by looking at the proof of Theorem 4.1.7, one sees that $\nu(\mathrm{Pic}^{g-1}(X)) = \nu(\mathrm{Pic}^{g+1}(X))$ is nothing else than the cardinality of the image of c_{g-1}.

Let F be a branch of W_{g-1} and let $\mathcal{O}_X(D)$ be its general element, where $D = p_1 + \cdots + p_{g-1}$. Then there is a branch $\tilde{F}$ of $\mathrm{Pic}^{g-1}(X) \simeq \mathrm{Pic}^{g+1}(X)$ containing F, namely the branch of $\mathrm{Pic}^{g+1}(X)$ whose general point is $\mathcal{O}_X(D + p + p^S)$, where $p \in X$ is a general point. It is clear that if F, F' are distinct branches of W_{g-1}, then $\tilde{F}$, $\tilde{F}'$ are also distinct. Indeed the map c takes different values on the general points of F and F' (see the proof of Theorem 4.1.7, and hence also on the general points of $\tilde{F}$ and $\tilde{F}'$.

Therefore:

(i) If $0 < \nu(X) < g+1$, there is a 1-1 correspondence between the branches of W_{g-1} and those of $\mathrm{Pic}^{g-1}(X)$ containing them, since the two sets of branches are both bijective to the image of c. Therefore we get

$$\nu(W_{g-1}) = \nu(\mathrm{Pic}^{g-1}(X))$$

according to Theorem 4.1.7.

(ii) If $\nu(X) = g+1$ each branch of $\mathrm{Pic}^{g-1}(X)$ contains the corresponding component of W_{g-1} except for the component $c^{-1}(1, \dots, 1)$, which clearly contains no effective divisor. Hence one has

$$\nu(W_{g-1}) = \nu(\mathrm{Pic}^{g-1}(X)) - 1$$

which also fits with Theorem 4.1.7.

Using the same argument as in the proof of Theorem 4.2.3, one proves that each branch of W_{g-1} contains exactly 2^{g-1} real odd theta-characteristics. Indeed, keeping the notation of the proof of Theorem 4.2.3, it suffices to observe that two values of σ corresponding to the same values of the b_i's, clearly belong to the same branch of $J(X)$ since they can be connected with a path in $J(X)$ consisting of real points. Hence the corresponding theta-characteristics belong to the same branch of $\mathrm{Pic}^{g-1}(X)$, and therefore of W_{g-1}.

In conclusion, using again Theorem 4.2.3 one can analyse the different situations which can occur in relation with the possible values of the invariants $a(X)$ and λ. We summarize the results in the following theorem.

Theorem 4.2.6. *Let (X, S) be a general, smooth, projective curve of genus g with a real structure. Then the following assertions hold.*

(i) *If $a(X) = 1$ every branch of W_{g-1} contains 2^{g-1} odd real theta-characteristics, whereas the corresponding branch of $\mathrm{Pic}^{g-1}(X)$ contains 2^{g-1} even real theta-characteristics, for a total of 2^g real theta-characteristics.*

(ii) *If $a(X) = 0$ and $\lambda \neq 0$ there exists a unique branch U of W_{g-1} which does not contain odd theta-characteristics, while all the others contain 2^{g-1} odd theta-characteristics and the corresponding branches of $\mathrm{Pic}^{g-1}(X)$ contain 2^{g-1} even real theta-characteristics, for a total of 2^g real theta-characteristics. The branch U is contained in a branch of $\mathrm{Pic}^{g-1}(X)$ containing only 2^g even real theta-characteristics. Moreover one has $U = c^{-1}_{|W_{g-1}}(1, \dots, 1)$.*

(iii) *If $a(X) = 0$ and $\lambda = 0$ then every branch of W_{g-1} contains 2^{g-1} odd theta-characteristics and the corresponding branches of $\mathrm{Pic}^{g-1}(X)$ contain 2^{g-1} even real theta-characteristics, for a total of 2^g real theta-characteristics. The branch $c^{-1}(1, \dots, 1)$ of $\mathrm{Pic}^{g-1}(X)$ contains only 2^g even real theta-characteristics.*

All the assertions in the theorem above are an easy consequence of the previous results except for the assertion in (ii) concerning the characterization of U as $c^{-1}_{|W_{g-1}}(1, \dots, 1)$. This is a non trivial result which has been first proved by Comessatti in [C5]. For a different proof see [GrHa].

Remark 4.2.7. Going back to Remark 4.1.5, we are now ready to see that, if $\nu(X) = 0$ and g is odd, then there is a line bundle $\mathcal{L} \in \mathrm{Pic}^0(X)^S - \mathrm{Pic}^0(X)^+$ such that $\mathcal{L}^{\otimes 2} \simeq \mathcal{O}_X$. Indeed in this case we have $\nu(\mathrm{Pic}^0(X)) = 2$ (see Theorem 4.1.7, (iii)), hence $\mathrm{Pic}^0(X)^+$ is the connected component of the origin of $\mathrm{Pic}^0(X)^S$. There are $4 \cdot 2^{g-1} = 2^{g+1}$ real theta-characteristics, hence this is the number of line bundles $\mathcal{L}$ on X such that $\mathcal{L}^{\otimes 2} \simeq \mathcal{O}_X$. Therefore 2^g of such line bundles sit on $\mathrm{Pic}^0(X)^+$ and 2^g on $\mathrm{Pic}^0(X)^S - \mathrm{Pic}^0(X)^+$.

4.3. Examples

In this section we apply the previous results to two particular cases: real hyperelliptic curves and real plane curves. Both examples are quite classical (see [Kl], [C6]; for a modern reference, see [GrHa]). More examples, e.g., trigonal curves, will be found in [GrHa]. However the interested reader should consider that only very few questions about linear systems on real curves have been thoroughly studied. If one takes the book [ACGH] and starts looking for (suitable) extensions of the results there to the real case, he will find soon so many (interesting) question which could provide work for more than a life!

A. Real hyperelliptic curves. A curve X over $\mathbb{C}$ of genus $g \geq 2$ is called *hyperelliptic* if there exists a line bundle $\mathcal{L}$ in $\mathrm{Pic}^2(X)$ such that: $h^0(X, \mathcal{L}) = \dim_{\mathbb{C}} H^0(X, \mathcal{L}) = 2$. We briefly recall some well known results on complex hyperelliptic curves (see [ACGH]).

(i) If X is hyperelliptic then $K = \mathcal{L}^{\otimes(g-1)}$, where, as usual, K is the canonical line bundle of X.

(ii) The *canonical map*

$$\phi_K: X \to \mathbb{P}(H^0(X, K)^*) \simeq \mathbb{P}^{g-1}$$

of X is $2-1$ onto a rational normal curve Y.

(iii) The map

$$f: X \to \mathbb{P}(H^0(X, \mathcal{L})^*) \simeq \mathbb{P}^1$$

is the 2-sheeted cover of $\mathbb{P}^1$ ramified at $2g+2$ points corresponding to the map ϕ_K in (ii). Accordingly there is an involution on X exchanging points of X which are mapped by f to the same point. This is called the *hyperelliptic involution* of X.

(iv) Let R be the ramification divisor of f on X and B be the branch divisor on $\mathbb{P}^1$, which has degree $2g+2$. The theta-characteristics of X are of the form

$$\mathcal{A} = \mathcal{L}^{\otimes(m-1)} \otimes \mathcal{O}_X(D)$$

where $m \geq 0$ and D is a subdivisor of R. One has $h^0(X, \mathcal{A}) = m$ and $2(m-1) + deg D = g - 1$. The above representation for $\mathcal{A}$ is unique except in the case $m = 0$ where D may be interchanged with $R - D$.

Suppose now that X has a real structure (X, S). By (i), (ii) and (iii), $\mathcal{L}$ is the only element in $\mathrm{Pic}^2(X)$ such that $h^0(X, \mathcal{L}) = 2$. Hence $\mathcal{L} \in \mathrm{Pic}^2(X)^S$.

Lemma 4.3.1. *In the above setting we have:*

(i) *If either $X_S \neq \emptyset$ or if g is even then the map $f: X \to \mathbb{P}^1$ defined in* (iii) *above is real.*

(ii) *If $X_S = \emptyset$ and $g \geq 3$ is odd then X is either a 2-sheeted cover of $\mathbb{P}^1_{\mathbb{R}}$ or a 2-sheeted cover of the imaginary conic $u^2 + v^2 = -1$, the 2-to-1 maps being defined over the reals.*

Proof. (i) In this case the canonical map $\phi_K: X \to \mathbb{P}^{g-1}$, which is defined over $\mathbb{R}$, has image which is a rational curve Y with a real structure. If $X_S \neq \emptyset$, then $Y_S \neq \emptyset$ and therefore $Y \simeq \mathbb{P}^1$. If g is even, the degree of Y is $g - 1$ which is odd. Hence again $Y_S \neq \emptyset$, and we conclude as before.

In case (ii) the curve Y, which is rational, is either isomorphic to $\mathbb{P}^1_{\mathbb{R}}$ or to the imaginary conic $u^2 + v^2 = -1$ (see Examples 2.1.1, (ii) and 2.2.4, (i)). If $\mathcal{L} \in \mathrm{Pic}^2(X)^+$ then X may be represented as a 2-sheeted cover of $\mathbb{P}^1$ over $\mathbb{R}$, if $\mathcal{L} \notin \mathrm{Pic}^2(X)^+$ then X may be represented over $\mathbb{R}$ as a 2-sheeted cover of the imaginary conic. □

The above lemma may be rephrased by saying that in all cases, but when $X_S = \emptyset$, $g \geq 3$ and $\mathcal{L} \notin \mathrm{Pic}^2(X)^+$, the curve X may be represented by a real equation of the form

$$y^2 = F(x) \qquad (\alpha)$$

where F is a real polynomial of degree $2g + 2$ in x. In the remaining case X is given by a pair of real equations

$$u^2 + v^2 = -1, \quad y^2 = F(u, v) \qquad (\beta)$$

where F is a real polynomial of degree $g + 1$ in (u, v).

Proposition 4.3.2. *Let X be a real hyperelliptic curve such that either $X_S \neq \emptyset$ or g is even. The invariant $\nu(X)$ may assume all values between 0 and $g + 1$ for such a curve. Furthermore if f induces a degree 2 map of X_S onto $\mathbb{P}^1_{\mathbb{R}}$, then $\nu(X) \equiv g - 1$* (mod 2).

Proof. Let X be represented by an equation as in (α). Then $F(x) = 0$ represents the branch divisor B on $\mathbb{P}^1$, which is reduced, i.e., F has no multiple roots. If $F(x) = 0$ has real solutions then their number is even, say $2m$. Therefore $\mathbb{P}^1_{\mathbb{R}}$ may be divided into $2m$ real intervals such that $F(x) > 0$ in m of these intervals whereas $F(x) < 0$ in the others. Accordingly X_S has exactly $\nu(X) = m$ connected components corresponding to the intervals where $F(x) > 0$.

The case in which f induces a degree 2 map of X_S onto $\mathbb{P}^1_{\mathbb{R}}$ clearly corresponds to the fact that $F(x) = 0$ has no real solution. If this happens and $F(x) > 0$ for all $x \in \mathbb{P}^1_{\mathbb{R}}$, then the map

$$f: X_S \to \mathbb{P}^1_{\mathbb{R}} \simeq S^1$$

is an unramified double covering of S^1. If X_S is connected then g is even. This follows from the fact that $\pi: X \to \mathbb{P}^1_{\mathbb{C}}$ is ramified at $2g + 2$ points so that $\mathbb{P}^1_{\mathbb{R}} \subset \mathbb{P}^1_{\mathbb{C}}$ winds around an odd number of ramification points, and this number equals $g + 1$. In fact the branch points come in pairs of conjugate points. If X_S is disconnected then, for the same reason, g is odd.

If $F(x) = 0$ has no real roots and $F(x) < 0$ for all x, then $X_S = \emptyset$, i.e., this case does not occur in our hypotheses. □

Proposition 4.3.3. *Let X be a real hyperelliptic curve with* $a(X) = 0$. *Then either* $\nu(X) = g + 1$ *or*

$$\nu(X) = \begin{cases} 1 & \text{if } g \equiv 0 \pmod 2, \\ 2 & \text{if } g \equiv 1 \pmod 2. \end{cases}$$

Proof. Clearly $X_S \neq \emptyset$. Therefore X is given by an equation of the form (α). If all the roots of $F(x)$ are real, i.e., $F(x) = 0$ has $2(g+1)$ real solutions, then $\nu(X) = g + 1$ and $a(X) = 0$ (see Proposition 2.7.10). On the other hand, $F(x)$ cannot have n real roots with $0 < n < 2g + 2$. In fact let p be any point of X and let p' be the conjugate point in the hyperelliptic involution. We can connect p and p' by a path entirely contained in $X - X_S$, by lifting to X a path in $\mathbb{P}^1$ which starts and ends at $f(p)$, winds around some complex branch point and never meets the intervals of $\mathbb{P}^1_{\mathbb{R}}$ where $F(x) \geq 0$. This shows that $X - X_S$ is connected thus contradicting the assumption $a(X) = 0$. Actually this argument shows that $F(x)$ has to be positive on $\mathbb{P}^1_{\mathbb{R}}$, which corresponds to the case in which $f: X_S \to \mathbb{P}^1_{\mathbb{R}}$ is surjective of degree 2. Hence we conclude by the argument of the proof of Proposition 4.3.2. □

Remark 4.3.4. If X has equation of the form (α) and if π is the hyperelliptic involution, then the antiinvolution $S' = S \circ \pi$ determines a new real structure (X, S'), with real model

$$y^2 = -F(x).$$

If both real structures have real points, i.e., if F has real roots, then it is clear by the above discussion that the number of branches of the two real structures is the same. If $\nu(X) = 0$ then the number of branches of (X, S') is 1 or 2 according to whether g is even or odd. Thus S' induces the real complementary structure on $J(X)$. This follows from the fact that π induces the multiplication by -1 on $J(X)$. The same thing happens if X has odd genus g and real equation of the form (β). The real model with the complementary real structure is of the same type.

B. Real plane curves. Let X be a smooth plane curve with real structure (X, S). Then X_S consists of $\nu(X)$ disjoint branches which are circles in $\mathbb{P}^2_{\mathbb{R}}$, each one homeomorphic to S^1. We will call any such circle an *oval* if it is homotopic to zero in $\mathbb{P}^2_{\mathbb{R}}$, a *pseudoline* if it represents the non-trivial primitive class in $\pi_1(\mathbb{P}^2_{\mathbb{R}}) \simeq \mathbb{Z}_2$. An oval C disconnects $\mathbb{P}^2_{\mathbb{R}}$: the two components of the complement $\mathbb{P}^2_{\mathbb{R}} - C$ are homeomorphic to a disk and to a Moebius strip. A pseudoline instead does not disconnect $\mathbb{P}^2_{\mathbb{R}}$.

Two pseudolines intersect at an odd number of points, in particular their intersection is always not empty. Every oval intersects any other circle (oval or pseudoline) at an even number of points. Therefore:

(i) for a smooth real curve X of even degree the real locus X_S consists only of ovals;

(ii) if the degree of X is odd then X_S contains exactly one pseudoline.

Two ovals C_1 and C_2 are said to be *nested* if one of them lies in the interior of the other, where the *interior* of the oval C is the component of the complement $\mathbb{P}^2_{\mathbb{R}} - C$ homeomorphic to a disk.

One main, classical question in the topology of real plane curves is the nesting of their ovals. This question became well-known due to its inclusion by Hilbert in his famous list as the sixteenth problem: find all topological types of smooth, real, plane curves of degree 6. A complete answer to this question was classically known only for curves of degree $d \leq 5$. In the late sixties D. A. Gudkov [Gu] provided a complete answer also for $d = 6$. O. Viro [Vi] solved in 1980 also the case $d = 7$. The case $d \geq 8$ is still open.

Here we only make a few remarks relating the nesting of ovals with the invariants $\nu(X)$ and $a(X)$ for the first values of the degree d.

Case $d = 1$. In this case X is a line and $g = 0$. Therefore there is only one possibility, namely

$$\nu(X) = 1,\ a(X) = 0,\ X \simeq \mathbb{P}^1_{\mathbb{R}}.$$

Case $d = 2$. Here X is a conic and $g = 0$. There are two possibilities, namely either

$$\nu(X) = 0,\ a(X) = 1,\ X \simeq \mathbb{O} \tag{1}$$

where $\mathbb{O}$ is the imaginary conic with equation $u^2 + v^2 = -1$, or

$$\nu(X) = 1,\ a(X) = 0,\ X \simeq \mathbb{P}^1_{\mathbb{R}}. \tag{2}$$

Case $d = 3$. Here X is an elliptic curve and $g = 1$. There are again two possible configurations, namely either

$$\nu(X) = 1,\ a(X) = 1,\ \lambda = 1,\ \text{ and } X_S \text{ consists of one pseudoline} \tag{1'}$$

or

$$\nu(X) = 2,\ a(X) = 0,\ \lambda = 0,\ \text{ and } X_S \text{ consists of one pseudoline and one oval.} \tag{2'}$$

Case $d = 4$. If X is a plane quartic then X is the canonical model of any non-hyperelliptic curve of genus 3 (see [ACGH]). The real odd theta-characteristics correspond to real lines which are tangent at two real points i.e., to *bitangent lines*. If C_1 and C_2 are two ovals of X_S, then there is a real line tangent to each C_i if and only if C_1 and C_2 are not nested (see [GrHa, 7.2]).

Note also that by Bezout's theorem the nesting of ovals can only occur when $\nu(X) = 2$.

According to our results from §4.2 one easily sees that, for $\nu(X) \neq 2$ the branches of X_S, which are all ovals, give rise to the following possibilities:

$$\nu(X) = 0, a(X) = 1, \quad \text{4 real bitangent lines,}$$

$$\nu(X) = 1, a(X) = 1, \quad \text{4 real bitangent lines,}$$

$$\nu(X) = 3, a(X) = 1, \quad \text{16 real bitangent lines,}$$

$$\nu(X) = 4, a(X) = 0, \quad \text{28 real bitangent lines.}$$

Note that, according to Corollary 4.2.6, if $\nu(X) = 3$ the 16 bitangents are subdivided in four groups of four, and the bitangents of each group have a different behaviour according to the number of contact points which they have with ovals of the curve. The same can be said in case $\nu(X) = 4$, where the bitangents are subdivided in seven groups of four.

Finally we are left with the case $\nu(X) = 2$. Then we claim that if $a(X) = 0$ the ovals are nested, whereas if $a(X) = 1$ the ovals are not nested. From $\nu(X) = 2$ and $g = 3$ we get $\lambda = 2$ (see Theorem 4.1.11, (i)). Therefore by Corollary 4.2.6, the component $U = c^{-1}(1, 1)$ of $\mathrm{Pic}^2(X)$ contains no odd real theta charactesristics. Hence there are no bitangent lines with the two contact points each on one of the two ovals C_1 and C_2 of X_S, which implies that C_1 and C_2 are nested.

If $a(X) = 1$ then again from Corollary 4.2.6 it follows that there are four real bitangent lines to C_1 and C_2, the two branches of X_S, and therefore the ovals are not nested.

4.4. Moduli spaces and the theorem of Torelli

The classical theorem of Torelli states that there is a 1 to 1 correspondence between isomorphism classes of compact Riemann surfaces and their Jacobians considered as principally polarized abelian varieties (see §4.1). Here we consider the case when X has a real structure (X, S). By results in §4.1, the topological type of (X, S) is described in terms of the genus g, of the number of branches $\nu = \nu(X)$ and of ϵ, where $\epsilon = (o)$ or $\epsilon = (d)$ marks the type, orthosymmetric or diasymmetric, of the Jacobian. Moreover $\epsilon = (o)$ if and only if $a(X) = 0$ (see Theorem 4.1.11).

We let $\mathcal{M}(\mathbb{R})_g$ be the set of equivalence classes of real curves of genus g and $\mathcal{M}_{g,\nu,\epsilon}$ be the set of equivalence classes of curves of genus g with a real structure (X, S), with characters $\nu = \nu(X)$ and ϵ. Then $\mathcal{M}(\mathbb{R})_g = \cup \mathcal{M}_{g,\nu,\epsilon}$. To every real curve (X, S) which corresponds to a point in $\mathcal{M}_{g,\nu,\epsilon}$ we associate its Jacobian $J(X)$ which is a real principally polarized abelian variety of type (λ, ϵ), i.e., it corresponds to a point of $\mathcal{A}_{g,(\lambda,\epsilon)}$ (see §3.5).

Thus we obtain the so called *Torelli mapping*

$$\tau \colon \mathcal{M}(\mathbb{R})_g \to \mathcal{A}(\mathbb{R})_g$$

(for the definition of $\mathcal{A}(\mathbb{R})_g$ see again §3.5). According to results by Seppälä and Silhol [SS] (see also [Sp1,2] for further information about moduli spaces), the set $\mathcal{M}(\mathbb{R})_g$ has a natural structure of a real analytic space of pure dimension $3g-3$, if $g \geq 2$, of dimension 1 if $g = 1$, which makes it a coarse moduli space for real structures on curves of genus g. Its connected components are the subspaces $\mathcal{M}_{g,\nu,\epsilon}$. The Torelli map is then a morphism of real analytic spaces, which can be decomposed in analytic mappings between the various components

$$\tau\colon \mathcal{M}_{g,\nu,\epsilon} \to \mathcal{A}_{g,(\lambda,\epsilon)}$$

where

$$\lambda = g+1-\nu \quad \text{if } \nu \neq 0$$

$$\lambda = g \quad \text{if } \nu = 0 \quad \text{and} \quad g \equiv 0 \pmod 2$$

$$\lambda = g-1 \quad \text{if } \nu = 0 \quad \text{and} \quad g \equiv 1 \pmod 2$$

Examples 4.4.1. (i) If $g = 0$ then $\mathcal{M}(\mathbb{R})_0$ consists of two points corresponding to $\mathcal{M}_{0,0,o}$ and to $\mathcal{M}_{0,1,o}$. Since $\mathcal{A}(\mathbb{R})_0$ consists of a single point, the map τ is, in this case, not injective, hence the theorem of Torelli does not hold.

(ii) If $g = 1$ then $\mathcal{M}(\mathbb{R})_1$ consists of three components $\mathcal{M}_{1,0,o}$, $\mathcal{M}_{1,1,d}$ and $\mathcal{M}_{1,2,o}$ (see Proposition 2.7.10 and Theorem 4.1.11). We have already discussed this case in the Example 3.5.3. The Torelli mapping yields isomorphisms

$$\tau\colon \mathcal{M}_{1,0,o} \to \mathcal{A}_{1,(0,o)},$$

$$\tau\colon \mathcal{M}_{1,1,d} \to \mathcal{A}_{1,(1,d)},$$

$$\tau\colon \mathcal{M}_{1,2,o} \to \mathcal{A}_{1,(0,o)}.$$

Therefore when $g = 1$ the Torelli mapping is also not injective.

(iii) In general, for all $\nu \neq 0$ the component $\mathcal{M}_{g,\nu,\epsilon}$ of the moduli space $\mathcal{M}(\mathbb{R})_g$ is sent by τ to the component $\mathcal{A}_{g,(g+1-\nu,\epsilon)}$. The component $\mathcal{M}_{g,0,o}$ is sent to $\mathcal{A}_{g,(g,o)}$ if $g \equiv 0 \pmod 2$, and to $\mathcal{A}_{g,(g-1,o)}$ if $g \equiv 1 \pmod 2$. The component $\mathcal{A}_{g,(g,o)}$ however contains also the image of $\mathcal{M}_{g,1,o}$, while the component $\mathcal{A}_{g,(g-1,o)}$ contains the image of $\mathcal{M}_{g,2,o}$. Despite the negative cases pointed out in the above examples, there is a real version of the theorem of Torelli, due to Silhol (see [S2], [GrHa]).

Theorem 4.4.2. *The Torelli mapping τ is injective as soon as $g \geq 2$.*

Proof. Let (X, S) and (X', S') be real curves such that $\tau(X, S) = \tau(X', S')$. By Torelli's theorem for complex curves (see [ACGH, VI 3]), this implies that X and X' are isomorphic over $\mathbb{C}$. By Theorem 2.6.4, there is a bijection between real structures on X modulo the equivalence relation, and the elements of the cohomology group $H^1(G, \mathrm{Aut}(X))$. Let $J(X)$ be the Jacobian with its principal polarization, let $\mathrm{End}_0(J(X))^*$ be the group of invertible endomorphisms of $J(X)$ fixing the polarization and let $\mathrm{Aut}_0(J(X))$ be the group of automorphisms of $J(X)$ fixing the polarization. The universal property of the Albanese variety (see Theorem 1.4.1), and the fact that $J(X) \simeq \mathrm{Alb}(X)$ imply the existence of a natural homomorphism

$$\phi\colon \mathrm{Aut}(X) \to \mathrm{End}_0(J(X))^*.$$

For $g \geq 2$ this map is injective. According to (iii) of Corollary 1.3.2, we have

$$\mathrm{Aut}(J(X)) \simeq J(X) \times \mathrm{End}(J(X))^*$$

where we recall that the product structure on the right hand side is given by

$$(y, f) \cdot (y', f') = (y + f(y'), f \circ f').$$

Since the translations fix any polarization on an abelian variety, we accordingly have

$$\mathrm{Aut}_0(J(X)) \simeq J(X) \times \mathrm{End}_0(J(X))^*$$

and therefore we have the injective homomorphism

$$\iota\colon f \in \mathrm{End}_0(J(X))^* \to (0, f) \in \mathrm{Aut}(X).$$

Hence we have the injective homomorphism

$$\psi = \iota \circ \phi\colon \mathrm{Aut}(X) \to \mathrm{Aut}_0(J(X))$$

It is now easy to see that the induced map ψ_* on cohomology (see Theorem 2.6.3)

$$f_*\colon H^1(G, \mathrm{Aut}(X)) \to H^1(G, \mathrm{Aut}_0(J(X))$$

is injective, which, in view of Theorem 2.6.4 concludes the proof of the theorem. □

It seems interesting to us to conclude this section by describing, following Comessatti [C2, II] but without giving proofs, the image of the Torelli mapping τ in the case $g = 2$.

The components $V_1 = \mathcal{M}_{2,0,o}$ and $V_2 = \mathcal{M}_{2,1,o}$ of $\mathcal{M}(\mathbb{R})_2$, both of dimension 3, are mapped by τ to the 3-dimensional subvariety $\mathcal{A}_{2,(2,o)}$ of $\mathcal{A}_2(\mathbb{R})$. The images of V_1 and V_2 are the two open subsets U_1 and U_2 of $\mathcal{A}_{2,(2,o)}$ which correspond to values $\delta > \frac{1}{16}$ and $\delta < \frac{1}{16}$, δ being the invariant defined in the statement of Theorem 2.8.5 (recall also Proposition 3.1.6). U_1 and U_2 are interchanged under the natural involution $\gamma := \gamma_{2,(2,o)}$ on $\mathcal{A}_{2,(2,o)}$, as defined in Example 3.5.3, which sends a principally polarized real abelian variety to its complementary variety. Jacobians exchanged by γ correspond to $\mathbb{C}$-isomorphic curves of the form $y^2 = F(x)$ and $y^2 = -F(x)$, where F has no real roots. If F is everywhere positive, these curves corresponds to elements in V_1 and V_2 respectively.

The set $\mathcal{A}_{2,(2,o)} - (U_1 \cup U_2)$ is the set of abelian varieties corresponding to the value $\delta = \frac{1}{16}$. These are products of complex conjugate elliptic curves, and can be considered as Jacobians of degenerate curves of genus two, formed by the two elliptic curves glued together at conjugate points. These curves appear as limits either of curves belonging to $\mathcal{M}_{2,(0,o)}$ or of curves in $\mathcal{M}_{2,(1,o)}$, as the reader will see after one moment of reflection.

4.5. Singular curves

In this section we briefly discuss the extension of some of the results of §4.1 to any real curve, not necessarily smooth neither projective (see [PW1,2]). We will consider an algebraic curve X with a fixed real structure (X, S). We will consider two topological invariants of the set X_S of real points of X, equipped with the euclidean topology, i.e., $\nu(X) = \dim_{\mathbb{Z}_2} H^0(X_S, \mathbb{Z}_2)$ and $l(X) = \dim_{\mathbb{Z}_2} H^1(X_S, \mathbb{Z}_2)$. If X is smooth and projective then $l(X) = \nu(X)$, while in general $\nu(X)$ represents the number of connected components of X_S and $l(X)$ the number of *loops* of X_S (see Remark 4.2.5.)

Let X_0 be a real model of X. The assumption that the complex curve X is irreducible is usually expressed by saying that the real curve X_0 is *geometrically irreducible*. Note that $\mathrm{Pic}(X_0) \simeq \mathrm{Pic}(X)^+$. The following proposition follows from Proposition 2.7.4 and the results of §4.1.

Proposition 4.5.1. *Let X_0 be a geometrically irreducible, smooth, projective curve over $\mathbb{R}$, which is the real model of the real structure (X, S). Let $\nu = \nu(X)$ be the number of real components of X and let g be its genus. Then* $\mathrm{Pic}(X_0) \simeq \mathrm{Pic}(X)^+$ *is a subgroup of* $\mathrm{Pic}(X)$ *and, as an abelian group, one has*

$$\mathrm{Pic}(X_0) \simeq \mathbb{Z} \times (\mathbb{R}/\mathbb{Z})^g \times \begin{cases} (\mathbb{Z}_2)^{\nu-1} & \text{if } \nu \neq 0, \\ 0 & \text{if } \nu = 0. \end{cases}$$

Thus $\mathrm{Pic}(X_0) \otimes \mathbb{Z}_2 \simeq (\mathbb{Z}_2)^\nu$ *if $\nu \neq 0$, while* $\mathrm{Pic}(X_0) \otimes \mathbb{Z}_2 \simeq \mathbb{Z}_2$ *if $\nu = 0$.*

Remark 4.5.2. If $X_S = \emptyset$, then as we observed in Remark 4.1.6, there is the short exact sequence (4.1.5) which splits when g is odd and does not split when g is even. In the latter case the map d is just the degree map modulo 2. We want to make some more comments on this.

We first observe that (4.1.5) is a particular case of an exact sequence, valid for any real variety X, which can be deduced from the Leray spectral sequence associated to the structure morphism $f: X_0 \to \mathrm{Spec}(\mathbb{R})$ in the following way. Consider the étale topologies on both X_0 and $\mathrm{Spec}(\mathbb{R})$ (see [Mi] for definitions and basic properties of étale topology and étale cohomology). Let $\mathbb{G}_m$ be the multiplicative sheaf on X_0, i.e., $\mathbb{G}_m(U) = \mathcal{O}_U^*$ for every open U in X_0. Then we get the following Leray spectral sequence:

$$H^p_{\mathrm{et}}(\mathrm{Spec}(\mathbb{R}), R^q f_* \mathbb{G}_m) \Rightarrow H^{p+q}_{\mathrm{et}}(X_0, \mathbb{G}_m).$$

If $\xi = \mathrm{Spec}(\mathbb{R})$, the stalk of the sheaf $R^q f_* \mathbb{G}_m$ over a geometric point $\bar{\xi}$ is $H^q(X, \mathbb{G}_m)$.

Let G be, as usual, the Galois group of $\mathbb{C}$ over $\mathbb{R}$. The étale cohomology on $\mathrm{Spec}(\mathbb{R})$ coincides with Galois cohomology, i.e., with G-cohomology as defined in

§2.6 (see [Mi]). Therefore in the above spectral sequence the E_2-terms are given by

$$E_2^{0,1} = H^0(\mathbb{R}, H^1(X, \mathbb{G}_m)) = \mathrm{Pic}(X)^G,$$
$$E_2^{1,0} = H^1(\mathbb{R}, f_*\mathbb{G}_m) = H^1(G, \Gamma(X, \mathcal{O}_X)^*),$$
$$E_2^{2,0} = H^2(\mathbb{R}, f_*\mathbb{G}_m) = H^2(G, \Gamma(X, \mathcal{O}_X)^*).$$

From the above spectral sequence we get the following exact sequence:

$$\begin{aligned} 0 \to H^1(G, \Gamma(X, \mathcal{O}_X)^*) &\to \mathrm{Pic}(X_0) \to \mathrm{Pic}(X)^G \\ &\to H^2(G, \Gamma(X, \mathcal{O}_X)^*) \to H^2_{\mathrm{et}}(X, \mathbb{G}_m). \end{aligned} \tag{4.5.4}$$

If X is projective then $\Gamma(X, \mathcal{O}_X)^* \simeq \mathbb{C}^*$ hence in (4.5.4) we have $H^1(G, \mathbb{C}^*) = 0$ and

$$H^2(G, \mathbb{C}^*) \simeq H^2_{\mathrm{et}}(\mathrm{Spec}(\mathbb{R}), \mathbb{G}_m) \simeq \mathrm{Br}(\mathbb{R})$$

where $\mathrm{Br}(\mathbb{R})$ is the *Brauer group* of the field $\mathbb{R}$, which is isomorphic to $\mathbb{Z}_2$, since it is generated by the class of $\mathbb{R}$ and by the class of the quaternions, see [Se1]. So, in the case X_0 is a projective real variety, the exact sequence (4.5.4) becomes the following one:

$$0 \to \mathrm{Pic}(X_0) \to \mathrm{Pic}(X)^G \to \mathbb{Z}_2 \to H^2_{\mathrm{et}}(X, \mathbb{G}_m).$$

If X_0 is either an affine real variety or a real curve then the group $H^2_{\mathrm{et}}(X_0, \mathbb{G}_m)$, which is usually called the *cohomological Brauer group*, is isomorphic to the usual Brauer group $\mathrm{Br}(X_0)$ (see [Mi]). In the case of a real curve X_0 with no real points, i.e., such that $X_S = \emptyset$, then $\mathrm{Br}(X_0) = 0$: this has been proven by Witt (see [Wi]) in the case of a smooth projective real curve and has then been extended to any real curve in [PW]. Hence the exact sequence (4.5.4) gives back the exact sequence (4.1.5) for any real curve.

If $X_S = \emptyset$, the G-module structure of $\mathrm{Pic}^0(X)$ depends upon the parity of the genus g. Using definitions and calculations in Example 2.6.1, (iii), and the results of §§4.1-4.2, we find that

$$H^1(G, \mathrm{Pic}^0(X)) \simeq \begin{cases} 0 & \text{if } g \text{ is even,} \\ \mathbb{Z}_2 & \text{if } g \text{ is odd,} \end{cases}$$

and as a subgroup of $\mathrm{Pic}(X)$, we have

$$\mathrm{Pic}(X)^G \simeq \begin{cases} \mathbb{Z} \oplus (\mathbb{R}/\mathbb{Z})^g \oplus \mathbb{Z}_2^{\nu-1} & \text{if } \nu = \nu(X) \neq 0, \\ \mathbb{Z} \oplus (\mathbb{R}/\mathbb{Z})^g & \text{if } \nu = 0 \text{ and } g \text{ is even,} \\ 2\mathbb{Z} \oplus (\mathbb{R}/\mathbb{Z})^g \oplus \mathbb{Z}_2 & \text{if } \nu = 0 \text{ and } g \text{ is odd,} \end{cases}$$

(see the proof of Theorem 4.1.7). On the other hand one can see that if g is even then $\mathrm{Pic}(X) \simeq A \times \mathbb{Z}$ as a G-module, but if g is odd then $\mathrm{Pic}(X)$ does not split as a G-module. Notice the contrast with the sequence (4.1.5), which does not split when g is even, but splits when g is odd.

From Proposition 4.5.1 one easily deduces the following result.

Corollary 4.5.5. *Let X_0 be a geometrically irreducible smooth real curve which is the real model of the real curve (X, S) having $l(X)$ loops. Then one has*

$$\operatorname{Pic}(X_0) \otimes \mathbb{Z}_2 \simeq \mathbb{Z}_2^{l(X)}$$

with the only exception when X is a geometrically irreducible projective curve with $X_S = \emptyset$. In that case one has

$$\operatorname{Pic}(X_0) \otimes \mathbb{Z}_2 \simeq \mathbb{Z}_2.$$

More generally, if X is any smooth real curve with $l(X)$ loops, let $E(X)$ denote the number of irreducible algebraic components X_i of X which are projective and for which $(X_i)_S$ is empty. Then one has

$$\operatorname{Pic}(X_0) \otimes \mathbb{Z}_2 \simeq \mathbb{Z}_2^{l(X)+E(X)}.$$

In order to extend the computation of $\operatorname{Pic}(X_0) \otimes \mathbb{Z}_2$ to any real curve, we consider étale cohomology on X_0 with coefficients in the sheaf μ_n of n-th roots of the unity. For $n = 2$, since $\mathbb{R}$ contains a primitive square root of 1, we have an isomorphism between the sheaf μ_2 and the constant sheaf $\mathbb{Z}_2$ on X_0. Hence multiplication by 2 on $\mathcal{O}_X^*$ yields a short exact sequence of sheaves on X_0,

$$0 \to \mathbb{Z}_2 \to \mathbb{G}_m \to \mathbb{G}_m \to 0.$$

Associated to the above sequence there is a *Kummer sequence* for étale cohomology (see [Mi]) which, using the isomorphism $H^2(X, \mathbb{G}_m) \simeq \operatorname{Br}(X_0)$ gives rise to the following exact sequence

$$0 \to \operatorname{Pic}(X_0) \otimes \mathbb{Z}_2 \to H^2_{\text{et}}(X_0, \mathbb{Z}_2) \to \operatorname{Br}(X_0) \to 0 \tag{4.5.6}$$

The étale cohomology groups of a real curve may be computed using the following result, for the proof of which we refer to [PW].

Theorem 4.5.7. *Let X_0 be a real curve which is the real model of a real structure (X, S). Then:*

(i) $H^0_{\text{et}}(X_0, \mathbb{Z}_2) \simeq \mathbb{Z}_2^a$ *where a is the number of connected components of X;*

(ii) $H^1_{\text{et}}(X_0, \mathbb{Z}_2) \simeq (H^0(X_0, \mathcal{O}_X^*) \otimes \mathbb{Z}_2) \oplus ((\operatorname{Pic} X_0)_2 \otimes \mathbb{Z}_2)$;

(iii) $H^2_{\text{et}}(X_0, \mathbb{Z}_2) \simeq \mathbb{Z}_2^{\nu(X)+l(X)+E(X)}$;

(iv) $H^i_{\text{et}}(X_0, \mathbb{Z}_2) \simeq \mathbb{Z}_2^{\nu(X)+l(X)}$ *for every $i \geq 3$.*

The group $\operatorname{Br}(X_0)$ is isomorphic to $\mathbb{Z}_2^{\nu(X)}$. This result, which was originally proved by Witt (see [Wi]) for smooth projective curves, has been extended in [PW1, (3.6)] to any real curve. Therefore from the exact sequence (4.5.6) one gets the following result which computes $\operatorname{Pic}(X_0) \otimes \mathbb{Z}_2$ for any real curve.

Theorem 4.5.8. *Let X_0 be a real curve which is the real model of a real structure (X, S). Then*

$$\mathrm{Pic}(X_0) \otimes \mathbb{Z}_2 \simeq \mathbb{Z}_2^{l(X)+E(X)}.$$

Example 4.5.9. If X_0 is the projective curve defined by $xy(x^2 + y^2) = 0$, then $l(X) = 2$ and $E(X) = 1$. Thus $\mathrm{Pic}(X_0) \otimes \mathbb{Z}_2 \simeq \mathbb{Z}_2^3$.

References

[ACGH] E. Arbarello, M. Cornalba, Ph. Griffiths, J. Harris, Geometry of algebraic curves, vol. I, Grundlehren Math. Wiss. 267, Springer-Verlag 1985.

[Ar] E. Artin, Ueber Zerlegung definiter Funktionen in Quadrate 1927 Abh. Math. Sem. Hamburg 5 (1927), 100–115.

[BCR] J. Bochnack, M. Coste, M. F. Roy, Géométrie algébrique réelle, Ergeb. Math. Grenzgeb. (3) 12, Springer-Verlag 1987.

[BS] A. Borel, J. P. Serre, Théorème de finitude en cohomologie galoisienne, Comment Math. Helv. 39 (1965), 111–164.

[Ca] H. Cartan, Quotients d'un espace analytique par un groupe d'automorphismes, Symp. in Honour of S. Lefschetz, Princeton Univ. Press 1957.

[Ch1] S. Cherubino, Sulle varietà abeliane reali e sulle matrici di Riemann reali, mem. I e II, 1922–1923 Giornale di Matem. di Battaglini 60–61 (1922–1923), 65–94, 47–68.

[Ch2] S. Cherubino, Sul problema della normalizzazione nella teoria generale delle varietà abeliane reali, Atti Ist. Veneto 88 (1929).

[Ch3] S. Cherubino, Sulla normalizzazione delle matrici di Riemann reali, Atti Ist. Veneto 89 (1930).

[Ci] C. Ciliberto, Endomorfismi di jacobiane, Rend. Semin. Mat. e Fis. Milano 59 (1989), 213-242

[CP] C. Ciliberto, C. Pedrini, Comessatti and Real Algebraic Geometry, Atti Convegno "Algebra e Geometria 1860–1940: il contributo italiano", Rend. Circ. Mat. Palermo (II) 36 (1995), 71–102 .

[C1] A. Comessatti, Intorno ad un nuovo carattere delle matrici di Riemann, Mem. R. Accad. d'Italia VII (1936), 81–129.

[C2] A. Comessatti, Sulle varietà abeliane reali I e II, Ann. Mat. Pura Appl. 2–4 (1924–1926), 67–106, 27–71.

[C3] A. Comessatti, Sopra certe trasformazioni dei periodi normali, Atti Ist. Veneto 83 (1924), 735–750.

[C4] A. Comessatti, Sulle trasformazioni involutorie delle varietà algebriche, Atti Ist. Veneto 85 (1926), 471–494.

[C5] A. Comessatti, Complementi al problema dei gruppi semicanonici reali, Rend. Circ. Mat. Palermo 48 (1924), 1–31.

[C6] A. Comessatti, Osservazioni sulle curve iperellittiche reali, 1928 Boll. Un. Mat. Ital. 7 (1928), 3–7.

[C7] A. Comessatti, Problemi di realità per le superficie e varietà algebriche, Atti Convegno Volta (1940), 3–29.

[GH] Ph. Griffiths and J. Harris, Principles of Algebraic Geometry, J. Wiley and Sons, New York 1978.

[GrHa] B. Gross and J. Harris, Real Algebraic Curves, Ann. Sci. École Norm. Sup. Paris 14 (1981), 157–182.

[Gu] D. A. Gudkov, The topology of real projective algebraic manifolds, Russian Math. Surveys 29 (1974), 1–79.

[Hr] R. Hartshorne, Algebraic Geometry, Grad. Texts in Math. 52, Springer-Verlag, 1977

[H] J. Huisman, Real abelian varieties with complex multiplication, Ph. D. thesis, Vrije Univ. of Amsterdam 1992.

[HS] P. J. Hilton, U. Stammbach, A Course in Homological Algebra, Grad. Texts in Math. 4, Springer-Verlag 1971.

[HW] P. J. Hilton, S. Wylie, Homology Theory, Cambridge Univ. Press 1960.

[Kl] F. Klein, Ueber Realitätsverhältnisse bei der einem beliebigen Geschlechte zugehörigen Normalcurve der φ, Math. Ann. 42 (1892), 1–29.

[Kn] M. Knebusch, On algebraic curves over real closed field I and II, Math. Z. 150/151 (1976), 49–70, 189–205.

[K] V. A. Krasnov, Harnack–Thom inequalities for mappings of real algebraic varieties, Math. USSR Izvestiya 22 (1984), 247–275.

[LB] H. Lange and Ch. Birkenhake, Complex Abelian Varieties, Grundlehren Math. Wiss. 302, Springer-Verlag 1992.

[M] D. Mumford, Abelian Varieties, Oxford Univ. Press 1970.

[Mi] J. S. Milne, Etale cohomology, Princeton Univ. Press 1980.

[PW] C. Pedrini and C. Weibel, Invariants of real Curves, Rend. Sem. Mat. Univ. Politecnico Torino 49 (1991), 139–173.

[Ra] M. Raimondo, Introduzione allo studio delle varietà abeliane reali, preprint 1981, Seminario dell'Istituto Matematico Genova (136).

[R] D. Robinson, A Course in the Theory of Groups, Graduate Texts in Math. 80, Springer-Verlag 1982.

[Sp1] M. Seppälä, Computation of period matrices of real algebraic curves, Discrete Comput. Geom. 11 (1994), 65–81.

[Sp2] M. Seppälä, Computation conformal geometry, preprint 1994.

[SS] M. Seppälä and R. Silhol, Moduli Spaces for Real Algebraic Curves and Real Abelian Varieties, Math. Z. 201(1989), 151–165.

[Se1] J.-P. Serre, Cohomologie Galoisienne, Lecture Notes in Math. 5, Springer-Verlag 1965.

[Se2] J.-P. Serre, Géométrie algébrique et géométrie analytique, Ann. Inst. Fourier 6 (1956), 1–42.

[Sm] G. Shimura, On the real points of an arithmetic quotient of a bounded symmetric domain, Math. Ann. 215 (1975), 135–164.

[S1] R. Silhol, Real Algebraic Varieties, Lecture Notes in Math. 1392, Springer-Verlag 1989.

[S2] R. Silhol, Real abelian varieties and the theory of Comessatti, Math. Z. 181 (1982), 345–364.

[Vi] O. Viro, Curves of degree 7, curves of degree 8 and Ragsdale conjecture, Soviet Math. Dokl. 22 (1980), 566–569.

[Wh] G. Weichold, Ueber symmetrische Riemann'sche Flächen und die Periodicitätsmoduln der zugehörigen Abel'schen Normalintegrale der ersten Gattung, Z. Math. Phys. 28 (1883), 321–352.

[W1] A. Weil, The field of definition of a variety, Amer. J. Math. 78 (1956), 509–524.

[W2] A. Weil, Zum Beweis des Torellischen Satzes, Nachr. Akad. Wiss. Göttingen Math.-Phys. Kl. II (1957), no. 2, 33–53.

[Wi] E. Witt, Zerlegung reeller algebraischer Funktionen in Quadrate. Schiefkörper über reellen Functionenkörper J. Reine Angew. Math. 171 (1934), 4–11.

Dipartimento di Matematica, Università di Roma "Tor Vergata", Viale della Ricerca Scientifica, 00133 Roma, Italy
e-mail: `ciliberto@mat.utovrm.it`

Dipartimento di Matematica, Università di Genova, Via Dodecaneso 35, 16146 Genova, Italy
e-mail: `pedrini@dima.unige.it`

Mario Raimondo's contributions to real geometry

Alberto Tognoli

Some historical remarks

Here I try to sketch some relevant events in the study of real algebraic geometry of this century. A. Comessatti dedicated a large part of his mathematical work to the study of real algebraic geometry. He had a deep knowledge of complex algebraic geometry and he made an extensive use of the concept of complexification. The results obtained by Comessatti are numerous and deep. The techniques used are those of the Italian school of algebraic geometry. In some area (e.g., the theory of Abelian varieties) he reached very complete and difficult results. For these reasons the study of Comessatti's works is difficult, and only very recently "modern proofs" became available.

The decline of the Italian school of algebraic geometry induced a slowing down of research in real algebraic geometry. The following two crucial articles can be considered as the beginning of the modern real algebraic geometry: the paper "Real algebraic manifolds" (1952) by J. Nash and "On periodic points" (1965) by M. Artin and B. Mazur. In these articles new problems are studied and the authors begin the study of real algebraic functions (usually called "Nash functions").

Recently the study of real algebraic geometry became more systematic, the following tools, among others, were used:

1) complexification;

2) the theory of schemes;

3) the theory of Nash functions.

Each of these tools has some advantages and some disadvantages. Complexification allows us to use the information given by the theory of complex algebraic geometry and from this to deduce properties of the real part. Moreover this method saves the geometric feeling of the problem. One of the disadvantages is that the complexification depends on the embedding of the real variety in $\mathbb{R}^n$. Only the germ of the complexification near the real part is canonically defined.

So geometric objects (algebraic vector bundles etc.) defined on the real algebraic variety V cannot in general be extended to the complexification $\tilde{V}$, but only to an open neighbourhood.

The theory of schemes overcomes some of these disadvantages. In fact the affine scheme associated to the ring Γ_V of regular functions on V is intrinsically defined and it is easier to extend geometical objects from V to $\operatorname{Spec}\Gamma_V$. In the category of affine schemes we have a nice cohomological theory ("Theorems A and B" for coherent sheaves), but this is not sufficient to ensure similar results for the real part because $\operatorname{Spec}\Gamma$ is not a Hausdorff space. So no reasonable cohomological result is known for real algebraic varieties.

One of the disadvantages of the use of the scheme theory is the loss of the geometric intuition and the impossibility of the use of transcendental methods.

The theory of Nash functions seems to be one of the most powerful instruments for the understanding of the features of real algebraic varieties. In fact the implicit function theorem holds in the class of Nash functions and the locus of zeroes of a global Nash function is a "sheet" of an algebraic variety. So this class of functions seems to be large enough to overcome the lack of the implicit function theorem and small enough to study only the analytic components of real algebraic varieties. And in fact several interesting results can be obtained using this class of functions. Unluckily the theory of Nash functions seems to be very difficult. No reasonable cohomology theory is available (also in the complex setting) and many central problems about the existence of global Nash functions are still open.

From the study of the basic properties of the real algebraic varieties a number of "strange" facts appeared. We shall list some of them.

For coherent sheaves defined over a real affine variety Theorem A does not hold.

All projective real algebraic varieties are affine, so it seems that in some sense non affine real algebraic varieties are "pathological".

Initially no "natural" family of non affine real varieties was known.

The cohomology groups $H^1(\mathbb{R}^n, \mathcal{O}_{\mathbb{R}^n})$, where $\mathcal{O}_{\mathbb{R}^n}$ is the sheaf of germs of rational functions is different from zero, if $n > 1$. So no local triviality holds for the cohomology theory.

Similar facts hold for the sheaf of real or complex Nash functions.

This was the basic setting of a part of the theory and these were some of the main tools that were developed to try to study real algebraic geometry.

A short review of Mario's work in real algebraic geometry

One of the first natural problems in the study of the algebraic geometry over a non algebraically closed field K was the characterization of the ring Γ_V of the regular functions over an affine variety. Such a characterisation allows one to state a duality between the category of affine varieties over K and a category of rings and hence to use scheme theory. This characterization was obtained in a paper by M. G. Marinari and M. Raimondo in 1979 (see [1]).

As remarked before, one of the most natural problems was to try to define a sort of canonical complexification. This seemed to be an easy problem and several attempts were made. This problem was solved negatively in two interesting papers "Affine curves over an algebraically non closed field" by M. G. Marinari, F. Odetti, and M. Raimondo and "On complete intersection over an algebraically non closed field" by M. G. Marinari and M. Raimondo. In these articles it is shown that, also in the case of the curves, no reasonable notion of canonical complexification can be given.

Only one example of a non affine real algebraic variety was known in 1978. This example was found by M. Galbiati using a construction of H. Hironaka. At that time it was also known that for some "bad" algebraic vector bundles $F \to V$ defined over an affine real variety V the set of algebraic global sections does not generate the fiber at any point. In a paper of 1979 by M. G. Marinari and M. Raimondo it is proved that the total space of a real algebraic vector bundle $F \to V$, defined over an affine variety V, is affine if and only if the set of algebraic sections generates the stalk in any point. This result shows that the two pathologies are strictly correlated and it provides a large set of "natural" non affine real algebraic varieties. In the same paper it is also proved that $\dim H^1(\mathbb{R}^n, \mathcal{O}_{\mathbb{R}^n}) = \infty$ if $n > 1$. This result shows that in this setting we have no hope of having cohomological finiteness results.

M. Raimondo published several articles on the theory of Nash functions. He attacked most of the interesting open problems. Here we give a list of some of the principal results proved by M. Raimondo.

He found criteria on U, to ensure that the ring of global Nash functions on U is Noetherian. He treated this problem also in the singular case.

He found criteria to ensure that the natural embedding $\Gamma(\mathcal{N}_U) \to \Gamma(\mathcal{A}_U)$ of the ring of Nash functions into the ring of analytic functions is faithfully flat.

He studied normalization in the analytic and Nash setting and he proved that in several cases they coincide.

I wish to point out two crucial results in this area. In a joint paper with F. Mora (1983) he solved, under the hypothesis of normality of the locus of zeroes, the problem of showing that if a Nash function f is a product of two analytic functions, $f = f_1 f_2$, then there exist two Nash functions h_1, h_2 such that $f = h_1 \cdot h_2$ and $\{f_i = 0\} = \{h_i = 0\}$. This is a partial solution of a crucial problem that is still open. No essential improvement of this result is known up to now.

I believe that M. Raimondo was the first to understand that also for Nash functions it was important to study the complex case. In 1985 he published an article "Sur l'anneau des fonctions de Nash complexes" where he proved several interesting properties of complex algebraic functions and of the cohomology theory of their sheaves. I believe that this article is very important.

Another nice paper was published in 1987 together with E. Fortuna and S. Łojasiewicz. The main result is that a complex analytic germ which is semialgebraic is also a complex Nash germ.

Finally I shall recall a paper where M. Raimondo gave a modern exposition of several results of A. Comessatti about abelian varieties. In this exposition it is a hard and difficult work to understand the old papers of Comessatti and I regret the excess of rigour of Mario. His rigorous style obstructed the circulation of these notes and this is a pity. But this rigorous style had the result that Mario never faced silly or easy problems.

References

[1] M. G. Marinari, M. Raimondo, Properties of the regular functions ring on affine varieties defined over any field, Rend. Sem. Mat. Univ. Politec. Torino 37, 1 (1979), 63–70.

[2] M. G. Marinari, M. Raimondo, Fibrati vettoriali su varietà algebriche definite su corpi non algebricamente chiusi, Boll. Un. Mat. Ital. A 16 (1979), 128–136.

[3] M. G. Marinari, F. Odetti, M. Raimondo, Affine curves over an algebraically non-closed field, Pacific J. Math. 107 (1983), 179–188.

[4] M. G. Marinari, M. Raimondo, On complete intersections over an algebraically non closed field, Canad. Math. Bull. 29 (1986), 140–145.

[5] M. G. Marinari, M. Raimondo On complete intersection real curves, Rocky Mountain J. Math. 14 (1984), 919–920.

[6] M. Raimondo, On the global ring of Nash functions, Boll. Un. Mat. Ital. A 18 (1981), 317–321.

[7] M. Raimondo, An algebraic property for Nash rings, Ann. Univ. Ferrara Sez. VII (N.S.) 29 (1983), 21–28.

[8] M. Raimondo, Some remarks on Nash rings, Rocky Mountain J. Math. 14 (1984), 921–922.

[9] F. Mora, M. Raimondo, Sulla fattorizzazione analitica delle funzioni di Nash, Le Matematiche (Catania) 37, fasc. 2 (1982), 251–256.

[10] F. Mora, M. Raimondo, On noetherianness of Nash rings, Proc. Amer. Math. Soc. 90 (1984), 30–34.

[11] M. Raimondo, Sur l'anneau des fonctions de Nash complexes, C.R. Acad. Sci. Paris Sér. I Math. 301 (1985), 19–21.

[12] M. Raimondo, On normalization of Nash varieties, Rend. Sem. Mat. Univ. Padova 73 (1984), 137–145.

[13] E. Fortuna, S. Łojasiewicz, M. Raimondo, Algébricité de germes analytiques, J. Reine Angew. Math. 374 (1987), 208–213.

[14] M. Raimondo, Introduzione allo studio delle varietà abeliane reali, Sem. dell'Ist. di Mat. dell'Univ. di Genova 136 (1981), 1–42.

Dipartimento di Matematica, Università di Trento, v. Sommarive 14, 38050 Povo, Trento, Italy

Mario Raimondo's contributions to computer algebra*

Tomas Recio and Maria-Emilia Alonso

Personal remembrances

If I remember correctly, the first time that I ever met Mario Raimondo was June 25, 1979, at Bressanone, Italy, on the occasion of a meeting organized by Professor Enzo Stagnaro, a generic "Simposium di Geometria Algebrica." Mario was there presenting a paper entitled "Completion of affine curves defined over any field," jointly written with M. G. Marinari and F. Odetti. I was presenting a scientific communication about "Another Positivstellensatz in semialgebraic geometry." Although we had no previous knowledge of each other, we very soon became friends, as young people do when sharing interests on the same subject. But let me trace back the origin of our papers.

There was a line of research, arising from some works of Silhol[1] and Beretta and Tognoli[2] considering algebraic geometry in the general non-algebraically closed case, but there was in these papers always a stress on the case of real fields (in fact, the work of Galbiati and Tognoli[3] was also the origin of this generalization to non-algebraically closed fields). Behind this there was, of course, the work of Tognoli with Acquistapace and Broglia about real analytic spaces, extensively studied in the sixties and the seventies (see, for instance, the booklet of Tognoli[4], that has been the early syllabus for many real geometers). Mario's mathematical research started with his thesis (under the direction of S. Greco): "Sul gruppo di Picard di certi anelli di dimensione uno"[5] and continued with some works that, for various reasons, were

*The paper is written in cooperation with M. E. Alonso, but the personal memories and subjective opinions are exclusively those of the senior author, Tomas Recio.

[1] Géométrie algébrique sur un corp non algébriquement clos, Comm. Algebra 6 (1978), 1131–1155

[2] Some Basic Facts in Algebraic Geometry on a non Algebraically Closed Field, Ann. Scuola Norm. Sup. Pisa Cl. Sci. (4) 3 (1976), 341–359

[3] Alcune propietà delle varietà algebriche reali, Ann. Scuola Norm. Sup. Pisa Cl. Sci. (3) 27 (1973), 359–404

[4] Algebraic Geometry and Nash Functions, Institutiones Mathematicae 3, Academic Press, 1978

[5] Rend. Sem. Mat. Padova 48 (1973), 23–37

never published. Mario's paper in Bressanone therefore had its roots in a mixture of commutative algebra and analytic geometry, with a special interest in real geometry; a very personal combination that never left him.

On the other hand, I had become interested in real geometry starting also from the work on real analytic and semianalytic sets introduced by Łojasiewicz in the sixties, but in 1977 I had "discovered" Dubois' Nullstellensatz (done seven years earlier) and I was moving to a more algebraic setting (because the analytic case was too difficult for me). So there I was in Bressanone, presenting a variant of Stengle's Positivstellensatz in real algebraic geometry.

Therefore Mario and I had real geometry as a common interest. Moreover, we were, together with Maria Grazia Marinari and Carlos Andradas, who was my first student, the only people in the area of real geometry present in Bressanone. I was very frightened about my talk, for it was my first scientific experience outside Spain, but I was comforted by Mario's gentle and respectful mood towards me. Mario treated me as a colleague who works in a very difficult subject, while he spoke of his own work as being almost trivial or pretended that it was mainly the work of the co-authors. If you knew Mario you would realize how natural this was for him; with such a gentleman it was impossible not to be on good terms.

We met again at the 1983 AMS–NATO "Real Algebraic Geometry" meeting in Boulder, USA, of which I was co-organizer. Here he presented two papers, a joint paper with Maria Grazia Marinari and one by himself, devoted to the study of Nash functions on algebraic varieties defined on arbitrary fields. Professor Tognoli will detail this part of his work. After that time we met practically every year at least once: meetings on real geometry and quadratic forms in Oberwolfach, Luminy, Corvallis (USA), and Trento. Also we almost simultaneously started showing appreciation for the algorithmic aspects of real geometry: in 1987 we presented a joint paper in the AAECC-5 (Applied Algebra and Error Correcting Codes) meeting held in Menorca. That year we also attended a conference on computational aspects of geometry and topology that took place in Sevilla. In 1988 we got together in the course "Introduction to Computational Geometry" organized by Professor Ferro at the University of Catania.

Then in 1989 I visited him for the first time at his home university in Genoa, when attending the CoCoA II (Computer Commutative Algebra) meeting. We met again that year in Portland, Oregon, on the occasion of the ISSAC (International Symposium in Symbolic and Algebraic Computing), where we both presented papers for the first time in this well-known international series of meetings on computer algebra. That was the last time we saw each other; some health problems on my side during 1990 and later his mortal sickness obstructed our contacts.

During those ten years of scientific and human contact we had some time to develop—so to speak—bilateral relations. In 1987 Mario came to Santander to be part of the commission judging a dissertation presented by one of my students. The weather was absolutely horrible that January and I recall Mario enduring with a frozen (sic) smile the strong winds and the heavy hail while I, enthusiastic but rather impolite, gave him a guided tour to Santander beaches and other summer recreation places. A few months later—but surely not as a consequence of the winter charm of my home city—Mario and I initiated a scientific cooperation agreement between our two universities plus the Universities of Madrid and Pisa. This agreement, mainly devoted to the exchange of visitors in the fields of computer algebra and algebraic and analytic geometry, has continued to the present day. Alonso, Andradas, Cucker, González-Vega, Luengo, Mora, Marinari, Traverso, and Tognoli are some of the scientists that have benefited from this cooperation that Mario—reluctant as he always was to appear in the front page of any scientific proposal—had the generosity and the vision to start. Besides the expected effect of promoting the exchange of ideas between researchers from the two countries, this agreement also had the unexpected but desirable effect of strengthening the cooperation among scientists in their own universities. For instance, Mario started working as a team with Alonso (one of my former students) and Mora (one of his colleagues with whom he had previously had little scientific contact) and this team produced most of the computer algebra work we are going to describe below.

Mario was a "caballero" in the deep Spanish meaning of this word: generous and devoted to friendship, helpful and patient, elegant and charming. On the scientific side Mario has been the germ and support of a lot of good mathematical work besides his own. Many other mathematicians would have capitalized on it for their own benefit, but he always preferred to remain in a second place. That is maybe why he had everywhere so many friends.

The transition from real geometry to computer algebra in Mario Raimondo's work

For Mario and me, our joint paper "On the computation of local and global analytic branches of a real algebraic curve"[6] (cf. [1]) was also our first paper where algorithmics played a relevant role.

Since the well known work of Collins,[7] algorithmic methods for real algebraic and semialgebraic objects were present in computer algebra, mainly throughout the work of Collins and his students. A quick look at the book "Computer Algebra, Symbolic and Algebraic Computation"[8] shows that the development of this important area of computer algebra was, in 1982, essentially reduced to Collins' team, and somehow disconnected from the main areas of research in real algebra and geometry that had been arising in the late seventies and in which Mario's work was developing. For example, in the above-mentioned 1983 meeting in Boulder on "Ordered Fields and

[6]In what follows numbers in brackets refer to the attached list of Mario's publications in computer algebra.

[7]Quantifier Elimination for Real Closed Fields by Cylindrical Algebraic Decomposition, in: Lecure Notes in Comput. Sci. 33, pp. 134–163, Springer-Verlag 1975

[8]Buchberger, B., Collins, G., Loos, R. (eds), Computing, Suppl. 4, Springer-Verlag 1982

Real Algebraic Geometry" Arnon and McCallum, two students of Collins, presented a paper[9] in which they obtained independently the same result as two Italian researchers in real geometry, Gianni and Traverso[10]

Mario's paper "On the computation of local and global analytic branches of a real algebraic curve" [1] contributed to the fact that the year 1987 marked the inflexion point when people who had been working in real geometry started systematically working on the algorithmic applications of ten years of theoretical developments. The authors of the paper were directly influenced by the beautiful ideas behind the work of Coste and Roy, "Thom's lemma, the coding of real algebraic numbers and the topology of semialgebraic sets"[11], that opened the possibility of computing with algebraic numbers in the real closure of a not necessarily archimedean ordered field, and that was the door to computer algebra also for several other mathematicians from real geometry.

Using his knowledge of real geometry and many other ideas and techniques of computer algebra in complex algebraic geometry (like Gröbner and standard basis), Mario communicated, in the three subsequent years 1988–1990, several results regarding the computation of analytical and topological features of algebraic objects in the two leading international conferences in computer algebra, namely AAECC and ISSAC. His last contribution in this field was to the EUROGRAPHICS meeting in October 1991, three months before his premature death.

Raimondo's work in computer algebra

As we have mentioned before, because of his previous work on Nash functions, Mario was mainly interested in the computation of the analytic (local) structure of real algebraic sets. Thus, in [1], the main contribution is the improvement of Roy's algorithm ("Computation of the Topology of a Real Curve"[12]) for the computation and combinatorial description of the local and global analytic branches of a real plane algebraic curve. A second major contribution of this paper, and one which is greatly due to Mario's ability to clarify a tangled exposition, is the detailed account of the main techniques that at that time were being developed for dealing with real algebraic numbers. In this respect the paper was pioneering (only rough sketches of the main tools had been published before this paper) and a useful reference for later works.

A natural continuation of this paper was to extend it to the case of two dimensional

[9]A polynomial time algorithm for the topological type of a real algebraic curve, Proc. of the meeting "Ordered Fields and Real Algebraic Geometry", Rocky Mountain J. Math. 14 (1984)

[10]Shape determination for real curves and surfaces, Publ. 23, Dipto. di Matematica, Univ. Pisa, July 1983

[11]J. Symbolic Comput. 5 (1988), 121–129

[12]Proc. of the Conference "Computational Geometry and Topology and Computation in Teaching Mathematics", Sevilla 1987, to appear in Astérisque

semialgebraic sets in the plane, which was done in [2]. Here the testing of signs of polynomials on sample points corresponding to two-dimensional cells—as is required in the CAD algorithm of Collins for quantifier elimination—is replaced by the computation of signs at suitable two-dimensional points of the real spectrum of the plane. A way to codify these points combinatorially is also developed.

Another natural continuation of [1] was the study of the local structure of algebraic sets (curves and surfaces) in three dimensional space. This is essentially more difficult than the planar case. Mario's approach was to study first the classical case of points in surfaces in which one variable can be solved as a fractional power series on the other two variables, that is the case of quasi-ordinary points. Thus in [3] he presented an algorithm for computing the analytic branches of quasi-ordinary singularities of surfaces. This involved giving a constructive version of the Abhyankar–Jung theorem, because the starting point for his algorithm could be any regular polynomial, of order d with respect to one variable, and satisfying that the discriminant locus with respect to this variable is a normal crossing. He constructed, then, the branch expansion.

In this setting, if one tries to translate the classical Newton–Puiseux procedure from the plane curve case one sees that it can be reasonably attempted if the Newton polyhedron has special properties, as happens in the particular case of the so called *ν-quasi-ordinary* polynomials[13].

Even if one considers only this case it turns out that, when performing the classical transformation of the Newton–Puiseux method, the ν-quasi-ordinary property is not preserved. This led him to develop a method for manipulating the Newton polyhedron in which changes of variables involving power series were essential. The intrinsic need to compute using power series comes from the fact that parametrizations of the branches of the surface can have an infinite number of terms between two significative exponents (namely the so called *distinguished pairs*[14]) that correspond to "features" of the Newton polyhedron similar to the ones of the polygon for the case of planar curves.

Therefore, as a technical device, he developed an original method to compute in a symbolic way with algebraic power series, using as main tool the *Mora Tangent Cone Algorithm*[15]. We shall comment below on this point, as Mario detailed the complete method in further papers ([4], [5], [6], [7]). In [3] he just gives a glimpse of the method, pursuing it only to solve the computations required by his algorithm for the branch description of quasi-ordinary singularities.

An important application of this algorithm is that it enables the computation of all the *distiguished pairs* of a singularity of this type, and therefore the computation of the topological and equisingularity type of a complex surface, locally at such a point[16].

---

[13]cf. Luengo I., A new proof of the Jung–Abhyankar Theorem, J. Algebra 85 (1983), 399–409

[14]cf. Lipman J., Quasi-Ordinary Singularities of Surfaces in $\mathbb{C}^3$, Proc. Symp. Pure Math. 40 (1983), 161–172

[15]Mora F., A Constructive characterization of Standard Bases, Boll. Un. Mat. Ital. D 2 (1983), 41–50

[16]cf. Lipman J., loc. cit.

cf. Gau, Y. N., Topology of the quasi-ordinary surface singularities, Topology 25 (1986), 495–519

As we have said, a complete method for computing with algebraic power series (of any number of variables and with coefficients in a computable field) is developed in [4][17] and [5]. The work in [3] suggested that a more suitable way of manipulating algebraic power series was to consider so called *Locally Smooth Systems* (*LSS* for short). These are polynomial systems in n variables plus s parameters satisfying at the origin the conditions of the Implicit Function Theorem (also called Dini's theorem in some European countries) and therefore defining uniquely an n-tuple of algebraic power series in the s parameters. Moreover, it is known that each algebraic power series is a component of an n-tuple solution of such a system[18]. In [5], Mario showed that the smallest subring of the ring of power series generated by all polynomials in the s parameters and by the n-tuple of solutions of a given LSS has "good behaviour" from the point of view of local computational algebra. In fact, his main achievement in [5] is to obtain, in this model of computation, an effective version of Weierstrass' Preparation and Division Theorems, Noether's Normalization Theorem and an algorithm for standard basis computation, for elements and ideals of these subrings. Moreover, no previous symbolic method was known for working systematically in this setting of algebraic power series.

The next issue he studied dealt with the size of the objects computed by the above mentioned algorithms. Given an algebraic power series f there can be many LSS's such that f is one component of its solution n-tuple. For every LSS let us consider the number of variables involved and the product of the degrees of the polynomials. Then from all LSS's describing f one selects first the LSS's having the smallest number of variables $\lambda(f)$, and among them the ones of smallest degree $\delta(f)$. In this model of computation the pair $(\lambda(f), \delta(f))$ is defined to be the complexity of f (cf. [7]). Another complexity measure for algebraic power series can be deduced from the work of Ramanakoraisina[19]: it is more classical to work with algebraic power series considering their minimal polynomial and Taylor expansion; therefore it is also natural to associate as complexity measure of an algebraic series the degree of the minimal polynomial and some measure of the cost of distinguishing one of the roots at the origin from the others; for instance, at least the multiplicity of the polynomial at the origin. In [7] both measures of complexity are compared, and it is shown for example that the LSS-complexity gives a simply exponential bound for the complexity of the output of the Weierstrass Preparation Theorem as a function of the complexity of the input. No similar estimates have been directly obtained for the other (classical) complexity measure (but the comparison made in [7] between the two complexities allows then to establish such an estimate).

In [6] a general scheme for computing several ideal-theoretic decomposition algorithms in the local case (local polynomial rings and algebraic power series) is exhibited, adapting the ideas of the algorithms used for the polynomial case[20] to the computa-

[17] A preliminary version of [5]

[18] cf. Artin M., Mazur, B., On periodic points, Ann. of Math. 81 (1965), 82–99

[19] Ramanakoraisina, R., Complexité des fonctions de Nash, Comm. Algebra 17 (1989), 1395–1406

[20] cf. Gianni P., Trager B., Zacharias G., Gröbner bases and primary decomposition of polynomials

tional model developed in the previous papers for the local case. Moreover, in the same paper, a new algorithm for computing the radical of an ideal in the polynomial ring was also introduced. As in previous algorithms, some components of the ideal of maximal dimension are isolated by extension-contraction. Here the novelty is that the other components are recovered not by enlarging the basis of the ideal, which introduces spurious components, but by ideal quotienting, thus avoiding the introduction of spurious components and even removing embedded ones. It is probably the best algorithm available at present for computing the radical of an ideal.

The last paper [8], revisits the topic of the computation of the local analytic structure of curves, but in n-dimensional space. Here Mario gives an algorithm to parametrize space curves avoiding the use of plane projections.

In the case of space curves, to be sure that one makes the Newton–Puiseux transformation on the "whole equations" one has to consider a standard basis of the ideal of the curve with respect to some preassigned weights. Then the coefficients of the parametrizations corresponding to these powers of the parameter appear as solutions of a 0-dimensional system in several variables, given by a Gröbner basis (as a result of the standard basis computation). This allows Mario to manipulate them recursively, in order to find new coefficients and weights, and to proceed further with the algorithm by using some ideas on the *arithmetic* of algebraic numbers due to Mora and Traverso[21]. The search for the admissible weights—i.e., the initial exponents of the parametrization—is linked to the computation of the so called *Critical Tropisms* of the ideal of the curve[22]: roughly, the weights for which the standard basis changes. In Mario's paper [8] these are computed using some ideas from the analogous notion coming from Gröbner bases[23] and adapting an algorithm by Assi[24]. Also he formalizes in "modern language" some ideas from an old paper by MacMillam[25] which allows him to have a good estimate of the number of steps of the algorithm.

In summary, Mario's work in computer algebra was important in many respects. He contributed, from his knowledge of real analytic and algebraic geometry, to settling algorithmic issues in difficult and substantial areas, some of them classically studied from a non-algorithmic point of view (like the determination of the structure of real algebraic sets), others originally opened through his work (like the development of a computation model for series). His early death, in the climax of his creative period, is a loss that only the scientific bonds that he established with other colleagues (he

ideals, J. Symb. Comput. 6 (1988), 149–167.

cf. Gianni P., Mora T., Algebraic solutions of systems of polynomials equations using Gröbner bases, Proc. AAECC-5, Lecture Notes in Comput. Sci. 356, Springer-Verlag 1989

[21] Mora T., Traverso C., Linear Gröbner methods and "natural" representations of algebraic numbers, preprint, May 1992

[22] Lejeune-Jalabert M., Tessier B., Transversalité, polygone de Newton et installations, Astérisque 7–8 (1973), 75–119

[23] Mora, T., Robbiano, L., The Gröbner Fan of an Ideal, J. Symb. Comput. 6 (1988), 183–208

[24] Assi, A., Standard Bases and Flatness, preprint, Université J. Fourier de Grenoble, 1991

[25] A method for determining the solutions of a system of analytic functions in the neighborhood of a branch point, Math. Ann. 72 (1912), 180–202

most often worked in a team, and as the focus of the team) can help to diminish, both through the continuation of the work he started, and through the imitation, by all of us, of his human and scientific example.

List of Mario Raimondo's publications in computer algebra

[1] On the computation of local and global analytic branches of a real algebraic curve (with F. Cucker, L. M. Pardo, T. Recio, M. F. Roy), in: Lecture Notes in Comput. Sci. 356, pp. 161–182, Springer-Verlag 1989.

[2] The Computation of the Topology of a Planar Semialgebraic Set (with M. E. Alonso), Rend. Sem. Mat. Univ. Politec. Torino, 46 (1988), 327–342.

[3] An Algorithm on Quasi-Ordinary Polynomials (with M. E . Alonso and I. Luengo), in: Lecture Notes in Comput. Sci. 357, pp. 59–73, Springer-Verlag 1989.

[4] Computing with Algebraic Series (with M. E. Alonso and T. Mora), Proc. ISSAC '89, ACM, New York 1989.

[5] A Computational Model for Algebraic Power Series (with M. E. Alonso and T. Mora), J. Pure Appl. Algebra 77 (1992), 1–38.

[6] Local Decomposition Algorithms (with M. E. Alonso and T. Mora), in: Lecture Notes in Comput. Sci. 508, pp. 208–221, Springer-Verlag 1990.

[7] On the Complexity of Algebraic Power Series (with M. E. Alonso and T. Mora), in: Lecture Notes in Comput. Sci. 508, pp. 197–207, Springer-Verlag 1990.

[8] Local Parametrization of Space Curves at Singular Points (with M. E Alonso, T. Mora and G. F. Niesi), Proc. EUROGRAPHICS '91. Sta. Margherita Ligure, 1991.

Departamento de Matemáticas, Estadística y Computación, Universidad de Cantabria, 39071 Santander, Spain
e-mail: `recio@matsun1.unican.es`

Departamento de Algebra, Facultad C.C. Matemáticas, Universidad Complutense, 28040 Madrid, Spain
e-mail: `mariemi@sunal1.mat.ucm.es`